Common Structures

(a) Octahedral MG/TM clusters generally have 26/86 valence electrons.
(b) Tetrahedral MG/TM clusters generally have 20 (16)/60 (56) valence electrons. The bracketed counts are less common.
(c) Triangular TM carbonyl clusters usually have 48 valence electrons.

The transition metals are associated with the following numbers of valence electrons:

Sc, Y, La	3
Ti, Zr, Hf	4
V, Nb, Ta	5
Cr, Mo, W	6
Mn, Te, Re	7
Fe, Ru, Os	8
Co, Rh, Ir	9
Ni, Pd, Pt	10
Cu, Ag, Au	11

Hence Sc is described as d^3, Cr as d^6, Pt as d^{10}, etc.

The donor properties of some common ligands are:

Terminal Cl, OR, SR	1
PR_3, and all CO	2
μ_2-Cl, μ_2-OR, μ_2-SR	3
μ_3-Cl, μ_3-OR, μ_3-SR	5

μ_2 = edge bridging μ_3 = face bridging

Throughout the text we use the convention that upper-case letters represent particular irreducible representations and symmetry labels in lower-case letters denote molecular orbitals which form bases for such representations.

Prentice Hall Advanced Reference Series

Physical and Life Sciences

**Prentice Hall
Inorganic and Organometallic Chemistry Series**

RUSSELL N. GRIMES, *Series Editor*

MINGOS AND WALES   *Introduction to Cluster Chemistry*

Forthcoming Books in this Series (tentative titles)

MARTELL AND HANCOCK   *Metal Complexes in Aqueous Solutions*

WILLIAMS   *Organic Superconductors*

Introduction to Cluster Chemistry

D. MICHAEL P. MINGOS
University of Oxford
and
DAVID J. WALES
University of Cambridge

Prentice Hall, Englewood Cliffs, New Jersey 07632

Library of Congress Cataloging-in-Publication Data

Mingos, D. M. P.,
  Introduction to cluster chemistry / D. Michael P. Mingos and David J. Wales.
    p.   cm. -- (Prentice Hall advanced reference series. Physical and life sciences) (Prentice Hall inorganic and organometallic chemistry series)
  Includes bibliographical references.
  ISBN 0-13-479049-9
  1. Metal crystals. 2. Metal-metal bonds. 3. Transition metal compounds.   I. Wales, David J.   II. Title.   III. Series.
  IV. Series: Prentice Hall inorganic and organometallic chemistry series.
  QD921.M645   1990
  541.2′24--dc20   90-30253
                                                                    CIP

Editorial/production supervision
  and interior design: Elaine Lynch
Cover design: Ben Santora
Manufacturing buyer: Kelly Behr

Prentice Hall Advanced Reference Series

Prentice Hall Inorganic and Organometallic Chemistry Series

© 1990 by Prentice-Hall, Inc.
A Division of Simon & Schuster
Englewood Cliffs, New Jersey 07632

All rights reserved. No part of this book may be reproduced, in any form or by any means, without permission in writing from the publisher.

Printed in the United States of America
10 9 8 7 6 5 4 3 2 1

ISBN 0-13-479049-9

PRENTICE-HALL INTERNATIONAL (UK) LIMITED, *London*
PRENTICE-HALL OF AUSTRALIA PTY. LIMITED, *Sydney*
PRENTICE-HALL CANADA INC., *Toronto*
PRENTICE-HALL HISPANOAMERICANA, S.A., *Mexico*
PRENTICE-HALL OF INDIA PRIVATE LIMITED, *New Delhi*
PRENTICE-HALL OF JAPAN, INC., *Tokyo*
SIMON & SCHUSTER ASIA PTE. LTD., *Singapore*
EDITORA PRENTICE-HALL DO BRASIL, LTDA., *Rio de Janeiro*

To our parents and to Stacey and Caroline

Contents

FOREWORD xiii
PREFACE xv

1 A SURVEY OF CLUSTER CHEMISTRY 1

1.1 Introduction 1
1.2 Thermochemical Aspects 6
1.3 Kinetic Factors 10
1.4 General Survey of Cluster Types 13
1.5 Multiple Bonding 26
1.6 Lanthanide and Actinide Cluster Chemistry 31
1.7 π-Acid Clusters 32
1.8 Phosphine Cluster Compounds 46
1.9 Hydrido Cluster Compounds 49
1.10 Other Ligands 55
1.11 Main Group Cluster Compounds 55
1.12 Clusters in Molecular Beams 64
Exercises 66
References 67

2 CLOSED-SHELL ELECTRONIC REQUIREMENTS FOR CLUSTER COMPOUNDS 72

2.1 Introduction 72
2.2 The Effective Atomic Number Rule 73

	2.3	Localized and Delocalized Representations 81	
	2.4	Face-Localized Bonding Schemes 99	
	2.5	Octahedral Halide Clusters 100	
	2.6	The Styx Method 103	
	2.7	Electron-Rich Cage Molecules 105	
	2.8	Metal Carbonyl Clusters in Molecular Beams 108	
	2.9	Summary 112	
		Exercises 113	
		References 115	

3 INTRODUCTION TO TENSOR SURFACE HARMONIC THEORY 117

3.1 Introduction   117
3.2 Some Helpful Analogies   117
3.3 Introduction to Spherical Harmonics   123
3.4 σ Cluster Orbitals   126
3.5 π Cluster Orbitals   132
3.6 The TSH Pairing Principle   145
3.7 δ Orbitals   148
3.8 Summary
   Appendix: Proof of the Pairing Principle   151
   Exercises   152
   References   153

4 CLUSTERS WHERE RADIAL BONDING PREDOMINATES 154

4.1 Introduction   154
4.2 Theoretical Considerations   155
4.3 Oblate and Prolate Clusters   163
4.4 Clusters with Interstitial Atoms   168
4.5 Multispherical Clusters   179
   Exercises   180
   References   180

5 CLUSTERS WHERE σ AND π INTERACTIONS ARE IMPORTANT   182

5.1 Introduction   182
5.2 Bonding in the Boron Hydrides   183
5.3 Shapes and Patterns in Boranes and Carboranes   186
5.4 The Pairing Principle   192
5.5 *nido* and *arachno* Clusters   195
5.6 Three-Connected Clusters   198
5.7 Four-Connected Clusters   200
5.8 Transition Metal Clusters   201
5.9 Symmetry-Forced Deviations in Electron Count   207
5.10 Partial Involvement of the Tangential Orbitals   212
Exercises   216
References   217

6 SKELETAL REARRANGEMENTS IN CLUSTERS   218

6.1 Introduction   218
6.2 The Diamond-Square-Diamond Rearrangement   219
6.3 The Conservation of Orbital Symmetry   220
6.4 TSH Analysis of Skeletal Rearrangements   222
6.5 Application to Boranes   227
6.6 Single Edge-Cleavage Processes   229
6.7 The SDDS Mechanism   231
6.8 Transition Metal Clusters Conforming to the PSEPT   232
6.9 Radially Bonded Metal Clusters   234
6.10 Clusters with Capping Ib Metal Atoms   236
6.11 Platinum and Palladium Clusters   236
6.12 Rearrangements of the Ligand Shell   240
6.13 Summary   245
Exercises   245
References   246

7 CONDENSED AND HIGH-NUCLEARITY CLUSTER COMPOUNDS   249

7.1 Introduction   249
7.2 Geometric and Theoretical Aspects   250

7.3	Edge-Sharing Condensed Clusters	256
7.4	Face-Sharing Clusters	260
7.5	The Capping Principle	263
7.6	High-Nuclearity Spherical Clusters	274
	Exercises	282
	References	284

8 CLUSTERS WHERE δ ORBITALS MUST BE CONSIDERED EXPLICITLY 285

8.1	Introduction	285
8.2	Octahedral Clusters	286
8.3	*nido* and *arachno* π-Donor Clusters	291
8.4	Three-Dimensional Linked Clusters	294
8.5	π-Donor Clusters with Interstitial Atoms	298
8.6	Columnar Clusters	304
	Exercises	309
	References	310

INDEX 313

Information Boxes

1. *Synthesis*

SYNTHESIS OF HALIDE CLUSTERS	**19**
IRON-MOLYBDENUM-SULFUR "CUBANE" CLUSTERS	**26**
REDOX PROPERTIES OF IRON-SULFUR CLUSTERS	**27**
SYNTHESIS OF METAL CARBONYL CLUSTERS	**35**
SYNTHESIS OF HETEROMETALLIC CLUSTERS	**39**
SYNTHESIS OF CLUSTERS WITH INTERSTITIAL ATOMS	**44**
SYNTHESIS OF GOLD AND PLATINUM PHOSPHINE CLUSTERS	**49**
SYNTHESIS AND PHYSICAL PROPERTIES OF POLYHEDRAL BORANES	**56**
SYNTHESIS AND PROPERTIES OF BORON HALIDE POLYHEDRAL MOLECULES	**57**
SYNTHESIS, COLORS, AND BOND LENGTHS OF "NAKED" CLUSTERS	**59**

2. Structural and Chemical Properties

DELTAHEDRA	61
HYDROCARBON POLYHEDRA	63
TETRAHEDRAL ELECTRON-PRECISE MOLECULES	64
BOND LENGTHS IN POLYHEDRAL BORANES AND CARBORANES	87
CHARACTERIZATION OF SODIUM CLUSTERS USING ESR	170
MÖSSBAUER DATA	172
CLUSTERS AND COLOR	202
INFRARED DATA FOR CARBONYL, CARBIDO, AND HYDRIDO CLUSTERS	205
LOCATION OF HYDRIDO LIGANDS	208
MASS SPECTROMETRY FOR CHARACTERIZATION OF MOLECULAR CLUSTERS	211
NMR AND INFRARED DATA FOR POLYHEDRAL BORANE ANIONS $B_n H_n^{2-}$	228
HIGH-RESOLUTION SOLID STATE NMR USING MAGIC ANGLE SAMPLE SPINNING	235
NMR STUDIES OF PLATINUM CLUSTERS: THE USEFULNESS OF ^{195}PT ISOTOPOMER EFFECTS	237
^{13}C NMR DATA FOR METAL CARBONYL CLUSTERS	242
PARAMAGNETIC TRANSITION METAL CARBONYL CLUSTERS	268
LUMINESCENCE AND REDOX PHOTOCHEMISTRY OF $[M_6 X_{14}]^{2-}$ CLUSTERS (M = Mo or W; X = Cl or Br)	290
COLORS OF π-DONOR CLUSTERS	292
ALKOXIDE CLUSTERS: A BRIDGE BETWEEN π-DONOR AND π-ACCEPTOR LIGANDS	295
CHEMICAL SHIFTS OF INTERSTITIAL ATOMS	299

Foreword

Inorganic chemistry today enjoys a level of interest and activity that is without precedent, at least in modern times. It is striking to note how many of the widely publicized spectacular advances in science and technology in recent years involve inorganic materials or reactions: high-temperature superconductivity, the success of drugs such as *cis*-$Pt(NH_3)_2Cl_2$ ("cisplatin"), and the strong shift of emphasis in molecular biology toward metal-centered activity. In space exploration, the fantastic pictures and data relayed from the Voyager 2 encounter with Neptune and Triton, and the earlier ones from Saturn, Jupiter, and Uranus, remind us that the solar system is a vast cauldron of inorganic chemistry, some of it undoubtedly exotic (for example, the storms of Jupiter and the strange properties of martian soil which at first seemed to suggest the possibility of living organisms). Then there are the burgeoning fields of inorganic polymers, low-dimensional electrical conductors, ceramics, metal reagents for stereoselective synthesis, radiotherapy and immunotherapy, and of course the exploding area of transition-metal organometallic chemistry. And we could go on.

Reflecting this interest, the Inorganic Division of the American Chemical Society has in recent years consistently attracted the largest number of contributors and papers of all the 30-odd divisions of that organization to its programs at national and international ACS meetings. This ferment of research activity has generated a vastly increased volume of excellent published research, as reflected in the establishment of several first-rate new journals (*Organometallics*, *Chemistry of Materials*, etc.), the expansion of old ones, and many monographs, serial publications, and review volumes dedicated to inorganic chemistry. The present series is a new venture whose intent is to provide high-quality books on frontier topics of inorganic and organometallic chemistry, written or edited by major contributors in these areas. With this goal in mind, I have been acutely aware of the importance of the first volumes in the series in setting a tone, style, and level of quality and timeliness to be emulated in subsequent books. This inaugural contribution by Mingos and Wales deals, fittingly, with the centrally important topic of molecular clusters and is the first truly comprehensive, unified book-length treatment of this subject from the perspective of modern bonding theory.

As the authors point out in their preface, clusters have become important in a variety of fields which range across the traditional boundaries of chemistry, from organic molecules such as prismane, dodecahedrane, and the postulated "Buckminsterfullerene," to enzyme centers, solid state materials, gas-phase clusters, novel organometallics, metal clusters, and of course the polyhedral boranes and carboranes and their many thousands of metallo derivatives. This volume represents an ambitious attempt to provide a solid theoretical basis for understanding cluster structure, bonding, and reactivity, and it serves as an ideal launching pad for what we hope will be a significant new series.

Russell N. Grimes
Series Editor

PRENTICE HALL INORGANIC AND ORGANOMETALLIC
CHEMISTRY SERIES

Charlottesville, Virginia

Preface

Plato and his school first emphasized the states of matter rather than the atomistic approach of Democritus. So much did the symmetry of simple geometric shapes impress the ancient Greeks that they constructed their theories of nature around them. Earth, air, fire, and water were associated with the cube, tetrahedron, octahedron, and icosahedron, and the cosmos was represented by the dodecahedron. Although these theories have long been discarded, the structures which captivated Plato retain their fascination, and in recent years chemists have synthesized compounds based upon all the regular polyhedra, as well as a great many more which are less symmetrical. The intrigue of the Platonic solids has not diminished over the intervening millennia.

The word "cluster" was first coined by Prof. F. A. Cotton in the early 1960s to describe compounds containing metal-metal bonds and has since evolved to include other species with element-element bonds such as main group polyhedra and any molecule containing metal-metal bonds. Contributions to this exciting area of chemistry have been made by workers in various fields. Organic chemists have synthesized a wide range of alicyclic hydrocarbons culminating with dodecahedrane, $C_{20}H_{20}$. Inorganic chemists have synthesized a multitude of borane, metalloborane, and organometallic clusters including tetrahedral, octahedral, and icosahedral geometries. Solid state chemists have shown that octahedral clusters may be linked to form infinite lattices, some of which exhibit superconductivity. Bioinorganic chemists have shown that the electron transfer properties of some ferredoxin proteins depend upon the redox properties of iron-sulfur "cubane" clusters. Most recently, physical chemists have developed molecular beam techniques for studying clusters in the gas phase. This book will touch on all these areas in an attempt to show how some simple theoretical ideas can help us to find patterns amid this panoply of structure.

The framework for most of our discussions is Stone's tensor surface harmonic (TSH) theory, and we have endeavored to give an introduction to this method suitable for both advanced undergraduates and workers in the field, who are assumed to have some knowledge of the elementary use of group theoretical methods in chemistry. We have therefore included exercises at the end of each chapter as well as numerous references to the original literature. Many of the exercises have been taken from the recent literature. There are two reasons for this: first, to demonstrate the application of the theories developed in the book to current chemical problems; and second, to provide an entry for students into the literature.

We have aimed throughout to give clear explanations with a few specific examples, rather than to give a full account of all the molecules which have been treated with this method. Particular attention is paid to electron counting rules for clusters, for which the TSH theory provides general and satisfying descriptions. A treatment of skeletal rearrangements in clusters is also given as an example of the general utility of this theory. Illustrative examples of synthesis, characterization, and properties of clusters are also provided within our context of a structural approach, but we have by no means attempted to provide an exhaustive treatment of such aspects. We are particularly conscious of our inability to give a complete account of cluster reactivity in such an introductory text.

Mike Mingos *David Wales*

ACKNOWLEDGMENT

D.J.W. expresses his gratitude to Anthony Stone for introducing him to tensor surface harmonic theory.

Introduction to Cluster Chemistry

1

A Survey of Cluster Chemistry

1.1 INTRODUCTION

During the last 20 years we have witnessed the emergence of a major new area of inorganic chemistry based on compounds with metal-metal bonds. In the early 1960s few examples of such molecules were known, and more often than not they were chance products of reactions aimed at synthesizing alternative mononuclear products.[1] In those early days the relatively unsophisticated X-ray equipment and computing facilities and the many crystallographic disorder problems which were encountered meant that the characterization process was far from straightforward. The elucidation of the formative structural examples in the 1960s is a tribute to the skills and tenacity of crystallographers such as Dahl.[2] Since those days more predictable synthetic routes have been pioneered by Chini[3] and Lewis[4] and have led to the characterization of compounds containing up to 50 metal atoms. In addition, synthetic routes to heterometallic compounds have been developed by many research workers, with the contributions of Stone[5] and Vahrenkamp[6] being particularly noteworthy.

The development of more efficient spectroscopic (particularly infrared and nmr) and diffraction techniques has greatly speeded up the characterization of cluster compounds in the solid and solution states and ensured that the field reached maturity quickly.[7] In addition, a preliminary understanding of the bonding in these compounds has arisen from a combination of theoretical ideas derived from molecular orbital theory[8] and the application of spectroscopic techniques such as photoelectron spectroscopy.[9]

Although many of the structures observed in metal cluster chemistry resembled those which had been observed in main group cluster chemistry (particularly the boranes), the two fields developed separately. However, the synthesis of clusters with both transition metal and main group atoms demonstrated the futility of such rigid demarcation lines between the disciplines. The development of a common theoretical framework[8] also helped to break down these barriers.

The field of metal cluster chemistry has reached a particularly exciting point in its development where the sizes of the clusters being characterized are sufficiently large for meaningful comparisons of physical and chemical properties to be made with both the bulk metal and metal surfaces. The late Earl Muetterties was particularly influential in popularizing the idea that metal cluster compounds would serve as effective models for metal surfaces, which play such an important role in heterogeneous catalytic processes.[10]

A degree of control having been developed in this area of chemistry during the last 25 years, this is an apt moment to analyze some of the basic factors which influence the occurrence, stability, and structural variety shown by compounds with metal-metal bonds.

A metal cluster compound has been defined[11] as a group of two or more metal atoms where direct and substantial metal-metal bonding is present. This has proved to be a reasonable working definition, although the phrase "direct and substantial metal-metal bonding" has proved to be more ambiguous than originally envisaged, because the lengths of metal-metal bonds vary over a considerable range even for compounds with the same formal bond order. Table 1.1 summarizes the metallic radii of the transition metals derived from crystallographically determined distances in the solid state.[12] Twice this radius should provide a benchmark for the presence of substantial metal-metal bonding. For example, for molybdenum a metal-metal distance of the order of 2.80 Å should indicate the presence of significant metal-metal bonding, but the following series of compounds, which all formally have a bond order of 1, have metal-metal bond lengths which vary by as much as 0.7 Å:

$[Mo_2(CO)_6(C_5H_5)_2]$	3.24 Å
$[Mo_2(CO)_4(C_5H_5)_2(Et_2C_2)]$	2.98 Å
$[Mo_6Cl_8]^{4+}$	2.62 Å
$[ZnMo_3O_8]$	2.52 Å

Therefore, even metal-metal bond lengths which exceed the sum of the covalent radii by 15% are formally considered as a bond, although antiferromagnetic interactions of the spins rather than a strong covalent interaction may actually provide a better description of some such "bonds."

TABLE 1.1 METALLIC RADII FOR 12-COORDINATION (Å)

Li	Be												
1.57	1.12												
Na	Mg	Al											
1.91	1.60	1.43											
K	Ca	Sc	Ti	V	Cr	Mn	Fe	Co	Ni	Cu	Zn	Ga	Ge
2.35	1.97	1.64	1.47	1.35	1.29	1.37	1.26	1.25	1.25	1.28	1.37	1.53	1.39
Rb	Sr	Y	Zr	Nb	Mo	Tc	Ru	Rh	Pd	Ag	Cd	In	Sn
2.50	2.15	1.82	1.60	1.47	1.40	1.35	1.34	1.34	1.37	1.44	1.52	1.67	1.58
Cs	Ba	La	Hf	Ta	W	Re	Os	Ir	Pt	Au	Hg	Tl	Pb
2.72	2.24	1.88	1.59	1.47	1.41	1.37	1.35	1.36	1.39	1.44	1.55	1.71	1.75

4f elements: Ce (1.82)–Lu (1.72), but Eu (2.06), Yb (1.94)

5f elements: Th Pa U Np Pu
1.80 1.63 1.56 1.56 1.64

This does raise the interesting question of why metal-metal distances show a much greater variation than those observed between light main group atoms, such as carbon, and indeed greater than those associated with metal-ligand bonds. The metal-metal bonds in a cluster are generally weaker than the metal-ligand bonds. The thermodynamic justification for this statement will be given in a subsequent section. It is therefore very likely that the potential energy surface associated with deforming a metal-metal bond is softer than those associated with the other bonds in the molecule. It has been calculated that for a first-row transition metal carbonyl cluster the metal-metal force constant is approximately 0.6 mdyn/Å and the metal-carbon force constant about 2.0 mdyn/Å. Assuming harmonic potentials, a uniform expansion of M—M by 0.1 Å requires 2.6 kcal/mol and an expansion of M—CO by 0.03 Å requires 1.8 kcal/mol. The metal-metal bond is therefore the most deformable bond in the structure, and the small energy changes associated with a lengthening of the metal-metal bond can be compensated for by more favorable bonding and nonbonding steric interactions for the ligands.[13] In addition, as Woolley[14] has pointed out, the transition metals' valence d, s, and p orbitals behave quite differently in a metal-metal bond. The d orbitals are rather contracted and do not give a large overlap integral, but can result in a substantial stabilizing effect. In contrast, the s and p valence orbitals are rather diffuse and despite their superior overlap give rise to a repulsive interaction because they impinge upon the cores of the neighboring metal atoms. Therefore, relatively minor changes in the contributions of the metal d, s, and p orbitals to the metal-metal bond can result in large changes in the equilibrium metal-metal distances.

The occurrence of "direct" metal-metal bonding associated with the above definition of clusters can be problematical when the metal atoms are also connected by bridging ligands. Since the orbitals associated with the metal-metal bond have exactly the same symmetry characteristics as some linear combinations of orbitals of the bridging ligand atoms, it is not possible to dissect the problem using either symmetry arguments or crude molecular orbital calculations. Either accurate calculations or spectroscopic measurements are required to justify whether the metal-metal bond is making a major contribution to the total energy, or whether it represents a weak antiferromagnetic coupling of electron spins. Much time has been spent on sophisticated theoretical calculations in order to resolve this problem.[15] The general conclusion is that metal-metal bonding in the latter case does not contribute substantially to the total energy of the molecule, but a formal metal-metal bond designation is retained in order to maintain a connection with the simplified bonding schemes based on the inert gas rule (see Chapter 2). This formalism is a useful one for comparing the metal-metal bond lengths in pairs of compounds with very similar bridging groups. For example, in the following pairs of molecules the metal-metal separation increases substantially on reducing the formal bond order from 1 to 0.[16] (The notation μ_2 signifies that the ligand bridges two metal atoms.)

Compound	Formal bond order	Metal-metal distance (Å)
$[Cr(C_5H_5)(\mu_2\text{-}SR)(NO)]_2$	1	2.95
$[Fe(C_5H_5)(\mu_2\text{-}SR)(CO)]_2$	0	3.39
$[Co(C_5H_5)(\mu_2\text{-}PR_2)]_2$	1	2.56
$[Fe(C_5H_5)(\mu_2\text{-}PR_2)]_2$	0	3.36

Nevertheless, the metal-metal separation for a series of closely related compounds can vary over a wider range than that associated with a change in bond order, because of the different electron donating capabilities of the bridging ligands; e.g., the distances are 2.72 Å for $[Fe(SR)(NO)_2]_2$ and 3.05 Å for $[FeI_2(NO)_2]_2$. Consequently, when the bridging groups and the formal bond orders are varied in a series of molecules, a very complex and rather impenetrable distribution of bond lengths results.

The bonding in a series of related nitrosyl and carbonyl bridged compounds has been analyzed carefully by Fenske; the results are summarized below:[17]

Compound	Formal bond order	Metal-metal distance (Å)
$[Fe(C_5H_5)(NO)]_2$	2	2.363
$[Co(C_5H_5)(CO)]_2$	2	2.338
$Co_2(C_5H_5)_2(CO)(NO)$	1.5	2.370
$[Co(C_5H_5)(CO)]_2^-$	1.5	2.372
$[Co(C_5H_5)(NO)]_2$	1	2.372
$[Ni(C_5H_5)(CO)]_2$	1	2.357

Fenske has noted that the dominant bonding interactions occur between the bridging ligands (NO or CO) and the metals—not between the metal atoms. For the carbonyls, the lone-pair interactions with the metal orbitals are particularly important, but for nitric oxide the antibonding π^* orbitals are lower-lying and make a larger contribution. Since the latter stabilize orbitals which are metal-metal antibonding, the metal-metal distance is longer in those compounds with bridging nitrosyls than those with bridging carbonyls with the same formal bond order.

Attempts have also been made to view the electron density associated with the metal-metal bond directly by accurate X-ray diffraction studies, but the results have been rather disappointing. Most experimental studies measure the standard deformation electron density, $\Delta\rho(\mathbf{r})$, which is defined as the molecular electron density minus the electron density of the "promolecule" made up by the superposition of the isolated, neutral, and spherically averaged ground-state atoms. In those molecules which have been studied and are known to have a formal metal-metal bond, e.g., $[Mn_2(CO)_{10}]$, no bonding electron density is observed in the metal-metal region. For those with strong multiple bonds, e.g., $[Re_2Cl_8]^{2-}$, some electron density in this region is observed. In the former example it is difficult initially to reconcile the standard deformation studies with our usual concept of the covalent bond based on a detailed knowledge of the hydrogen molecule ion, H_2^+. Hall and Ruedenberg have independently proposed that this discrepancy arises because the choice of "promolecule" based on spherical atoms is inappropriate for metal-metal and related weak homopolar bonds. If the "promolecule" is based on atoms in their valence states, then the deformation electron density map does indeed show a buildup of electron density in the metal-metal region.[18-19]

1.2 THERMOCHEMICAL ASPECTS

Only a limited number of thermochemical measurements (using microcalorimetry) have been reported on the transition metal carbonyl clusters, $[M_m(CO)_n]$. The enthalpy of disruption, ΔH_D, referring to the following process:[20]

$$M_m(CO)_n(g, 298\ K) \longrightarrow m M(g, 298\ K) + n\ CO(g, 298\ K),$$

can be calculated from the following primary thermodynamic data:

$$\Delta H_D = m\ \Delta H_f^\circ[M, g] + n\ \Delta H_f^\circ[CO, g] - \Delta H_f^\circ[M_m(CO)_n, g].$$

[M, g] refers to metal atoms in their ground states rather than their valence states. If the structures of the clusters are described in terms of the number of localized metal—CO terminal bonds (T), metal—CO bridging bonds (B), and metal-metal bonds (M), then the mean bond enthalpy contributions $\bar{T}$, $\bar{B}$, and $\bar{M}$(cluster) summarized in Table 1.2 can be calculated. These are based on the usual assumptions that the bond enthalpy contributions $\bar{T}$, $\bar{M}$, and $\bar{B}$ are transferable and that bond length variations are averaged out. The cohesive energies of the metals can be used to estimate a mean metal-metal bond enthalpy contribution, $\bar{M}$(metal), for the bulk metal if the coordination numbers of the metals, n, are taken into account according to $\bar{M}$(metal) $= 2\Delta H_f^\circ[M, g]/n$. The relevant calculated values for $\bar{M}$(metal) are also summarized in Table 1.2.

Connor[20] has proposed that $\bar{T}$, $\bar{M}$(cluster), $\bar{B}$, and $\bar{M}$(metal) can be related through the following empirical relationships:

$$2\bar{T} \approx 3\bar{M}(\text{cluster}) \approx 4\bar{B} \approx 2\bar{M}(\text{metal}).$$

The following broad generalizations can be drawn from the thermodynamic data:

(1) $\bar{M}$(cluster) and $\bar{M}$(metal) are related and follow the same general periodic trends. Therefore, it can be reasonably assumed that the type

and strength of metal-metal bonding is comparable in clusters and the bulk metal.

(2) The $\bar{M}$(cluster) bond enthalpy contributions increase with increasing atomic number for a group of elements in the same column of the periodic table.

(3) There is insufficient information to comment on the variations across the transition series for $\bar{M}$(cluster), but it is established that $\bar{M}$(metal) approaches a maximum toward the center of the transition series where the d and s bands are half filled.

(4) The metal-metal bond enthalpy contributions not only are smaller than $\bar{T}$ for carbonyl clusters, but are also likely to be smaller than those for commonly occurring ligands in transition metal chemistry such as F, Cl, and OR. Table 1.3 gives some typical mean bond enthalpy data for such ligands.

Although both $\bar{M}$(cluster) and $\bar{M}$(metal) increase down a periodic group for the transition elements, this does not represent a general trend for all elements of the periodic table. For the pre- and post-transition elements, $\bar{M}$(metal) decreases down the group. This difference can be attributed to differences in the radial distribution functions of the relevant valence orbitals. For the first-row transition elements, the $3d$ orbitals are rather contracted and an increase in principal quantum number leads to better overlap between d orbitals on adjacent metal atoms. It has also been argued that relativistic effects[21] operating for the third-row transition elements leads to a contraction of the $6s$ valence orbitals and improved metal-metal bonding. For the post-transition elements the optimum overlap integrals between the s and p valence orbitals are achieved for the first-row elements, e.g., boron and carbon.

TABLE 1.2 STANDARD ENTHALPIES OF FORMATION (ΔH_f°) OF METAL CARBONYLS $M_m(CO)_n$ IN THE GAS PHASE: BOND ENTHALPY CONTRIBUTIONS ($\bar{T}, \bar{M}, \bar{B}$) FOR TERMINAL CARBONYLS, METAL-METAL BONDS, AND BRIDGING CARBONYLS WITH RESPECT TO THE ENTHALPY OF DISRUPTION (kJ/mol)[20]

	ΔH_f°	$\bar{T}$	$\bar{M}$	$\bar{B}$
[Mn$_2$(CO)$_{10}$]	−1598	100	67	
[Fe$_3$(CO)$_{12}$]	−1753	117	82	54
[Ru$_3$(CO)$_{12}$]	−1820	172	117	
[Os$_3$(CO)$_{12}$]	−1644	190	130	
[Co$_4$(CO)$_{12}$]	−1749	136	83	68
[Rh$_4$(CO)$_{12}$]	−1749	166	114	83
[Ir$_4$(CO)$_{12}$]	−1715	110	130	
[Rh$_6$(CO)$_{16}$]	−2299	166	114	83

TABLE 1.2 (Continued)
STANDARD ENTHALPIES OF FORMATION OF GASEOUS METAL ATOMS $\Delta H_f^\circ (M, g)$ AND ESTIMATES OF $\bar{M}$ BASED ON THE COORDINATION NUMBER IN THE METAL $(kJ/mol)^{20}$

Element	(Coordination number n^*)	$\bar{M}$	Element	(Coordination number)	$\bar{M}$	Element	(Coordination number)	$\bar{M}$
Sc	(12c)	63						
Ti	(12h)	78	Zr	(12h)	100	Hf	(12h)	104
V	(8)	129	Nb	(8)	181	Ta	(8)	197
Cr	(8)	99	Mo	(8)	164	W	(8)	213
Mn	(8)	71	Tc	(12h)	116	Re	(12h)	129
Fe	(8)	104	Ru	(12h)	108	Os	(12h)	132
Co	(12h)	71	Rh	(12h)	93	Ir	(12h)	112
Ni	(12c)	71	Pd	(12c)	62	Pt	(12c)	94
Cu	(12c)	56	Ag	(12c)	48	Au	(12c)	62

*12c = cubic close packed, 12h = hexagonal close packed, 8 = body-centered cubic.

$$\bar{M} = 2[\Delta H_f^\circ (M, g)]/n$$

TABLE 1.3 MEAN BOND DISSOCIATION ENTHALPIES $\bar{D}$(M—L), kJ/mol, FOR ML_n COMPOUNDS[20]

	Ti	Zr	Hf	Nb	Ta	Mo	W
CH_3	(260)	(310)	(330)		261		159
NMe_2	307	350	(370)		328		220
OPr^n	441			418	440		
OPr^i	444	518	534				
F	586	648	605	573	603	449	508
Cl	430	490	497	406	430	305	347

Figures in parentheses are based on estimates and extrapolations.[20]

Pr^i = isopropyl; Pr^n = n-propyl.

With regard to the more general problem of cluster synthesis and occurrence, the thermodynamic data provide some insight into the trends for the enthalpy changes for the following prototype reaction which leads to cluster formation from a mononuclear precursor:

$$nML_p \longrightarrow M_nL_m + (pn - m)L.$$

In general the reaction will have a negative enthalpy change only if $x\bar{M}(\text{cluster}) \geq y\bar{T}$, where x is the number of metal-metal bonds formed in the cluster M_nL_m and $y = pn - m$ is the number of ligands, L, lost in the above process. Since $x \sim y$ in the majority of realistic examples, e.g.,

$$3Fe(CO)_5 \longrightarrow Fe_3(CO)_{12} + 3CO,$$

it follows that the enthalpy will be negative only when $\bar{M} > \bar{T}$. In practice the positive entropy term associated with such reactions means that the free energy for cluster formation will be negative when $\bar{M}(\text{cluster})$ approaches $\bar{T}$. The entropic term also favors high-temperature aggregation reactions of the following type:

$$Os_3(CO)_{12} \xrightarrow{\text{heat}} Os_6(CO)_{18} + Os_4(CO)_{13} + Os_5(CO)_{16} \quad \text{etc.}$$

This is, of course, not a selective synthesis, and the products have to be separated by chromatography.

For a transition metal, this crude analysis suggests that cluster formation is favored for the third-row elements. Furthermore, the associated ligands should have the type of electronic properties which stabilize the low oxidation states associated with effective metal-metal bond formation, but should not have too large a $\bar{T}$ enthalpy contribution. In practice this means the presence of reasonable π-acid ligands such as CO, CNR, and PR_3 for metal cluster compounds with zero or negative oxidation states, and ligands such as S and Cl for metals formally in +2 to +4 oxidation states. For main group elements, cage and cluster compounds are favored for the lighter elements, where $\bar{M}$(cluster) is able to compete more effectively with $\bar{T}$. Therefore, although there are some examples of cluster compounds for the heavier post-transition elements, e.g., Sn, Ge, and Bi, the most extensive series of cage compounds occur for boron and carbon. Interestingly, for elements such as Sn, Ge, and Bi, where $\bar{M}$(cluster) is not large, the cluster compounds do not have ligands associated with them and thereby remove the competition between metal-ligand and metal-metal bond enthalpy contributions.

1.3 KINETIC FACTORS

In common with other aspects of transition metal and main group chemistry, kinetic factors also influence the isolation and chemical behavior of cluster compounds.[22] Even though the following type of reaction leading to cluster decomposition might be thermodynamically favorable,

$$M_n L_m \longrightarrow nM + mL,$$

it will have chemical significance only if the relevant ligand dissociation and metal aggregation processes have appreciable rates. Therefore, the isolation of cluster compounds which are thermally and chemically robust is favored by metal-ligand combinations which are dissociatively inert. Since complexes of the transition metals become progressively more inert as a periodic group is descended, the kinetic and thermodynamic effects reinforce each other to give a more extensive range of cluster compounds for the heavier transition metals. The following comparison between related iron and osmium carbonyl cluster compounds in their reactions with CO underlines how these kinetic and thermodynamic effects influence their relative stabilities and reactivities:

$$Fe_3(CO)_{12} + 3CO \xrightarrow{25°, 20 \text{ atmos.}} 3Fe(CO)_5,$$

$$Os_6(CO)_{18} + 3CO \xrightarrow{160°, 90 \text{ atmos.}} Os_6(CO)_{21}.$$

In the latter example, even under the forcing conditions which prevail, the cluster does not degrade to form a mononuclear complex, but manages to incorporate additional CO molecules by adopting a more open skeletal geometry. Kinetic effects also favor the isolation of cluster molecules with closed-shell electronic configurations and large HOMO-LUMO gaps in preference to molecules with unpaired electrons, which are generally kinetically labile.

The ligand can also influence the nuclearity and geometry of the cluster by exerting its steric requirements. Clearly, as the ligand size increases, the rates of reactions leading to metal-metal bond formation are reduced, and for a given ligand/metal ratio the higher-nuclearity clusters are destabilized by ligand-ligand repulsion effects. The following examples of cluster-forming reactions in gold cluster chemistry underline the importance of steric effects:[23]

$$[AuCl(PMe_2Ph)] \xrightarrow{\text{reducing agent}} [Au_{11}(PMe_2Ph)_{10}]^{3+}$$

$$[AuNO_3(PPh_3)] \xrightarrow{\text{reducing agent}} [Au_9(PPh_3)_8]^{3+}$$

$$[AuNO_3(PCy_2Ph)] \xrightarrow{\text{reducing agent}} [Au_6(PCy_2Ph)_6]^{2+}$$

The steric influence of the ligand has been estimated using an extension of Tolman's cone angle concept.[24] The "cluster cone angle" is defined in Figure 1.1, and some typical values for common ligands and metal nuclearities are given in Table 1.4. Since the cluster cone angle is defined from the origin of the cluster, the angle subtended by the ligand is dependent not only on the van der Waals radii of the ligand atoms but also on the metal nuclearity. The angle subtended at the center by two adjacent metal atoms imposes a theoretical upper limit on the number of ligands which can be accommodated around the cluster. For example, for a first-row transition metal cluster with a tetrahedral geometry, the theoretical upper limit for the number of carbonyl ligands is about 12. When this limit is exceeded, either a significant isotropic expansion of the metal tetrahedron occurs, e.g., $[Fe_3Cr(CO)_{14}]^{2-}$, or the metal cluster adopts a more open skeletal geometry; for example, in

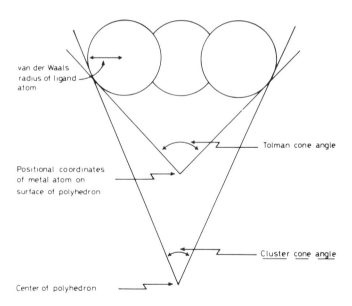

Figure 1.1 Definition of the "cluster cone angle" and comparison with the Tolman cone angle. The ligand dimensions are determined by the van der Waals' radii of the atoms.

TABLE 1.4 CLUSTER CONE ANGLES (°) FOR SOME COMMON METAL-LIGAND FRAGMENTS

	Tetrahedron	Octahedron	Icosahedron
Ideal cone angle	109.5	90.0	63.9
M(η-C$_5$H$_5$)[a]			
M—M = 2.50 Å	97	92	82
M—M = 2.90 Å	90	85	75
M(CO)$_3$[b]			
M—M = 2.50 Å	114	108	96
M—M = 2.90 Å	108	102	90
M(CO)$_2$[b]			
M—M = 2.50 Å	68 (99, 36)[c]	64 (94, 34)	54 (78, 28)
M(CO)[b]			
M—M = 2.50 Å	36	34	30
M—M = 2.90 Å	33	31	28
MCl[d]			
M—M = 2.50 Å	55	51	44
M—M = 2.90 Å	50	47	40

[a] C—C = 1.42 Å, C—H = 1.08 Å, M—C = 2.21 Å (M—M = 2.50 Å), 2.29 Å (M—M = 2.90 Å), van der Waals' radius hydrogen 1.00 Å.

[b] C—O = 1.18 Å, M—C = 1.82 Å (M—M = 2.50 Å), 1.95 Å (M—M = 2.90 Å), C—M = C = 90°, van der Waals' radius O = 1.40 Å.

[c] For the anisotropic M(CO)$_2$ fragment the values in parentheses refer to the extreme cone angles in the plane of the fragment and perpendicular to it. The latter has been assumed to be equal to the cone angle of M(CO). A mean value has also been estimated.

[d] M—Cl = 2.40 Å (M—M = 2.50 Å), 2.50 Å (M—M = 2.90 Å), van der Waals' radius Cl = 1.80 Å.

[Fe$_4$H(CO)$_{13}$]$^-$, the metal tetrahedron opens into a butterfly geometry and the CO takes up an unusual bridging coordination mode.[13] For third-row metal tetrahedra, the limit appears to be about 14 ligands; e.g., Os$_4$(CO)$_{14}$ has been isolated and shown to have a distorted tetrahedral geometry.[13]

Although the inertness of cluster compounds toward dissociation is an important consideration in their isolation, it has other consequences which require consideration. First, the development of catalysts based on metal cluster compounds requires systems where ligand dissociation and addition processes proceed rapidly, and this condition is not usually met in metal carbonyl clusters. Second, one can envisage a mechanism for cluster growth which has as its initial step ligand dissociation. The formation of metal-metal bonds from this unsaturated growth point leads to clusters of relatively low symmetry. Indeed, it results in growth patterns where the observed geometries of the clusters can be described in terms of vertex-, edge-, and face-sharing polyhedra. Some examples of these "condensed clusters" are illustrated in Figure 1.2, and the structural and bonding aspects of this type of cluster

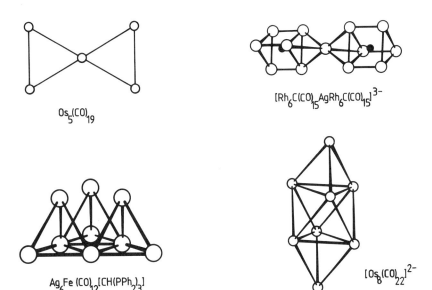

Figure 1.2 Some examples of condensed transition metal clusters.

compound will be discussed in Chapter 7. Of course, these cluster compounds can subsequently lose ligand molecules and undergo skeletal rearrangements to give more compact and spherical clusters.

For the pre- and post-transition elements the trends in chemical lability are quite different, and in general the compounds of heavier elements are more susceptible to nucleophilic or electrophilic attack. Thus it is the lighter members of the group which are kinetically more inert to substitution reactions. Once more, therefore, the kinetic and thermodynamic effects reinforce each other to lead to a more extensive range of cluster compounds for the lighter elements and in particular boron and carbon. The heavier posttransition metal clusters are very susceptible to nucleophilic attack if they are cationic (e.g., Bi_9^{5+}) and electrophilic attack if they are anionic (e.g., Sn_9^{2-}) and therefore have to be handled in systems either of low nucleophilicity or electrophilicity, respectively.[25]

1.4 GENERAL SURVEY OF CLUSTER TYPES

The elements which form either molecular clusters or compounds based on infinite chains of cluster units are indicated on the periodic table shown in Figure 1.3. The range of elements displaying this property is impressive; only the lanthanides and actinides and later main group elements are poorly represented. Figure 1.3 gives a periodic basis for the classification of cluster types.

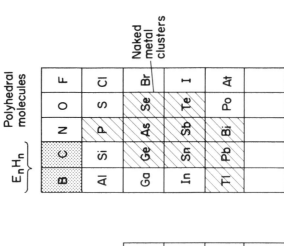

Figure 1.3 Major classes of cluster-forming elements.

1.4.1 Alkali and Alkaline-Earth Metals

Lithium forms a range of compounds where the metal atoms are separated by distances comparable with or shorter than those observed in the bulk metal (3.14 Å). Some examples are illustrated in Figure 1.4.[26] In these compounds the metal-metal bonds are spanned by bridging ligands and it is difficult to separate out unambiguously the contributions from metal-metal and metal-ligand bonds, since the bonding electrons are quite extensively delocalized. The bonding is best described in terms of three-center two-electron bonds and related schemes.

In Figure 1.4a the two lithium atoms are bridged by bicyclo-[1,1,0]butan-1-yl groups and the bonding is satisfactorily described in terms of three-center two-electron Li—C—Li bonds. The presence of the tetramethylethylenediamine (TMEDA) ensures that all four valence orbitals of lithium are used for bonding. The bonding in Figure 1.4a clearly resembles that in Al_2Me_6. In the absence of ligands such as TMEDA, the lithiums relieve their co-ordinative unsaturation by oligomerizing. For example, $[Li_4Me_4]$ is a tetrahedral tetramer (Figure 1.4b) with the methyl groups forming four-center two-electron bonds in the faces. In contrast, $[Li_6(C_6H_{11})_6]$ is a hexamer in the solid state (Figure 1.4c) with the lithium atoms forming a flattened chair and the cyclohexyl ligands bridging three adjacent metal atoms. Similar bonding schemes have been proposed for $[Li(N=CBu^t)]_6$ by Snaith and Wade.[27]

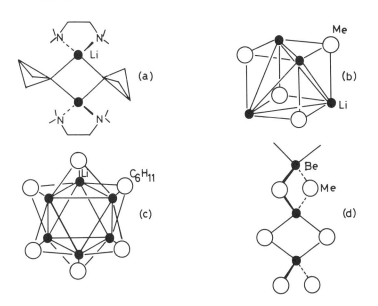

Figure 1.4 Examples of lithium and beryllium organometallic clusters: (a) $Li_2(TMEDA)_2 (C_4H_5)_2$; (b) Li_4Me_4; (c) $Li_6(C_6H_{11})_6$ and (c) $[BeMe_2]_n$.

Beryllium and magnesium form infinite chains of atoms linked by alkyl and aryl ligands (see Figure 1.4d), and the bonding can be described in terms of three-center two-electron bonds. This polymerization provides a satisfactory way of utilizing all the valence orbitals of the central atom. The heavier elements of group IIa and the group IIb metals do not form compounds where metal-metal bonding contributes to a significant degree.

For the heavier alkali metals, cluster compounds based on octahedra have been identified as a result of the elegant work of Simon and his coworkers.[28-29] In particular the "suboxides" Rb_9O_2 and $Cs_{11}O_3$ have been shown to contain centered M_6O octahedra linked by sharing one or two faces, respectively. The structures of these clusters and the packing of the latter in the solid state are illustrated in Figure 1.5. The O—M distances are close to those expected for M^+O^{2-}, and the ionic character of the metal atoms is reflected in short intracluster metal-metal distances (3.54 to 4.03 Å in Rb_9O_2 and 3.67 to 4.31 Å in $Cs_{11}O_3$). The intercluster distances are comparable to those found in metallic rubidium and cesium. Simon has proposed that these species are best formulated as $[Rb_9O_2]^{5+}$ and $[Cs_{11}O_3]^{5+}$ and thereby satisfy the inert gas configurations at the oxygen atoms (see Chapter 2). The remaining five electrons are donated to a conduction band which is both intra- and intercluster metal-metal bonding, and hence these compounds are metallic conductors.

When additional alkali metal atoms are added, we find compounds of the type

$Rb_9O_2 \cdot Rb_3$, $Cs_{11}O_3 \cdot Cs$, $Cs_{11}O_3 \cdot Cs_{10}$, $Cs_{11}O_3 \cdot Rb$, and $Cs_{11}O_3 \cdot Rb_7$,

whose structures can be interpreted in terms of sphere packings if the entire clusters are regarded as spheres.

The exposure of Na^+-zeolite-Y to a low concentration of sodium vapor results in brightly colored samples whose electron spin resonance spectrum suggests the presence of Na_4^{3+} clusters (see the information box on page 171).

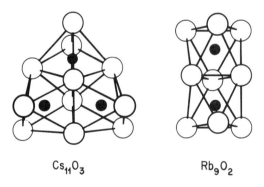

$Cs_{11}O_3$ Rb_9O_2

Figure 1.5

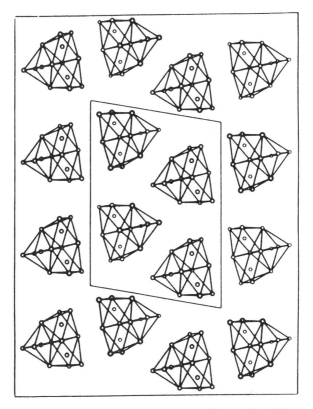

Figure 1.5 (*Contd.*) The structures of $Cs_{11}O_3$ and Rb_9O_2 and a solid state packing diagram for the former. Reproduced with permission from Simon, A., *Angew. Chem., Int. Ed.*, **27**, 159 (1988).

Presumably these highly reactive clusters are kinetically stabilized by being trapped in cavities of the correct dimensions in the zeolite cage.[30] (See also Chapter 4.) The cavities in zeolite structures have been used to stabilize clusters of a wide range of other metals.

1.4.2 Transition Metals

For the transition metals in their bulk states, maximum binding energies are achieved at the center of the transition series corresponding to half filling of the *d* and *s* bands. In molecular clusters, the ligands modify the electronic configuration of the metals in an attempt to emulate this situation. Therefore, for the earlier transition metals, π-donor ligands contribute extra electrons, and for the later transition metals π-acid ligands remove some of the excess electron density. The latter species are described in Section 1.7, and π-donor ligand clusters are discussed below and in Chapter 8.

Cluster compounds of the earlier transition metals are generally associated with ligands such as O^{2-}, S^{2-}, Cl^-, Br^-, I^-, and OR^-, and the metals usually have formal oxidation states of $+2$ or $+3$. The metal skeletal geometries are based on either metal triangles, e.g., $[Re_3Cl_9L_3]$, or octahedra, e.g., $[Mo_6Cl_8L_6]^{4+}$ and $[Ta_6Cl_{12}L_6]^{2+}$ (L represents a suitable donor ligand.) These important structural types are illustrated in Figure 1.6. In $[Mo_6Cl_8L_6]^{4+}$ the chloro bridges are located on the faces of the octahedron and are formally described as five-electron donors, whereas in $[Ta_6Cl_{12}L_6]^{2+}$

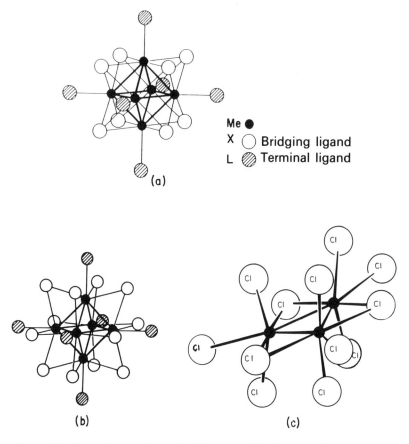

Figure 1.6 The octahedral clusters $M_6X_8L_6$ and $M_6X_{12}L_6$, where X is a π-donor ligand are illustrated in (a) and (b). The structure of $[Re_3Cl_{12}]^{3-}$ is shown in (c).

Synthesis of Halide Clusters

Halide clusters generally require high-temperature solid state reactions. A sloping tube with a temperature gradient allows volatile oxidation products to flow back to the center of the reaction.

(1) **Reduction of high-oxidation halide:** The reducing agents commonly used are P, Al, and $SnCl_2$, which form reasonably volatile oxidation products under the high-temperature conditions used.

Examples	Reagents	
Nb_3Cl_8	$NbCl_5/Nb$	Temperature gradient 700–670°
Ta_6Cl_{15}	$TaCl_5/Al$	Temperature gradient 400–200°
Nb_6Br_{14}	$NbBr_5/Al$	
$[Mo_6Cl_8Cl_4]$	$MoCl_5/Mo$	Heat to form $MoCl_3$, then $MoCl_2$

(2) **Controlled thermal disproportionation:** This reaction frequently follows an initial reduction step, for example:

$$TaCl_5 \xrightarrow{Al} TaCl_4 + TaCl_3 + Ta_6Cl_{14}$$
$$\downarrow$$
$$Ta_6Cl_{14} + Ta$$

There are some direct interconversions achieved in this manner:

$$Nb_3Cl_8 \xrightarrow{heat} Nb_6Cl_{14}$$

$$WCl_4 \xrightarrow[\text{in vacuo}]{heat\ 500°} [W_6Cl_8Cl_4]$$

(3) **Controlled thermal decomposition:** This route is particularly useful for the heavier halides:

$$NbI_5 \xrightarrow[\text{vacuum 300–400°}]{heat} Nb_6I_{14}$$

$$W_6Br_{16} \xrightarrow[\text{vacuum}]{200°} W_6Br_{14} \xrightarrow[\text{vacuum}]{320°} W_6Br_{12}$$

Less general routes to these clusters include the reactions of oxides with halogens and halogen carriers, such as CCl_4 and oxidation of the metal with halogens.

The anionic clusters $[Re_3Cl_{12}]^{3-}$, $[Mo_6Cl_8Cl_6]^{2-}$, and $[Nb_6Cl_{12}Cl_6]^{4-}$ can be obtained from the polymeric halogen bridged clusters and $[NR_4]X$ salts usually in EtOH or from alkali metal salts in acid solutions:

$$[Re_3Cl_9] \xrightarrow[\text{conc HCl}]{CsCl} [Re_3Cl_{12}]^{3-}$$

$$[Mo_6Cl_8]Cl_4 \xrightarrow[\text{NEt}_4\text{Cl}]{EtOH,\ HCl} [Mo_6Cl_8Cl_6]^{4-}$$

$$[Nb_6Cl_{14}]8H_2O \xrightarrow[NEt_4Cl]{EtOH} [Nb_6Cl_{12}Cl_6]^{4-}$$

Neutral ligands can be added in the same manner:

$$[Re_3Cl_9] \xrightarrow[EtOH]{PR_3} [Re_3Cl_9(PR_3)_3]$$

Redox processes on the clusters in aqueous solution can lead to retention of the basic cluster:

$$[Nb_6Cl_{12}]^{2+} \xrightarrow[I_2]{Fe(III)} [Nb_6Cl_{12}]^{4+}$$

$$[Ta_6Cl_{12}]^{2+} \xrightarrow[Hg(II)]{Fe(III)} [Ta_6Cl_{12}]^{3+}$$

the chloro ligands span the 12 edges of the octahedron and function as three-electron donors.[31-35]

The triangular and octahedral metal cluster units give rise to complex three-dimensional networks either by forming halide bridges or by cluster condensation processes based on vertex, edge, and face sharing. Some examples of ligand bridged clusters are illustrated in Figure 1.7. In these examples the ligand bridges replace the terminal ligands on two adjacent clusters by utilizing more than one lone pair. A notation for describing these species is described in Chapter 8. An alternative type of cluster condensation process involving the clusters themselves is illustrated in Figure 1.8.[33]

A particularly interesting series of oligomers based on Mo_3X_3 fragments is illustrated in Figure 1.9. The final infinite polymeric product is $[Mo_3X_3]_n$, which has an infinite chain of face-sharing octahedra.[33] Fragments of the octahedral $[M_6X_8]$ and $[M_6X_{12}]$ clusters have also been observed for the earlier transition metals. For example, $[Mo_5Cl_{13}]^{2-}$ has a square pyramid of metal atoms and $[Mo_4(OPr^i)_8Br_4]$ and $[Mo_4I_{11}]^{2-}$ have butterfly geometries based on the removal of two adjacent vertices from the parent metal clusters (Pr^i is isopropyl). In contrast, $[Mo_4(OPr^i)_8Cl_4]$ has a square-planar arrangement of metal atoms. These clusters are illustrated in Figure 1.10.[36-39]

Metal clusters are a common feature of mixed metal oxides of molybdenum in which the molybdenum is present in a low formal oxidation state. The first reported example was $Zn_2Mo_3O_8$, which contains Mo_3O_{13} units with an Mo—Mo distance of 2.532 Å. More recent examples include $NaMo_4O_6$ with an infinite chain of fused edge-sharing Mo_6 octahedra. In $Mo_{40}O_{62}$, there are finite chains of four and five edge-sharing octahedra, and in $LaMo_5O_8$ pairs of octahedral Mo_{10} units are linked by additional metal-metal bonds.

The octahedral metal cluster unit can also be stabilized by other main group ligands such as S, Se, Te, and Sb, and the research groups of Sergent, Simon, and Schäffer have developed an extensive chemistry based on the

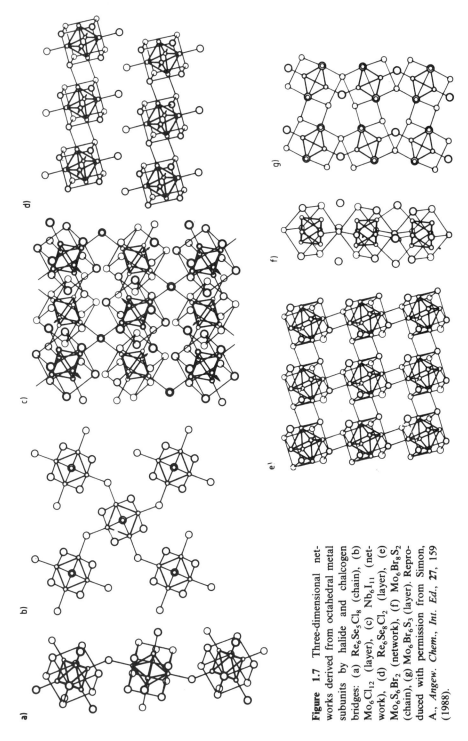

Figure 1.7 Three-dimensional networks derived from octahedral metal subunits by halide and chalcogen bridges: (a) $Re_6Se_5Cl_8$ (chain), (b) Mo_6Cl_{12} (layer), (c) Nb_6I_{11} (network), (d) $Re_6Se_8Cl_2$ (layer), (e) $Mo_6S_6Br_2$ (network), (f) $Mo_6Br_8S_2$ (chain), (g) $Mo_6Br_6S_3$ (layer). Reproduced with permission from Simon, A., *Angew. Chem., Int. Ed.*, **27**, 159 (1988).

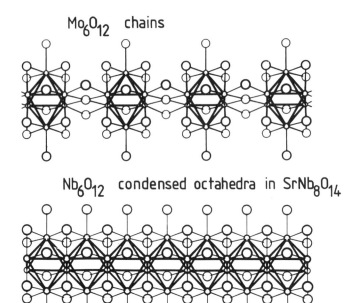

Figure 1.8 Comparison of a chain of condensed Mo_6O_{12} clusters in oxomolybdates (top) and the arrangement of discrete Nb_6O_{12} units in $SrNb_8O_{14}$. Reproduced with permission from Simon, A., *Angew. Chem., Int. Ed.*, **27**, 159 (1988).

linking of these octahedral units. The sulfido ligand is particularly important in this context because it leads to a series of molybdenum clusters with very interesting electrical conductivities and to iron clusters which have important biological implications.

In the Chevrel-Sergent compounds, $[M_6S_8]M'$ ($M' = Pb^{II}$, Cu^{II}, etc.), the central cluster moieties closely resemble those described earlier for $[Mo_6Cl_8]^{4+}$, but are connected in the lattice through metal-sulfur bonds to the countercation M', which is a post-transition metal.[31-34] The structure of one of these Chevrel-Sergent compounds is illustrated in Figure 1.11. In contrast to $[Mo_6Cl_8]^{4+}$, these $[Mo_6S_8]$ clusters do not have closed-shell electronic configurations, and there is unpaired electron spin density. The weak interactions between the paramagnetic cluster ions are mediated by interactions through the counter cations and lead to a band structure which supports metallic conduction. Furthermore, these compounds display superconducting properties at low temperatures and even in the presence of high magnetic fields.[40] The possibility of using the electrical and magnetic properties of clusters in the solid state has only recently been realized and should represent a major area of growth in future years.

Sulfido clusters also play an important role in certain proteins and enzymes. The iron-sulfur proteins constitute a large and important group of

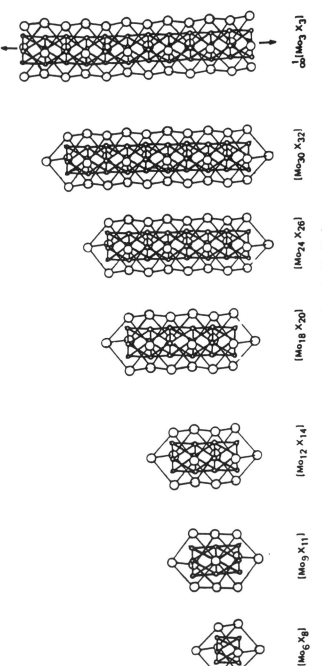

Figure 1.9 Clusters based on stacks of triangular Mo_3X_3 fragments.

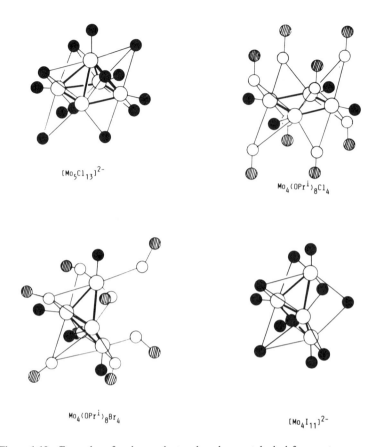

Figure 1.10 Examples of π-donor clusters based on octahedral fragments.

biological materials which are responsible for hydrogen uptake, ATP formation, pyruvate metabolism, nitrogen fixation, and photosynthetic electron transport. The 4Fe and 8Fe ferredoxins contain the Fe_4S_4 "cubane" cluster unit illustrated in Figure 1.12. This unit has been identified in both model systems and biological material by X-ray diffraction studies, and the geometries have been shown to be closely related. In both cases the clusters show significant distortions toward D_{2d} symmetry. In $[Fe_4S_4(SPh)_4]^{2-}$ this takes the form of two long and four shorter bond lengths, but in the reduced species $[Fe_4S_4(SPh)_4]^{3-}$ the Fe_4 core shows a more elongated D_{2d} geometry. Mössbauer and magnetic measurements both suggest an average oxidation state of 2.5 for the former compound. The dimeric anions $[Fe_6Mo_2S_8(SR)_9]^{3-}$ (R = Et, CH_2CH_2OH, or Ph) have the structure illustrated in Figure 1.12 containing two Fe_3MoS_4 "cubane" clusters linked by SR bridges. The molybdenum environment in these molybdenum-iron clusters is very similar to that proposed for this atom in the nitrogenase enzyme on the basis of EXAFS data. Mössbauer and esr studies have indicated that the cluster is best

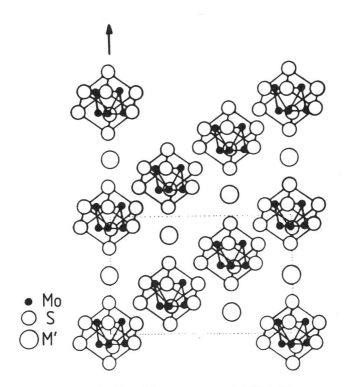

Figure 1.11 Structure of a Chevrel-Sergent compound, M′[Mo$_6$S$_8$].

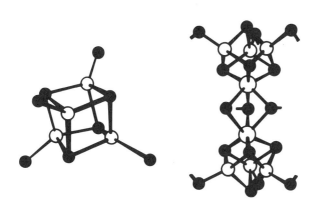

Figure 1.12 The structures of the "thiocubane" clusters [Fe$_4$S$_4$(SR)$_4$]$^{2-}$ and [Fe$_6$Mo$_2$S$_8$ (SR)$_9$]$^{3-}$. The metal atoms are shown as open circles.

Iron-Molybdenum-Sulfur "Cubane" Clusters

The model systems for ferredoxins and nitrogenase are synthesized quite simply by the following routes:

$$4FeCl_3 + \tfrac{1}{2}S_8 + 14LiSR \xrightarrow[25°C]{MeOH} Li_2[Fe_4S_4(SR)_4)] + 5RSSR + 12LiCl$$
black needles

$$3FeCl_3 + [MoS_4]^{2-} + 10NaOMe + 10RSH$$
$$\xrightarrow[25°C]{MeOH} [Fe_6Mo_2(SR)_8]^{3-} + [Fe_6Mo_2S_8(SR)_9]^{3-}$$
black air-sensitive crystals

It has also proved possible to obtain the $[Fe_4(SR)_4]^{2-}$ clusters directly from ferredoxins with solutions of SR^-.

The ion $[Fe_6Mo_2S_8(SR)_9]^{3-}(M^{3-})$ has been shown by cyclic voltammetry to undergo the following electrochemical processes in solution:

$$M^{2-} \longleftarrow M^{3-} \xrightleftharpoons{\text{reversible}} M^{4-} \xrightleftharpoons{\text{reversible}} M^{5-} \longrightarrow M^{6-} \longrightarrow M^{7-}$$

described in terms of a $2Fe^{II}Fe^{III}Mo^{IV}$ model with antiferromagnetic coupling between the unpaired spins on the iron atoms.[41–44]

The transition metals form a very extensive series of cubanelike clusters of the general type $[M_4S_4(\eta\text{-}C_5H_5)_4]$ which have been studied by a combination of spectroscopic and structural techniques.[45] (The notation $\eta\text{-}C_5H_5$ indicates that all the carbon atoms are equidistant from the metal atom.) Dahl has utilized detailed single-crystal X-ray crystallographic studies to trace the effect of electron addition on the metal-metal and metal-ligand bond lengths. These structural data therefore provide circumstantial evidence for the bonding character of particular molecular orbitals.

Recently Fenske et al. have synthesized a number of sulfido and selenido cluster compounds with terminal phosphine ligands. Examples of these compounds and some indication of their structures are given in Table 1.5. The most spectacular compound to arise from the work is $[Ni_{34}Se_{22}(PPh_3)_{10}]$, which has a central Ni_{10} pentagonal antiprism with bridging Se and $Ni(PPh_3)$ groups. This compound and some less complicated examples of such clusters are shown in Figure 1.13.[46–47]

1.5 MULTIPLE BONDING

An important general feature of modern inorganic chemistry has been the control of thermodynamic and kinetic conditions in order to isolate compounds with multiple bonds. In the transition metal series, multiple bonding

Redox Properties of Iron-Sulfur Clusters

The redox behavior of $[Fe_4S_4(SR)_4]^{n-}$ clusters is clearly important for understanding their role in iron-sulfur proteins. Cyclic voltammetry studies summarized here suggest that the hydrophobic environment around the cluster has a marked influence on the redox potentials. All three redox processes, $1-/2-$, $2-/3-$, and $3-/4-$, are observed to occur reversibly for the complexes 1 and 2 in DMSO solution whereas more limited redox behavior has been noted for the less shielded clusters. Steric effects and NH—S hydrogen bonding effects also influence the redox properties. The enclosed clusters (1) to (3) are also more stable toward oxidation by air. (Okuno, Y., et al., *J. Chem. Soc. Chem. Comm.* 1018 (1987).)

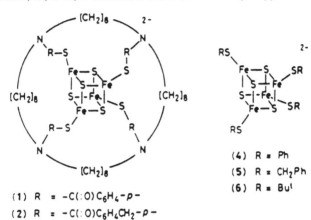

(1) R = $-C(:O)C_6H_4-p-$
(2) R = $-C(:O)C_6H_4CH_2-p-$
(3) R = $-C(:O)CH_2CMe_2-$

(4) R = Ph
(5) R = CH_2Ph
(6) R = Bu^t

REDOX POTENTIALS OF 4Fe–4S CLUSTERS IN DMSO SOLUTION[a]

Compound[b]	$1-/2-$	$E_{1/2}$/V $2-/3-$ $(E_{p,c} - E_{p,a})$, mV	$3-/4-$
(1)	−0.36	−0.85	−1.64
	(36)	(48)	(50)
(4)	—	−0.92	−1.70
	—	(74)	(134)
(2)	−0.35	−1.12	−1.86
	(70)	(20)	(30)
(5)	—	−1.14	—
	—	(70)	—
(3)	+0.25	−1.13	—
	(180)	(60)	—
(6)	−0.11	−1.40	—
	(46)	(70)	—

[a]Potentials vs. S.C.E. Electrochemical data were obtained for a cluster concentration of 1 mM in 0.2 M solutions of $Bu_4^nNBF_4$ using platinum as the working electrode and a saturated calomel electrode (S.C.E.) as the reference.
[b]Et_4N^+ salt.

TABLE 1.5 EXAMPLES OF NICKEL AND COBALT SULFIDO AND SELENIDO CLUSTER COMPOUNDS

	Structure
$Co_4Se_4(PPh_3)_4$	Distorted tetrahedron μ_3-Se ligands Co—Co 2.616–2.683 Å
$Co_6S_8(PPh_3)_6^+$ $Co_6Se_8(PPh_3)_6$	Octahedral cluster with μ_3-S or -Se ligands Co—Co 2.99–3.02 Å
$[Co_7S_6(PPh_3)_5Cl_2]$	Cube with one corner missing Co—Co 2.574–2.637 Å
$[Ni_8S_6(PPh_3)_6Cl_2]$	Highly distorted Ni_8 cube μ_4-S ligands Ni—Ni 2.66–2.70 Å
$[Ni_8S_5(PPh_3)_7]$	Two edge-to-edge coupled trigonal bipyramids
$Co_9Se_{11}(PPh_3)_6$	Face-shared octahedra μ_3- and μ_4-Se Co—Co 2.71–2.98 Å
$[Ni_9S_9(PEt_3)_6]^{2+}$	Face-shared octahedron
$[Ni_{34}Se_{22}(PPh_3)_{10}]$	Pentagonal antiprism which is coordinated by a μ_5-Se ligand at the terminal faces. Between the Ni_5 faces lies an intervening Ni_4 face. Five-connected Ni_8 cubes are thereby formed.

These examples have been taken from references 46 and 47.

is well established for V, Cr, Mo, W, Re, Ru, Os, and Rh, and some idea of the types of multiple bonds which can be formed and the range of bond lengths observed can be estimated from Figure 1.14.[11] A more detailed account of this important area of chemistry is to be found in recent books and reviews by Cotton and Walton.[11] For main group atoms the maximum formal bond order is limited to 3 because of the nodal characteristics of the p orbitals, but for transition metals formal bond orders of 5 can be achieved if the d orbitals are used to form σ, two π, and two δ bonds. Although there is some evidence for the formation of quintuple bonds for metal dimers in the gas phase, in molecular compounds the maximum bond order is generally 4 because one of the d orbitals is used in metal-ligand bonding.

The assignment of a formal bond order in a cluster rests on obtaining internal consistency from the following types of structural and spectroscopic data: bond lengths and conformations from single-crystal studies; ultraviolet photoelectron spectroscopy in combination with quantum mechanical calculations; magnetic anisotropy studies; and force constant data from infrared, Raman, and resonance Raman spectroscopy.

These multiply bonded compounds can form the basis of high-nuclearity clusters by condensation reactions which formally resemble cycloaddition processes. For example, McCarley's compound $[Mo_4Cl_8(PEt_3)_4]$ can be described in terms of the resonance form shown in Figure 1.15a and can thereby be related to the quadruply bonded dimers.[48] Similarly, $[Tc_6Cl_{12}]^{2-}$ can be described as the condensation product of three quadruply bonded units.[49–51] Finally it should be noted that the bonding in $[Re_3Cl_{12}]^{3-}$ can be described in terms of Re—Re double bonds around the metal triangle (Figure 1.15b).[11]

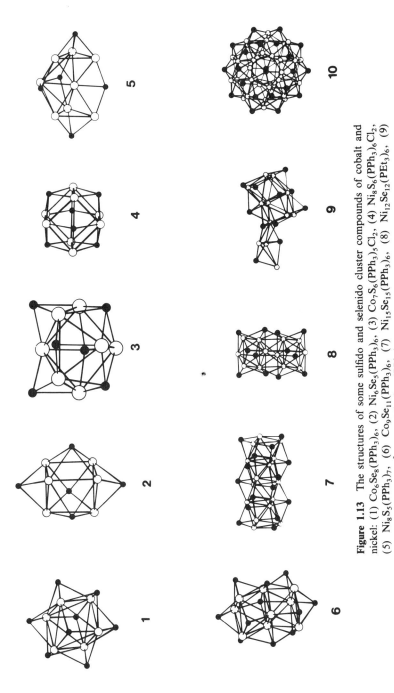

Figure 1.13 The structures of some sulfido and selenido cluster compounds of cobalt and nickel: (1) $Co_6Se_8(PPh_3)_6$, (2) $Ni_6Se_5(PPh_3)_6$, (3) $Co_7S_6(PPh_3)_5Cl_2$, (4) $Ni_8S_6(PPh_3)_6Cl_2$, (5) $Ni_8S_5(PPh_3)_7$, (6) $Co_9Se_{11}(PPh_3)_6$, (7) $Ni_{15}Se_{15}(PPh_3)_6$, (8) $Ni_{12}Se_{12}(PEt_3)_6$, (9) $[Ni_{12}Se_{11}(PPh_3)_8Cl]^{2+}$, (10) $Ni_{34}Se_{22}(PPh_3)_{10}$.

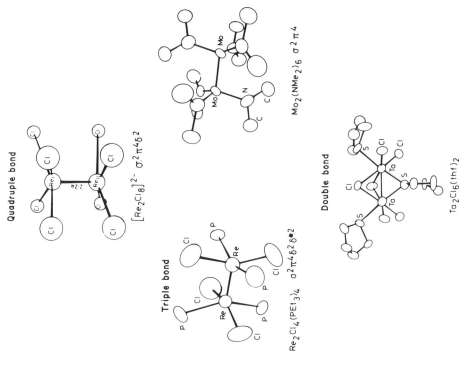

Figure 1.14 Ranges of metal-metal distances (Å) for metal-metal bonds of various bond orders formed by transition metals and some specific examples of compounds with multiple metal-metal bonds. A single dot means that only one such distance is known. The vertical arrows represent twice the Pauling metal bond radius. Adapted from Cotton, F. A., and Walton, R. A., *Multiple Bonds between Metal Atoms*, Wiley, New York, 1982.

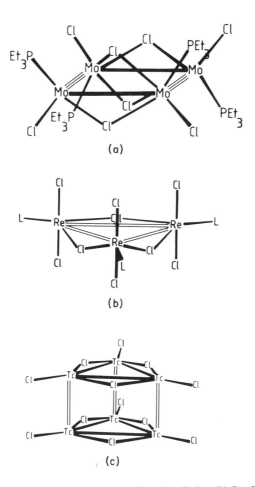

Figure 1.15 Multiple bond descriptions of (a) $Mo_4Cl_8L_4$, (b) $Re_3Cl_9L_3$, and (c) $Tc_6Cl_{12}{}^{2-}$.

1.6 LANTHANIDE AND ACTINIDE CLUSTER CHEMISTRY

The cluster chemistry of the lanthanide and actinide series is still at a very early stage in its development. There are no examples of molecular compounds for these elements, but investigation of the subhalides, notably by Simon and Corbett, has led to the characterization of infinite arrays of metal clusters. Some examples of these clusters are summarized in Table 1.6.[52-55]

TABLE 1.6 SOME EXAMPLES OF LANTHANOID CLUSTERS

Gd_2Cl_3	Parallel chains of trans edge-sharing octahedra with μ-Cl above all free edges (compare the M_6X_8 clusters) shared edges 3.37 Å long remainder 3.90 Å (cf. metal-metal distance of 3.57 Å in bulk)
Gd_2Cl_3N	Stabilization of Gd_2Cl_3 by two interstitial N atoms per octahedron. Gd—Gd very similar to Gd_2Cl_3.
$Gd_{10}Cl_{18}C_4$	Gd_6X_{12} dimers condensed via edge-sharing C_2
$Gd_{10}Cl_{17}C_4$	groups, which are parallel to each other and fill the
$Gd_{10}I_6C_4$	octahedral voids.

1.7 π-ACID CLUSTERS

The growth of the area of π-acid cluster chemistry has been particularly remarkable during the last 20 years and can be attributed in part to the belief that these compounds might provide a new generation of catalysts which would combine the high selectivities of mononuclear organometallic compounds and the high reactivities of metal surfaces.[56–57] The great majority of compounds of this type have carbonyl ligands coordinated to the metal atoms, although examples of compounds with NO, CNR, PR_3, and H ligands are also known. The structural variety shown by these clusters has proved to be truly remarkable, and examples of compounds with up to 50 metal atoms have been characterized by X-ray crystallographic techniques.[58] A systematic survey of the structural types will be given in subsequent sections dealing with the electronic structures of these compounds. Here the compounds will be classified according to ligand type.

1.7.1 Carbonyl Clusters

Carbon monoxide is easily the most important ligand for stabilizing cluster compounds of the later transition metals. Homonuclear cluster compounds are generally limited to the platinum metals, and the cluster chemistries of osmium, rhodium, and platinum are particularly well developed. The compounds can be either neutral or anionic, and the common structural building blocks are metal triangles, tetrahedra, and octahedra. Some of the simpler examples of cluster compounds of this type are given in Tables 1.7 and 1.8.

Interestingly, although the clusters adopt pseudospherical deltahedral geometries when the number of metal atoms is less than 7—i.e., when the structures are tetrahedra, trigonal bipyramids, and octahedra—for the higher-nuclearity examples condensed clusters rather than spherical deltahedra are

TABLE 1.7 EXAMPLES OF METAL CARBONYL CLUSTERS WITH UP TO SIX METAL ATOMS

Compound	Skeletal geometry	Other comments	No. of valence electrons	M—M (Å)
$Fe_3(CO)_{12}$	Triangular	Two μ_2-CO C_{2v} geometry	48	2.69 (×2), 2.51
$Ru_3(CO)_{12}$	Triangular	CO's all terminal D_{3h} geometry	48	2.85
$Os_3(CO)_{12}$	Triangular	D_{3h} geometry	48	2.88
$[Mn_3(CO)_{14}]^-$	Three-atom chain		50	
$Co_4(CO)_{12}$	Tetrahedral	Three μ_2-CO C_{3v} geometry	60	2.44–2.53
$Rh_4(CO)_{12}$	Tetrahedral		60	2.70–2.80
$Ir_4(CO)_{12}$	Tetrahedral	All CO's terminal. T_d geometry	60	2.68
$[Re_4(CO)_{16}]^{2-}$	Butterfly geometry	D_{2h} planar geometry	62	2.96–3.02
$Os_4S(CO)_{13}$	Butterfly geometry	C_{2v} geometry	62	
$Os_4(CO)_{16}$	Square-planar	D_{4h} geometry	64	2.99
$Os_5(CO)_{16}$	Trigonal bipyramid	D_{3h} geometry	72	
$[Ni_5(CO)_{12}]^{2-}$	Trigonal bipyramid	Elongated D_{3h} geometry	76	2.36–2.81
$Os_5S(CO)_{15}$	Square pyramidal	C_{4v} geometry	74	
$Co_6(CO)_{16}$	Octahedral		86	2.78
$Os_6H_2(CO)_{18}$	Capped square pyramidal	(C_s symmetry)	86	
$Os_6(CO)_{18}$	Bicapped tetrahedral	C_{2v} geometry	84	2.80
$Re_6(CO)_{18}(PMe_3)$	Trigonal prismatic		90	

TABLE 1.8 EXAMPLES OF METAL-CARBONYL CLUSTERS WITH MORE THAN SIX METAL ATOMS

Compound	Skeletal geometry	Other comments	No. of valence electrons	M—M (Å)
$Os_7(CO)_{21}$	Capped octahedral		98	2.86
$[Os_8(CO)_{22}]^{2-}$	Bicapped octahedral	*para*-caps	110	
$Pd_8(CO)_8(PMe_3)_7$	Bicapped octahedral	*meta*-caps	110	
$[Rh_9P(CO)_{21}]^{2-}$	Capped square antiprism	Interstitial P	130	
$[Rh_9(CO)_{19}]^{3-}$	Face-sharing octahedra		122	
$[Ni_9(CO)_{18}]^{2-}$	Trigonal prism and octahedron face sharing		128	2.44, 2.70
$[Os_{10}H_4(CO)_{24}]^{2-}$	Tetracapped octahedron		134	
$[Rh_{10}S(CO)_{22}]^{2-}$	Bicapped square antiprism	Interstitial S	142	
$[Rh_{11}(CO)_{23}]^{3-}$	Three face-sharing octahedra		148	
$[Rh_{13}H(CO)_{24}]^{4-}$	Centered cuboctahedron		170	2.80

Synthesis of Metal Carbonyl Clusters

There are four main routes for the synthesis of carbonyl clusters:

(1) **Thermal condensation:** Heating a saturated low-nuclearity carbonyl cluster in a high boiling solvent leads to carbonyl loss and oligomerization. The route is not very specific, and good separation techniques based usually on thin-layer chromatography are required.

$$Os_3(CO)_{12} \xrightarrow{200°} Os_6(CO)_{18}, Os_5(CO)_{16}, Os_6(CO)_{18}, Os_7(CO)_{21}, Os_8(CO)_{23}$$

In this reaction $Os_6(CO)_{18}$ is the major product (80%).

These reactions proceed more efficiently if the carbonyls are replaced by more labile CH_3CN ligands:

$$Os_3(CO)_{11}(CH_3CN) + H_2Os(CO)_4 \xrightarrow{\Delta} H_2Os_4(CO)_{13}$$

The labile complexes are preformed in the following manner:

$$Os_3(CO)_{12} + R_3NO \xrightarrow{CH_3CN} Os_3(CO)_{12-n}(CH_3CN)_n + nCO_2$$

(2) **Redox condensation:** The first example of a redox condensation was reported by Hieber and Schubert in 1965:

$$[Fe_3(CO)_{11}]^{2-} + Fe(CO)_5 \xrightarrow{25°C} [Fe_4(CO)_{13}]^{2-} + 3CO$$

It seems likely that the first step is a redox transfer giving $[Fe_3(CO)_{11}]^-$ and $[Fe(CO)_5]^-$ radicals, which are much more labile to nucleophilic addition and loss and allow CO loss and fragment addition.

These reactions often proceed at ambient temperatures and in high yield. Other examples include:

$$[Rh_5(CO)_{15}]^- + [Rh(CO)_4]^- \xrightarrow{25°\ THF\ 1\ atm\ CO} [Rh_6(CO)_{15}]^{2-} + 4CO$$

$$[Rh_6(CO)_{15}]^{2-} + [Rh(CO)_4]^- \xrightarrow{25°\ MeOH\ 1\ atm\ CO} [Rh_7(CO)_{16}]^{3-}$$

(3) **Base-promoted CO loss:** Base in alcoholic solution can frequently be utilized to cause CO loss and cluster condensation: e.g.,

$$[Rh_4(CO)_{12}] \xrightarrow[PrOH,\ H_2]{80°\ OH^-} [Rh_{13}(CO)_{25}H_3]^{2-}\ (50\%\ yield)$$

$$\xrightarrow[PrOH,\ N_2]{80°\ OH^-} [Rh_{15}(CO)_{27}]^{3-}\ (50\%\ yield)$$

(4) **Oxidative coupling of anionic species:** Oxidative coupling can be achieved by either using inorganic salts, e.g., Fe(III), or protonation, which leads to an unstable hydride species:

$$2[Ni_6(CO)_{12}]^{2-} + 2H^+ \xrightarrow[H_2O]{pH\ 4} [Ni_{12}(CO)_{21}H_2]^{2-} + 3CO\ (90\%\ yield)$$

observed. For example, with seven and eight metal atoms, capped octahedra are preferred to the pentagonal bipyramidal and dodecahedral deltahedral geometries. This behavior contrasts with that observed for the lighter main group elements such as boron and carbon, where the smaller clusters are rather unstable because of bond strain effects and larger icosahedral and cubic molecules are observed. This difference can be attributed to the different nodal characteristics of the primary valence orbitals. The s and p valence orbitals of the lighter main group atoms lead to a preference for bond angles close to the idealized tetrahedral angle, and considerable bond strain therefore arises in tetrahedral units which require bond angles of 60°. For the heavier main group elements, such as phosphorus and silicon, this is less problematical because of the higher proportion of p orbital character in the element-element bonds. *Ab initio* calculations on silicon clusters have also indicated a preference for capped structures rather than simple deltahedra.[59] For the transition metals the doubly noded character of the d orbitals permits the formation of low-nuclearity clusters without introducing any large bond strain effects. Therefore, homonuclear cluster compounds are generally based on triangles, tetrahedra, octahedra, and condensed clusters derived from them. Some examples of these clusters are illustrated in Figure 1.16.

This pattern based on the condensation of cluster fragments leads ultimately to the close-packed structures observed in the bulk metals. In particular the condensation of tetrahedral and octahedral units leads to the familiar cubic-close-packed and hexagonal-close-packed arrangements observed for the great majority of metals. In contrast, the lighter main group elements form structures which permit larger bond angles between adjacent atoms such as the diamond structure for carbon and linked icosahedra for boron.

In Table 1.9 examples of molecular carbonyl clusters which adopt

TABLE 1.9 EXAMPLES OF HIGH-NUCLEARITY CLUSTER COMPOUNDS

Compound	n_i	n_s	Structure
$[Au_9(PPh_3)_8]^+$	1	8	bcc
$[Au_{11}I_3(PPh_3)_7]$	1	10	bcc/icp
$[Au_{13}Cl_2(PMePh_2)_{10}]^{3+}$	1	12	icp
$[Pt_{19}(CO)_{22}]^{4-}$	2	17	icp
$[Rh_{22}(CO)_{35}H_{5-q-m}]^{q-}$	2	20	fcc/bcc
$[Au_{13}Ag_{12}Cl_6(PPh_3)_{12}]^{m+}$	3	22	icp
$[Pt_{26}(CO)_{32}]^{2-}$	3	23	hcp
$[Ni_{38}Pt_6(CO)_{48}H_{6-n}]^{n-}$	6	38	fcc
$[Pt_{38}(CO)_{44}H_m]^{2-}$	6	32	fcc

n_i = number of interstitial atoms

n_s = number of surface atoms

icp = icosahedral or fivefold symmetry packing arrangements

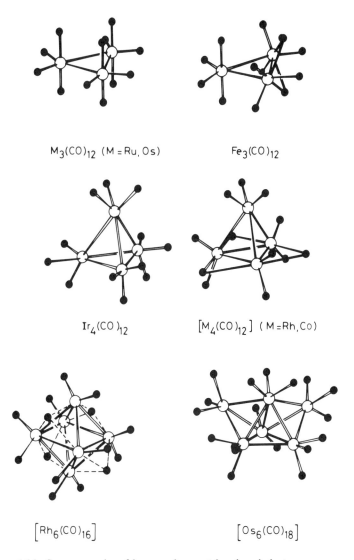

Figure 1.16 Some examples of homonuclear metal carbonyl clusters.

fragments of close-packed metallic structures are summarized. Interestingly, examples of face-centered cubic (fcc) packing, hexagonal close packing (hcp), and body-centered cubic (bcc) packing have been observed, and some examples are illustrated in Figure 1.17. In addition, there are examples of structures which are composites of these symmetrical alternatives. For example, $[Rh_{15}(CO)_{27}]^{3-}$ has a structure which is a composite of bcc and fcc. Furthermore, although it is not possible to generate an infinite structure based on

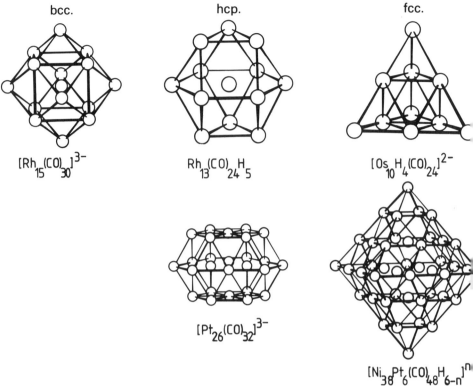

Figure 1.17 Examples of high-nuclearity metal carbonyl clusters which adopt close-packed structures.

fivefold symmetry and with regular translational symmetry, it is possible to get molecular structures with fivefold symmetry and reasonably close-packed arrangements of metal atoms. $[Pt_{19}(CO)_{22}]^{4-}$, which is illustrated in Figure 1.18, provides an example of this type of structure.[60]

Although metal clusters can condense to give pseudo-spherical close-packed structures, they can also condense in columns of triangles or squares.[61] Table 1.10 and Figure 1.19 give some examples of triangles or squares of metal atoms which are stacked either in an eclipsed or a staggered fashion to give columns of face-sharing trigonal prisms, octahedra, or antiprisms.

Studies of carbonyl clusters in solution have indicated that in many instances the carbonyl ligands migrate readily over the surface of the cluster. This is achieved by a concerted movement based on terminal-to-bridging exchange processes. This nonrigidity can be attributed to two important factors. First, the carbonyl ligand functions as a two-electron donating ligand in terminal and edge- and face-bridging situations, and therefore the change in coordination mode does not require the involvement of any additional ligand

Synthesis of Heterometallic Clusters

Until recently the majority of mixed metal clusters were not made by systematically devised syntheses, but rather by placing together relatively labile reagents containing different metal atoms, allowing them to react, and examining the reaction mixtures. The following important general classes of reaction have now been identified:

(1) **Pyrolysis:** Metal carbonyl dimers have been particularly useful precursors for the synthesis of mixed metal clusters, e.g.,

$$Fe_2(CO)_9 + Co(\eta\text{-}C_5H_5)(CO)_2 \xrightarrow[15\ min]{65°C\ petrol} Fe_2Co(\eta\text{-}C_5H_5)(CO)_9$$

(moderate yield).

Pyrolysis of clusters in the presence of monomers, dimers, or other clusters usually requires either more severe conditions or longer reaction times; e.g.,

$$Ru_3(CO)_{12} + Os_3(CO)_{12} \xrightarrow[CO,\ xylene]{175°C,\ 19\ h} Ru_2Os(CO)_{12} + RuOs_2(CO)_{12};$$

$$Ru_3(CO)_{12} + Pt(PMe_2Ph)_4 \xrightarrow[benzene]{25°C,\ 7\ days} Ru_2Pt(CO)_7(PMe_2Ph)_2.$$

(2) **Addition to coordinatively unsaturated molecules:** $Os_3H_2(CO)_{10}$ has hydrido-ligands doubly bridging one of the metal-metal edges and consequently has 46 valence electrons rather than the 48 generally associated with a metal triangle. This coordinative unsaturation has been utilized to add nucleophilic monomeric metal fragments; e.g.,

$$Os_3H_2(CO)_{10} + Pt(C_2H_4)(PPh_3)_2 \longrightarrow Os_3PtH_2(CO)_{10}(PPh_3)_2.$$

Stone and his coworkers* have very elegantly developed a range of less random methods for synthesizing mixed metal clusters based on the coordinative unsaturation of metal-carbyne, metal-carbene, and clusters with multiple metal-metal bonds. Examples are illustrated below.

(3) **Redox condensations:** The reactions of carbonyl metalates, $M(CO)_n^{x-}$, with a neutral metal carbonyl have been described as "redox condensation"

*Angew. Chem., Int. Ed., **23**, 89 (1984).

reactions by Chini and have been widely used for mixed metal cluster synthesis:

$$[W(\eta\text{-}C_5H_5)(CO)_3]^- + Fe_2(CO)_9 \xrightarrow[\text{THF}]{25°C,\ 0.5\ h} [W_2Fe_2(\eta\text{-}C_5H_5)_2(CO)_{10}]^{2-}.$$

Such procedures can be used to build up selectively rather complex mixed metal clusters; e.g.,

$$Fe_2Ru(CO)_{12} + [Co(CO)_4]^- \xrightarrow[\text{THF}]{\text{heat, 2 h}} [CoFe_2Ru(CO)_{13}]^-;$$

$$FeRu_2(CO)_{12} + [Co(CO)_4]^- \xrightarrow[\text{THF}]{\text{heat, 1.25 h}} [CoFeRu_2(CO)_{13}]^-.$$

(4) Addition of electrophilic capping groups: The $AuPPh_3^+$ fragment is readily generated from $Au(PPh_3)Cl$ and $TlPF_6$ in THF (tetrahydrofuran). The presence of an empty valence hybrid in this fragment has led to the suggestion that it can behave rather like H^+. Indeed, it is sufficiently electrophilic to act as a capping group for anionic and neutral clusters; e.g.,

$$Pt_3(CO)_3(PCy_3)_3 + [AuPPh_3]^+ \longrightarrow [Pt_3Au(CO)_3(PCy_3)_3(PPh_3)]^+;$$

$$[Os_{10}C(CO)_{24}]^{2-} + [AuPPh_3]^+ \longrightarrow [Os_{10}C(CO)_{24}AuPPh_3]^-.$$

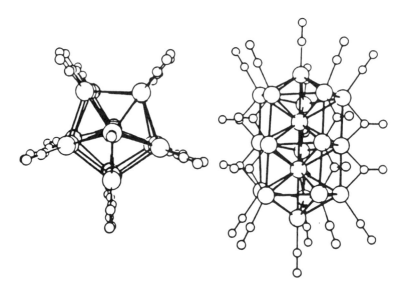

Figure 1.18 Two views of the structure of $Pt_{19}(CO)_{22}^{4-}$ (which has a fivefold rotation axis).

TABLE 1.10 EXAMPLES OF COLUMNAR METAL CARBONYL CLUSTERS

Compound	Description of structure
$[Pt_3(CO)_6]_n^{2-}$	$n = 2$–6, stacks of platinum triangles with a slight helical twist
$[Ni_9(CO)_{18}]^{2-}$	Trigonal prism and octahedron sharing a face
$[Rh_9(CO)_{19}]^{3-}$	Two face-sharing octahedra
$Rh_{12}(CO)_{25}$	Three face-sharing octahedra
$[Rh_{12}C_2(CO)_{24}]^{2-}$	Two square antiprisms sharing a common square face
$[Rh_{17}S_2(CO)_{32}]^{3-}$	Three stacked square antiprisms with interstitial S atoms in outer antiprisms and rhodium in the inner

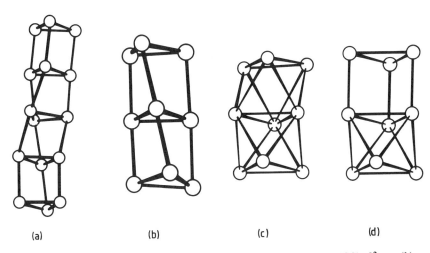

Figure 1.19 Examples of columnar clusters: (a) $[Pt_{15}(CO)_{30}]^{2-}$, (b) $[Pt_9(CO)_{18}]^{2-}$, (c) $[Rh_9(CO)_{19}]^{3-}$, (d) $[Ni_9(CO)_{18}]^{2-}$.

electrons. Contrast the chloride ligand, which changes from a one- (terminal) to a three- (edge-bridging) to a five- (face-bridging) electron donor. Second, the synergic bonding effects which operate in the alternative bridging modes of the carbonyl ligand lead to comparable stabilization energies and therefore a soft potential energy surface for the concerted transformations. The theoretical aspects of these terminal-bridging interchange processes have been discussed by several research workers,[62-64] who have attempted to rationalize why the ground-state structures of related pairs of molecules such as $[Fe_3(CO)_{12}]$ and $[Os_3(CO)_{12}]$, and $[Co_4(CO)_{12}]$ and $[Ir_4(CO)_{12}]$, have very different arrangements of carbonyl ligands (see Figure 1.16).

1.7.2 Carbonyl Clusters with Interstitial Atoms

The radius of the cavity at the center of a metal cluster is determined primarily by the nuclearity and geometry of the cluster. Table 1.11 summarizes the radius of the cavity for some common cluster geometries.[65] For an octahedral cluster the radius of the cavity is 0.414 times the metallic radius of the metal atoms on the surface of the cluster, and therefore only small main group atoms can be incorporated. In particular, H, C, and N atoms have been observed at the centers of octahedral clusters, and some specific examples are given in Table 1.12.[66-67] The calculated ratios of the covalent radius of the interstitial atom to the metallic radius of the surface atoms also given in Table 1.11 indicate that the effective radius of the interstitial atom is somewhat smaller than the covalent radius. Thus for cobalt, rhodium, and iridium clusters the radius ratios are 0.57 to 0.62, but interstitial carbon atoms are observed in octahedral clusters with a cavity radius ratio of 0.414.[68] The reasons for this are discussed in more detail in Chapter 8.

It seems impossible to incorporate nitrogen or carbon atoms into smaller clusters unless the metal adopts a more open skeletal arrangement. $[Os_4N(CO)_{12}]^-$ and $[Fe_4C(CO)_{12}]^{2-}$ provide examples of "butterfly" clusters with the main group atom located in a rather exposed position midway between the wingtip atoms. Similarly, in $[Ru_5C(CO)_{15}]$ the carbon atom is located in the base of a square pyramid of metal atoms (see Figure 1.20). In

TABLE 1.11 CAVITY RADII (IN UNITS OF r) FOR SOME COMMON CLUSTER GEOMETRIES WHERE THE METAL-METAL BOND LENGTHS ARE $2r$

Tetrahedron	0.225
Octahedron	0.414
Trigonal prism	0.528
Capped octahedron	0.592
Square antiprism	0.645
Dodecahedron	0.668
Cube	0.732
Tricapped trigonal prism	0.732
Icosahedron	0.902
Cuboctahedron	1.00
Anticuboctahedron	1.00

TYPICAL RATIOS OF COVALENT RADIUS OF INTERSTITIAL ATOM/METALLIC RADIUS OF SURFACE ATOM

Co—C	0.62	Rh—C	0.58
Rh—C	0.58	Rh—Si	0.87
Ir—C	0.57	Rh—Ge	0.91
		Rh—Sn	1.05

TABLE 1.12 EXAMPLES OF CARBONYL CLUSTERS WITH INTERSTITIAL C, N AND H ATOMS (I)

Compound	Interstitial atom	Description of structure	M—I (Å)
$[Os_{10}C(CO)_{24}H]^-$	H	Tetracapped octahedron with H in tetrahedral cavity	
$Fe_4C(CO)_{13}$	C	Butterfly geometry	1.94, 1.80
$[Fe_4N(CO)_{12}]^-$	N	Butterfly geometry	1.90, 1.77
$Fe_5C(CO)_{15}$	C	Square pyramid	1.89
$[Rh_{13}(CO)_{24}H_{(5-n)}]^{n-}$	H	Square-pyramidal cavity	
$[Fe_5N(CO)_{14}]^-$	N	Square-pyramidal cavity	
$[Fe_6C(CO)_{16}]^{2-}$	C	Octahedral	1.91
$[Ru_6N(CO)_{16}]^-$	N	Octahedral	
$[Co_6H(CO)_{15}]^-$	H	Octahedral	
$[Ni_{12}H(CO)_{21}]^{3-}$	H	Distorted octahedral	
$[Rh_6C(CO)_{15}]^{2-}$	C	Trigonal prism	2.13
$[Co_6N(CO)_{15}]^-$	N	Trigonal prism	1.94
$Re_7C(CO)_{21}$	C	Capped octahedron	2.13
$[Re_8C(CO)_{24}]^{2-}$	C	Bicapped octahedron	2.12
$[Os_{10}C(CO)_{24}]^{2-}$	C	Tetracapped octahedron	2.04
$[Co_8C(CO)_{18}]^{2-}$	C	Square antiprism	1.99
$[Rh_{10}PtN(CO)_{15}]^{3-}$	N	Distorted trigonal bipyramid of PtRh₄	1.92–2.12

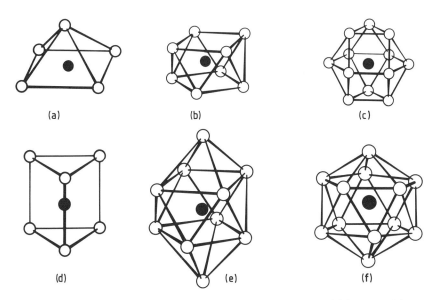

Figure 1.20 Examples of metal carbonyl clusters with interstitial atoms: (a) $Ru_5C(CO)_{15}$, (b) $[Co_8C(CO)_{18}]^{2-}$, (c) $[Rh_{13}H(CO)_{24}]^{4-}$, (d) $[Rh_6C(CO)_{15}]^{2-}$, (e) $[Rh_{10}S(CO)_{22}]^{2-}$, (f) $[Rh_{13}Sb(CO)_{27}]^{3-}$.

Sec. 1.7   π-Acid Clusters

Synthesis of Clusters with Interstitial Atoms

Carbido clusters

$[Fe_5C(CO)_{15}]$ was first obtained in 1962 as an unexpected and very minor side product from reacting $Fe_3(CO)_{12}$ and acetylenes at 90°. Subsequently the great majority of the carbido cluster compounds have been obtained by high-temperature reactions in inert solvents.

Examples

$$[Ru_3(CO)_{12}] \xrightarrow[6h]{Bu_2O\ 142°} [Ru_6C(CO)_{17}] (30\%\ \text{yield})$$

$$[Os_3(CO)_{11}py] \xrightarrow[pyrolysis]{vacuum} [Os_{10}C(CO)_{24}]^{2-} \qquad (py = pyridine)$$

Tracer experiments have suggested that the carbonyls are the source of C in these reactions.

For cobalt and rhodium, halogenated solvents have been used as alternative sources of "C" and generate the clusters in higher yields:

$$[Rh(CO)_4]^- + CCl_4 \xrightarrow[25°]{PrOH} [Rh_6C(CO)_{15}]^{2-} \quad (90\%\ \text{yield})$$

Nitrido clusters

NO and the azide ion (N_3^-) have proved to be the most efficient sources of interstitial nitrogen atoms:

$$[H_3Ru_4(CO)_{12}]^- + NO^+ \longrightarrow [HRu_4N(CO)_{12}]$$

$$Ru_3(CO)_{12} + N_3^- \longrightarrow [Ru_6N(CO)_{16}]^-$$

Phosphido clusters

The reaction of $[Rh(CO)_2(acac)]$ with PPh_3 in the presence of cesium benzoate in tetraethylene glycol dimethyl ether for 4 h at 400 atm CO and 140° to 160° gives $[Rh_9P(CO)_{21}]^{2-}$ in 80% yield.

Arsenido clusters

$[Rh_9As(CO)_{21}]^{2-}$ and $[Rh_{10}As(CO)_{22}]^{3-}$ were prepared similarly but PPh_3 was replaced by $AsPh_3$. The products are separated by fractional crystallization. Improved yields can be obtained by adding $CsBH_4$ to the reaction mixture.

Stibinido clusters

$[Rh_{12}Sb(CO)_{27}]^{3-}$ was synthesized in an analogous fashion, but PPh_3 was replaced by $SbPh_3$.

Sulfido clusters

$[Rh_{10}S(CO)_{22}]^{2-}$ was synthesized from $Rh_4(CO)_{12}$ and SCN^-.

the related [$Os_5C(CO)_{16}$] cluster, the carbido ligand is located at the center of a distorted trigonal bipyramid of metal atoms.[69] It is possible to incorporate the smaller hydrogen atom at the center of a tetrahedral cavity, and in the tetracapped octahedral cluster [$Os_{10}(CO)_{24}HC$]$^-$ the carbido ligand is located at the center of the metal octahedron with the hydrogen atom in a tetrahedral cavity associated with one of the capping osmium atoms.[70]

Higher-nuclearity clusters, such as [$Co_8C(CO)_{18}$]$^{2-}$ and [$Ni_8C(CO)_{16}$]$^{2-}$, are still capable of incorporating first-row atoms, but in addition the cavity is now sufficiently large to incorporate second-row atoms such as Si, P, and S.[71] The calculated radius ratios for Rh—C vs. Rh—Si given in Table 1.11 indicate an increase from about 0.58 to 0.87 when the first-row atom is replaced by a second-row atom. Examples of such cluster compounds are given in Table 1.13. Larger clusters are able to incorporate even larger atoms such as Sb in [$Rh_{12}Sb(CO)_{27}$]$^{3-}$ (calculated Sb/Rh radius ratio = 1.05).[72] Interestingly, although uncentered clusters with 8 to 12 metal atoms do not adopt the deltahedral geometries common to the borane anions, the clusters with interstitial atoms do adopt these geometries, e.g., capped and bicapped square-antiprismatic and icosahedral.

For cluster compounds with 12 or more atoms, the size of the central cavity is sufficiently large to allow the incorporation of another transition metal atom. This can be an identical atom, as in [$Rh_{13}(CO)_{24}H_3$]$^{3-}$, which has the centered anticuboctahedral geometry characteristic of hexagonal close-packed metallic lattices, or another atom, as in [$Rh_{12}Pt(CO)_{24}$]$^{4-}$.[73] Such clusters can therefore also be described as close-packed and have been discussed in some detail above. The radial metal-metal bond lengths are invariably shorter than the tangential metal-metal bond lengths and reinforce the suggestion that the effective radius of an interstitial atom is smaller than predicted from considerations of simple metallic radii.

TABLE 1.13 EXAMPLES OF TRANSITION METAL CARBONYLS WITH INTERSTITIAL MAIN GROUP ATOMS

Compound	Interstitial atom	Description of structure
$Ru_8P(CO)_{19}(\eta\text{-}CH_2C_6H_5)$	P	Square antiprism
[$Co_9Si(CO)_{21}$]$^{2-}$	Si	Capped square antiprism
[$Rh_9P(CO)_{21}$]$^{2-}$	P	
[$Rh_{10}P(CO)_{22}$]$^{3-}$	P	Bicapped square antiprism
[$Rh_{10}As(CO)_{22}$]$^{3-}$	As	
[$Rh_{10}S(CO)_{22}$]$^{2-}$	S	
[$Rh_{17}S_2(CO)_{32}$]$^{3-}$	S	Three square antiprisms sharing faces with S atoms in outer antiprisms
[$Rh_{12}Sb(CO)_{27}$]$^{3-}$	Sb	Icosahedron

The bonding between the interstitial atoms and the peripheral metal atoms (which is discussed in Chapter 8) is of a delocalized nature and is best described within a molecular orbital framework. This delocalized bonding is particularly effective and makes a significant contribution to the stability of the cluster. In general, cluster compounds with interstitial main group atoms are more thermally robust than corresponding clusters with no interstitial atoms.

There are also examples of "ethanido" cluster compounds where a C—C bond is retained with a variable length in the range 1.37 to 1.66 Å depending upon the molecule. Generally, such compounds are observed inside condensed clusters sharing a square face such as trigonal prisms and square antiprisms. Specific examples include $Rh_{12}C_2(CO)_{25}$ (C—C = 1.48 Å) and $Ni_{10}C_2(CO)_{16}^{2-}$ (C—C = 1.41 Å).[74] The formation of clusters with interstitial atoms is not limited to π-acid clusters, and there are may examples of π-donor clusters with interstitial Be, B, C, or N atoms.

1.8 PHOSPHINE CLUSTER COMPOUNDS

There are a large number of phosphine and phosphite cluster compounds which have been synthesized from the parent carbonyl clusters by substitution reactions. In general these compounds are isostructural with the parent compounds. There are very few examples of cluster compounds with only phosphine ligands. Indeed, only gold forms an extensive series of binary cluster compounds of the general types $[Au_m(PR_3)_m]^{x+}$ and $[Au_{m+1}(PR_3)_m]^{y+}$.[23] The latter contain an interstitial gold atom, and two distinct geometric classes of cluster compound have been identified. The spherical clusters have the peripheral atoms lying on the surface of a sphere and they are characterized by a total of $12m + 18$ valence electrons. The second class have the peripheral gold atoms lying on the surface of a torus and are characterized by a total of $12m + 16$ valence electrons. Examples of these gold cluster compounds are given in Table 1.14 and illustrated in Figure 1.21.[23]

There have been some reports of very high-nuclearity gold clusters,[75] such as

$$[Au_{55}(PPh_3)_{12}Cl_6],$$

but there have been no X-ray crystallographic data to confirm the structure. Teo has reported a mixed gold-silver cluster[76]

$$[Au_{13}Ag_{12}Cl_6(PPh_3)_{12}]^{m+}$$

which has a structure based on three icosahedra sharing pentagonal faces, as shown in Figure 1.22. This appears to be one of a more extensive series of silver-gold clusters based on linked icosahedra such as $Au_{18}Ag_{20}Cl_{14}(P(tolyl)_3)_{12}$ and $Au_{22}Ag_{24}Cl_{10}(PPh_3)_{10}$.

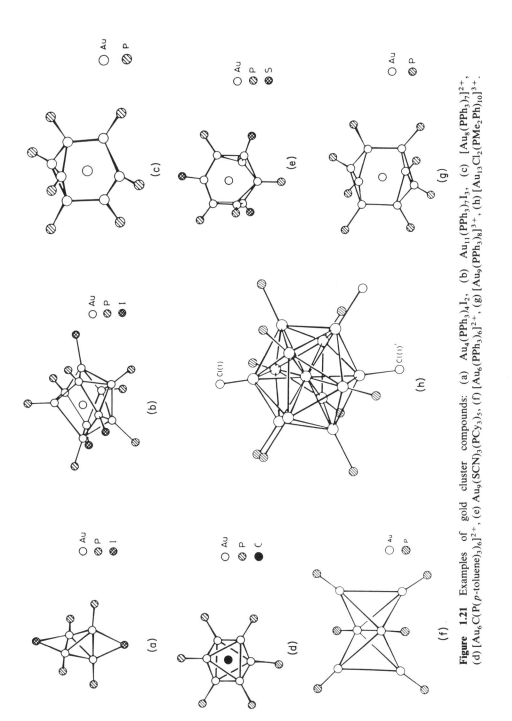

Figure 1.21 Examples of gold cluster compounds: (a) $Au_4(PPh_3)_4I_2$, (b) $Au_{11}(PPh_3)_7I_3$, (c) $[Au_8(PPh_3)_7]^{2+}$, (d) $[Au_6C(P(p\text{-toluene})_3)_6]^{2+}$, (e) $Au_9(SCN)_3(PCy_3)_5$, (f) $[Au_6(PPh_3)_6]^{2+}$, (g) $[Au_9(PPh_3)_8]^{3+}$, (h) $[Au_{13}Cl_2(PMe_2Ph)_{10}]^{3+}$.

TABLE 1.14 EXAMPLES OF GOLD CLUSTER CATIONS

Stoichiometry	Polyhedral geometry
Noncentered polyhedra	
$Au_4(\mu\text{-}I)_2(PPh_3)_4$	Tetrahedral
$[Au_6C(P(p\text{-}C_6H_4CH_3)_3)_6]^{2+}$	Octahedral
$[Au_7(PPh_3)_7]^+$	Pentagonal bipyramid
Condensed noncentered polyhedra	
$[Au_6(PPh_3)_6]^{2+}$	Edge-sharing bitetrahedral
$[Au_6(dppp)_4]^{2+}$	Di-μ-bridged tetrahedral
(dppp = bis(diphenylphosphino)propane)	
Toroidal centered clusters ($12n_s + 16$ electrons)*	
$[Au_8(PPh_3)_7]^{2+}$	Icosahedron missing five vertices
$[Au_9(P(p\text{-}C_6H_4CH_3)_3)_8]^{3+}$	Icosahedron missing four vertices
$[Au_9(P(p\text{-}C_6H_4OMe)_3)_8)]^{3+}$	Centered crown
$[Au_{10}Cl_3(PCy_2Ph)_6]^+$	Ring of six tetrahedra
Spherical centered clusters ($12n_s + 18$ electrons)*	
$[Au_8(PPh_3)_8]^{2+}$	Capped centered chair
$[Au_9(PPh_3)_8]^+$	Bicapped centered chair–cube distorted
$[Au_{11}I_3(P(p\text{-}C_6H_4F)_3)_7]$	Tridecahedron
$[Au_{13}Cl_2(PMe_2Ph)_{10}]^{3+}$	Icosahedron

*n_s = number of surface gold atoms.

Many of the remaining cluster compounds also have hydrido ligands coordinated to the metals; some examples of these clusters are illustrated in Figure 1.23. In some of these compounds it has not proved possible to grow sufficiently large crystals for neutron diffraction studies which would enable the hydrogen atom locations to be determined.[77] These hydrido cluster compounds are discussed in more detail in the subsequent section.

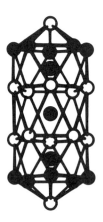

Figure 1.22 The structure of $[Au_{13}Ag_{12}Cl_6(PPh_3)_{12}]$; the gold atoms are shaded. For reasons of clarity the ligands have been omitted from the illustration.

Synthesis of Gold and Platinum Phosphine Clusters

In general, gold compounds have been made by the reduction of mononuclear gold phosphine compounds $AuX(PR_3)$ (X = halide, NO_3^-, etc.). The most commonly used reducing agent is $NaBH_4$, and some examples of its use are given below (Cy is cyclohexyl):

$$AuI(PPh_3) \xrightarrow{NaBH_4} Au_{11}I_3(PPh_3)_7$$

$$Au(NO_3)(PPh_3) \xrightarrow{NaBH_4} Au_9(PPh_3)_8(NO_3)_3$$

$$Au(SCN)(PCy_3) \xrightarrow{NaBH_4} Au_9(SCN)_3(PCy_3)_5$$

Recently B_2H_6 has been used instead to synthesize the very high-nuclearity gold cluster compound $Au_{55}(PPh_3)_{12}Cl_6$, which, although not structurally characterized, has been formulated on the basis of spectroscopic and electron microscopy data.

$Ti(\eta\text{-toluene})_2$ has also proved to be an effective reducing agent and has resulted in the synthesis of clusters which do not have triarylphosphine ligands such as $[Au_{11}(PMe_2Ph)_{10}][(BF_4)_3]$. Other gold clusters have been made by aggregation and degradation reactions:

$$[Au_9(PPh_3)_8]^{3+} + 2PPh_3 \longrightarrow [Au_8(PPh_3)_8]^{2+} + [Au(PPh_3)_2]^+$$

$$[Au_9(PPh_3)_8]^{3+} + I^- \longrightarrow Au_4I_2(PPh_3)_4$$

$$[Au_{11}(PMe_2Ph)_{10}]^{3+} + Cl^- \longrightarrow [Au_{13}Cl_2(PMe_2Ph)_{10}]^{3+}$$

$$[Au_9(PPh_3)_8]^{3+} + Cl^- \longrightarrow [Au_{11}Cl_2(PPh_3)_8]^+$$

Some properties of gold cluster compounds are summarized in Table 1.14.

The platinum hydrido clusters reported by Spencer et al. are derived from the reactions of "$Pt(PR_3)$" fragments generated *in situ* with molecular hydrogen, e.g.,

$$Pt(C_2H_4)_2(PBu_2^tPh) \xrightarrow[15°C,\ petrol,\ 5\ days]{H_2,\ 300\ atm} Pt_5H_8(PBu_2^tPh)_5 \quad \text{(brown crystals)}$$

$$Pt(C_2H_4)(PPr_2^iPh) \xrightarrow[16\ h]{H_2} Pt_4H_8(PPr_2^iPh)_4 \quad \text{(orange crystals)}$$

$$Pt(C_2H_4)_2(PBu_3^t) \xrightarrow[16\ h]{H_2} Pt_3H_6(PBu_3^t)_3 \quad \text{(yellow crystals)}$$

$$Pt_3H_6(PBu_3^t)_3 \xrightarrow[27°C,\ 24\ h]{C_2H_4,\ 1\ atm} Pt_4H_6(PBu_3^t)_4 \quad \text{(dark red crystals)}$$

1.9 HYDRIDO CLUSTER COMPOUNDS

Some examples of hydrido cluster compounds are summarized in Table 1.15. In the great majority of examples the hydrido ligands occupy bridging rather than terminal sites, and therefore the bonding in such compounds has to be discussed using multicentered bonding schemes. In edge-bridged hydrido

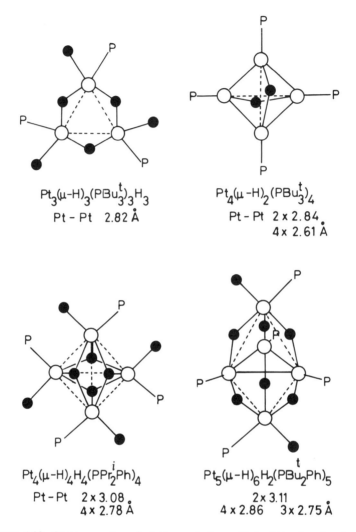

Figure 1.23 Some examples of platinum clusters with hydrido ligands whose precise locations have not yet been elucidated by neutron diffraction. The hydrogens are shown shaded.

clusters, the three-center two-electron bonding scheme illustrated in Figure 1.24 is appropriate and leads to a bonding molecular orbital which is both metal-hydrogen and metal-metal bonding. Photoelectron spectroscopy (see Figure 2.9 in Chapter 2) and molecular-orbital calculations have indicated that this orbital is more localized on the hydrogen atom than the metals[9,78] (again, see Figure 2.9 in Chapter 2). The three-center M—H—M bond can also be formally described as a protonation of a metal-metal bond and

TABLE 1.15   EXAMPLES OF HYDRIDO CLUSTER COMPOUNDS

Compound	Skeletal geometry	Location of hydrido ligands
$Re_3H_3(CO)_{12}$	Triangular	μ_2-
$Os_3H_2(CO)_{10}$	Triangular	$2\mu_2$-over one edge
$Re_4H_4(CO)_{12}$	Tetrahedral	μ_3-face-capping
$Ru_4H_4(CO)_{12}$	Tetrahedral	μ_2-edge-bridging
$Pt_4H_8(PPr^i_2Ph)_4$	Tetrahedral	H atom positions
$Pt_4H_2(PBu^t_3)_4$	Tetrahedral	inferred from
$[Pt_4H_2(PBu^t_3)_4]^{2+}$	Tetrahedral	nmr and X-ray work
$[Pt_4H_7PBu^t_3)_4]^+$	Butterfly	
$Pt_5H_8(PBu^tPh_2)_5$	Trigonal bipyramid	
$[Co_6H(CO)_{15}]^-$	Octahedral	μ_6-interstitial
$Cu_6H_6(PPh_3)_6$	Octahedral	
$Os_6H_2(CO)_{18}$	Octahedral	μ_3-face-capping
$[Os_6H(CO)_{18}]^-$	Octahedral	μ_6-interstitial

represented by the following valence bond canonical form:

$$\begin{array}{c} M \\ | \\ M \end{array} \longrightarrow H^+$$

This description leads to two important generalizations. First, the length of a hydrido-bridged metal-metal bond is longer than that of a comparable unbridged bond. This criterion is often used to locate hydrido ligands in clusters where only X-ray diffraction data are available.[79] Second, since the protonation process involves the formation of a dative bond from an existing metal-metal bond, the total number of valence electrons associated with the

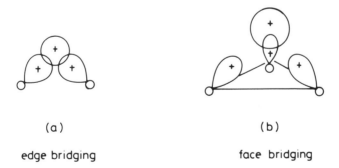

(a)
edge bridging

(b)
face bridging

Figure 1.24 Multicenter bonding schemes for hydrido ligands in clusters.

metal cluster remains unaffected and is identical to that of the parent cluster. For example, [Re₃H₃(CO)₁₂], which has each of the metal-metal bonds bridged by a hydrido ligand, has a total of 48 electrons and is therefore isoelectronic to the triangular [Os₃(CO)₁₂] cluster.

If two hydrido ligands bridge the same edge of the cluster, it is necessary to form two dative bonds to the protons, and formally this can be achieved only if the parent unprotonated cluster is formulated with a double bond. The relevant canonical forms for double and triple hydrido bridges are illustrated in Figure 1.25.

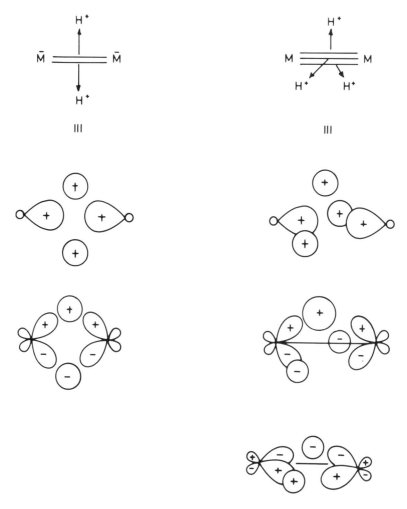

Figure 1.25 Localized bonding descriptions in doubly bridged, e.g., Re₂(CO)₈H₂, and triply bridged, e.g., [(triphos)₂Fe₂H₃]⁺, complexes and their relevant molecular orbital descriptions.

The metal-metal bonds in these multiply bridged compounds are therefore generally shorter than a conventional single metal-metal bond. Furthermore, the total number of valence electrons in the cluster is now related to that in the hypothetical multiply bonded cluster in its unprotonated form. For example, $[Os_3H_2(CO)_{10}]$, which has two hydrido ligands bridging the same edge of the triangle, has a total of 46 electrons corresponding to the doubly bonded hypothetical species $[Os_3(CO)_{10}]^{2-}$. Furthermore, the hydrido-bridged bond is 0.14 Å shorter than the unbridged metal-metal bonds.[80] The same principles apply to the doubly and triply bridged hydrido examples illustrated in Figure 1.25.

The localized view of the bonding presented above has to be modified to account for the occurrence of additional hydrido complexes which exceed the electron counts noted above. For example, $[(P_3)Co(\mu_2\text{-}H)_3Co(P_3)]^+$, where P_3 is a tridentate ligand, has a very similar structure to the corresponding iron compound illustrated above, but has an additional electron pair. Molecular orbital calculations have indicated that this electron pair partially occupies a degenerate pair of E'' symmetry molecular orbitals, which are nonbonding with respect to the bridging hydrogens but metal-metal antibonding.[81] This molecular orbital therefore not only accounts for the longer observed metal-metal bond (2.38 vs. 2.33 Å), but also for the observed paramagnetism.

A similar situation applies to the quadruply bridged hydrido dimer

$$[H_2(PPh_3)_2Re(\mu_2\text{-}H)_4Re(PPh_3)_2H_2],$$

which, if formulated as a protonated quadruple bond, would be expected to have a total of 28 valence electrons, but instead has 30. Molecular orbital calculations have indicated that the additional electron pair resides in an orbital of A_1 symmetry, which is only weakly metal-hydrogen antibonding and more strongly metal-metal antibonding. These examples have been discussed in some detail in order to emphasize that it is not always possible to rationalize the bonding in cluster compounds on the basis of localized bonding models.

In Figure 1.24 the molecular orbital scheme for a face-bridging hydrido cluster is illustrated. The most stable molecular orbital is bonding between the metal atoms and between the metal and hydrogen. The lowest unoccupied molecular orbitals are metal-hydrogen nonbonding and singly noded across the face of the tetrahedron. Therefore, the bonding in this situation can also be described in terms of a dative bond from a three-center metal-metal bonding orbital to a proton. Hence, in general the total number of valence electrons in the face-bridged hydrido clusters corresponds to that of the parent unprotonated cluster. For example, $[Rh_3H(\eta\text{-}C_5H_5)_4]$ has 48 valence electrons, which corresponds to the characteristic electron count for a triangular cluster. The tetrahedral clusters $[Co_4H_4(\eta\text{-}C_5H_5)_4]$ and $[Ru_4H_4(CO)_{12}]$ both have a total of 60 valence electrons, but in the first compound the hydrido ligands are face-bridging and in the second example edge-bridging. Therefore, in these examples the total number of valence electrons does not distinguish the alternative coordination modes. In tetrahedral clusters with 60 valence electrons the bonding can be described in terms of a localized bonding scheme based on two-center two-electron bonds along the edges of the tetrahedron.[82-83] Therefore the bonding in the edge- and face-bridged clusters can be described using the canonical forms illustrated in Figure 1.26. For the face-bridged cluster it is necessary to invoke resonance between the alternative dative bonds from the edges to reproduce the symmetry of the molecule.

$[Re_4H_4(CO)_{12}]$ has a total of only 56 electrons and therefore appears to be anomalous. However, the bonding in the parent hypothetical $[Re_4(CO)_{12}]^{4-}$ cluster can be represented by face-localized three-center metal-metal bonds. Donation of electron density from these face-localized metal-metal bonding orbitals to protons located on the tetrahedral faces gives a satisfactory account not only of the total number of valence electrons but also of the locations of the hydrido ligands.[82]

The conformations of hydrido cluster compounds are determined primarily by the formal electron configurations of the metal atoms. For example, all d^6 hydrido cluster compounds can be described in terms of edge- and face-sharing octahedra. Similarly, d^8 hydrido cluster compounds of rhodium and platinum have square-planar local geometries about the metal atoms. In a quadruply bonded rhenium complex the local ReH_4P_2 geometry is found to be square-antiprismatic.[84] (P_2 represents a bidentate ligand.)

In a previous section the possibility of locating the hydrido ligand in interstitial sites was noted. The related series of compounds $[Os_6H(CO)_{18}]^-$, $[Ru_6H(CO)_{18}]^-$, and $[Ru_6H_2(CO)_{18}]$ is particularly interesting because although the first and last examples have face-bridging hydrides, the middle example has an interstitial hydrogen.[85] Clearly the electronic and steric factors discriminating between the alternative coordination modes must be very subtle.

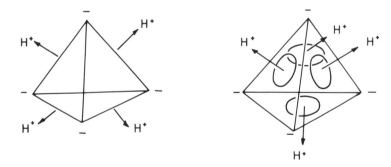

Figure 1.26 Localized bonding schemes for the tetrahedral hydrido clusters $Re_4H_4(CO)_{12}$ and $Ru_4H_4(CO)_{12}$.

1.10 OTHER LIGANDS

Cyclopentadienyl,[86] isocyanide,[87] sulfur dioxide,[88] and nitrosyl[89] metal cluster compounds are known, but there are far fewer examples than for carbon monoxide. For completeness, examples of such compounds are summarized in Table 1.16. In comparison with carbon monoxide ligated clusters, the cluster chemistries of these species have been neglected.

1.11 MAIN GROUP CLUSTER COMPOUNDS

The geometries of main group cluster compounds depend critically on the number of valence electrons available for skeletal bonding. Three-connected polyhedra require three hybrid orbitals each occupied by an unpaired electron to form localized two-center two-electron bonds along the edges of the

TABLE 1.16 EXAMPLES OF CYCLOPENTADIENYL, ISOCYANIDE, NITROSYL, AND SULFUR DIOXIDE CLUSTER COMPOUNDS

Compound	Skeletal geometry	Other comments	No. of valence electrons
Cyclopentadienyl			
$Rh_3(\eta-C_5H_5)_3H$	Triangular	μ_3-H	48
$Co_4H_4(\eta-C_5H_5)_4$	Tetrahedral	μ_3-H	60
$Ni_6(\eta-C_5H_5)_6$	Octahedral		90
Isocyanide			
$Pd_3(CNBu^t)_6$	Triangular		42
$Ni_4(CNBu^t)_7$	Tetrahedral		60
$Pt_7(CNR)_{11}(\mu\text{-}CNR{:}CN)$ R = xylyl	Di-edge-bridged trigonal bipyramid		96
Nitrosyl			
$Mn_3(NO)_4(\eta-C_5H_5)_3$	Triangular	$\mu_3(NO); 3 \times \mu_2(NO)$	48
$Fe_4S_4(NO)_4$	Tetrahedral		60
$[Fe_4S_3(NO)_7]^-$	C_{3v} distorted tetrahedron		66
Sulfur dioxide			
$Pt_3(\mu\text{-}SO_2)_3(PPh_3)_3$	Triangular	μ_2-SO_2	42
$Pt_5(CO)_3(SO_2)_3(PPh_3)_4$	Edge-bridged tetrahedron	μ_2-SO_2	70
$Rh_4(SO_2)_3(CO)_4(P(OPh)_3)_4$	Butterfly	μ_2-SO_2 and μ_3-SO_2	62
$Pd_5(SO_2)_4(PMe_3)_5$	Distorted trigonal bipyramid	μ_2-SO_2; $2 \times \mu_3$-SO_2	72

Synthesis and Physical Properties of Polyhedral Boranes

nido and *arachno* boranes

The higher boranes can be synthesized from B_2H_6, which is conveniently synthesized by the following route:

$$2NaBH_4 + I_2 \xrightarrow{\text{diglyme}} B_2H_6 + 2NaI + H_2$$

$$2B_2H_6 \xrightarrow[\text{5 days}]{\text{store under pressure } 25°} B_4H_{10} + H_2$$

- $250°/H_2$, 3-secs → B_5H_9
- $100\text{–}200°$, Me_2O catalyst → $B_{10}H_{14}$
- $+ B_2H_6$, hot/cold reactor → B_5H_{11}

Properties

Compound	m.p.	b.p.		Compound	m.p.	b.p.
B_2H_6	−165°	−93°	spontaneously inflammable	B_4H_{10}	−120°	18°
B_5H_9	−47	60		B_5H_{11}	−122°	65°
B_6H_{10}	−62	108		B_6H_{12}	−82°	111°
$B_{10}H_{14}$	99	213				

closo-**borane anions**

$$B_2H_6 + 2NaBH_4 \xrightarrow[180°]{NEt_3} Na_2B_{12}H_{12}$$

$$B_2H_6 + 2LiAlH_4 \xrightarrow{\sim 160°} Li_2B_{10}H_{10}$$

The borane anions are white crystalline solids soluble in polar organic solvents. They are thermally very stable (particularly $B_{12}H_{12}^{2-}$ and $B_{10}H_{10}^{2-}$ salts). Stable to nucleophiles but undergo electrophilic substitution reactions.

---

polyhedron. Clusters which satisfy this requirement are described as *electron-precise*.[90] If the vertex atoms have fewer electrons available for skeletal bonding, then it is necessary to utilize a delocalized bonding scheme in order to provide a viable bonding network. If the atoms lie approximately on a spherical surface, then the extent of delocalization is maximized for polyhedra which have triangular faces exclusively, i.e., deltahedra.[91–94] Deltahedra have the maximum number of edges for a given number of vertices, and therefore the electron density in the bonding regions comes under the influence of the largest possible number of nuclei and the extent of delocalization is

Synthesis and Properties of Boron Halide Polyhedra

The polyhedral boron halides are synthesized by the thermal decomposition of B_2Cl_4:

$$B_2Cl_4 \xrightarrow[\text{decomposition}]{\text{thermal}} \underbrace{B_8Cl_8}_{\text{purple}} + \underbrace{B_9Cl_9}_{\text{yellow}} + \underbrace{B_{10}Cl_{10}}_{\text{red}}$$

In contrast to the boron hydride anions they are highly colored; it has not been established whether this has its origins in charge transfer or skeletal excitations. Tetrahedral B_4Cl_4 has been synthesized from BCl_3 by radiofrequency discharge methods.

B_4Cl_4 is sufficiently volatile for its structure to be determined by electron diffraction techniques; B_8Cl_8 and B_9Cl_8 have been characterized by X-ray techniques. The relevant structural data are summarized below:

	B—B (Å)	B—Cl (Å)
B_4Cl_4 (X-ray) tetrahedron	1.69 (4)	1.70 (3)
	1.71 (4)	
B_4Cl_4 (electron diff.)	1.70 (1)	1.69 (1)
B_8Cl_8 dodecahedron	1.78–2.09 (4)	1.68–1.75 (3)
B_9Cl_9 tricapped trigonal prism	1.73–1.81 (1)	1.72–1.74 (1)

The occurrence of n skeletal electron pairs rather than $n + 1$ can be rationalized using the symmetry arguments given in Chapter 5.

---

maximized. Deltahedral main group clusters are therefore described as *electron-deficient*, because they do not have sufficient electrons to form localized bonds along all the edges of the cluster. Nevertheless, this is a misnomer, because the delocalized bonding molecular orbitals which are formed in deltahedral molecules are completely occupied and lead to a stable closed-shell electronic configuration. We will generally refer to such systems as "electron-delocalized."

If the vertex atoms have more electrons than those required to form three-connected polyhedra, then they are described as *electron-rich*.[90,95] The additional electrons must occupy antibonding skeletal molecular orbitals if the three-connected skeletal geometries are retained. This energetically unfavorable situation is avoided either by a symmetrical lengthening of all the bonds in the cluster or by a dramatic lengthening of one of the edges of the polyhedron. The former leads to a reduction of the antibonding character of the delocalized molecular orbitals, but retains the nodal characteristics. The latter corresponds to localization of the antibonding density in one specific bond, and substantial lengthening creates a pair of nonbonding electron pairs across the critical edge. The former distortion mode generally predominates, and consequently the majority of electron-rich clusters have either incomplete skeletal geometries or rings of atoms.

1.11.1 Electron delocalized clusters

The borane anions $B_n H_n^{2-}$ represent the most extensive series of electron-delocalized cluster compounds. They have the deltahedral geometries illustrated in Figure 1.27 and have $2n$ electrons involved in B—H bonding and $2n + 2$ occupying delocalized skeletal molecular orbitals. Since these polyhedra have $3n - 6$ edges, clearly the molecular orbitals cannot be localized onto all the edges of the polyhedron. However, we noted above that deltahedra maximize the number of nearest neighbor contacts within the cluster skeleton and are therefore most effective at delocalizing the electron density.

The heavier main group elements, such as Ge, Sn, Pb, and Bi, provide some examples of "naked" clusters, e.g., Pb_5^{2-}, Sn_5^{2-} (trigonal bipyramidal), and Ge_9^{2-} (tricapped trigonal prismatic). These are described as "naked"

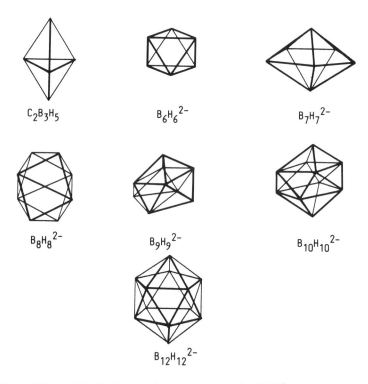

Figure 1.27 Deltahedral borane skeletons observed for $B_n H_n^{2-}$.

Synthesis, Colors, and Bond Lengths of "Naked" Clusters

Cationic clusters
Usually use superacid medium to limit the reactions of clusters with nucleophiles.

$$BiCl_3/AlCl_3 + Bi \xrightarrow[\text{fused 151°}]{\text{NaAlCl}_4} Bi_5(AlCl_4)_3$$
yellow

$$BiCl_3/AlCl_3 + \text{excess Bi} \xrightarrow[\text{fused 151°}]{\text{NaAlCl}_4} Bi_8(AlCl_4)_2 \quad (\text{Bi—Bi 3.10 Å; square antiprism})$$

$$Bi + BiCl_3 \xrightarrow[\substack{\text{cool slowly}\\\text{1–2 weeks 270°}}]{325°} (Bi_9^{5+})_2(BiCl_5^{2-})_4(Bi_2Cl_8^{2-}),$$
$(Bi_{24}Cl_{28}, \text{black})$

(Bi_9^{5+}: tricapped trigonal prism; 3.24 Å prism triangles; 3.74 Å between triangles; 3.10 Å to caps)

$$Te + AsF_5 \xrightarrow{AsF_3} Te_6(AsF_6)_4 \cdot 2AsF_3$$
brown

$$Se_8 + 5SbF_5 \xrightarrow[-23°]{SO_2} Se_8[Sb_2F_{11}]_2 \text{ (green) related compounds } Se_4^{2+} \text{ (yellow)}$$
Te_4^{2+} (bright red)

Anionic clusters
Cryptate polydentate ligand coordinates to the sodium ion and liberates cluster anion from alloy.

$$NaSn_{1.7} \xrightarrow[\text{ethylenediamine}]{\text{crypt}} [Na(\text{crypt})^+]_2[Sn_5^{2-}] \text{ analogous } [Pb_5^{2-}]$$
red   red

$$NaSn_{2.25} \xrightarrow[\text{ethylenediamine}]{\text{crypt}} [Na(\text{crypt})^+]_4[Sn_9]^{4-} \text{ analogous } [Pb_9]^{4-}$$
dark red   emerald green

Trigonal bipyramid M_5^{2-}:

    Axial bonds   2.85 Å (Sn); 3.00 Å (Pb)

    Equatorial bonds   3.10 Å (Sn); 3.24 Å (Pb)

Capped square antiprism Sn_9^{4-}: 2.95–2.01, except inner square (3.10–3.31 Å).

---

because they do not have any ligands bonded to the cluster vertices. All the above deltahedral molecules are described by the prefix *closo*, because the vertex atoms represent a complete set of deltahedral vertices.[25]

Boron also forms two other series of electron-delocalized clusters which are structurally related to the $B_nH_n^{2-}$ series. These compounds have additional hydrogen atoms in terminal and/or edge-bridging positions. The boron skeletons in the boranes can be represented as fragments of the deltahedra described above for $B_nH_n^{2-}$. In B_nH_{n+4} the boron atoms have a skeletal geometry based on a deltahedron but with one vertex (usually that of highest connectivity) missing. Some boranes are illustrated in Figure 1.28. The

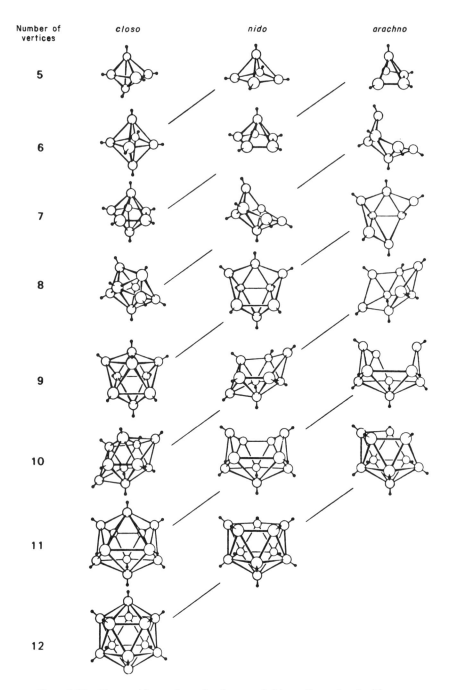

Figure 1.28 *Closo-*, *nido-*, and *arachno-*borane skeletons. Reproduced with permission from Rudolph, R. W., and Pretzer, W. R., *Inorg. Chem.*, **11**, 1974 (1972).

"nestlike" appearance of the boron skeletons leads to their description as *nido* deltahedra.

In B_nH_{n+6} the boron skeletons are related to those of the *closo*-boranes with two additional atoms, and the two missing vertices are generally the highest-connectivity vertex and an adjacent one. These more open skeletal frameworks have the appearance of a spider's web and are consequently described as *arachno* deltahedra. Therefore the boranes *closo*-$B_{n+2}H_{n+2}^{2-}$, *nido*-$B_{n+1}H_{n+5}$, and *arachno*-B_nH_{n+6} share the same deltahedral skeleton, although the latter have either one or two missing vertices. The situation is reminiscent of that in molecular compounds where the number of electron pairs decides the coordination geometry. For example, SF_6, BrF_5, and XeF_4 are all based on the octahedron because they have a total of six valence electron pairs around the central atom.

In these deltahedral clusters the molecules are sufficiently spherical that their wavefunctions are reasonably well described by descent in symmetry from the eigenfunctions of a free electron constrained to move on the surface of a sphere. This problem has a characteristic energy-level spectrum (as we shall see in some detail in following chapters), and the pattern is sufficiently well retained in the related *closo*, *nido*, and *arachno* clusters for them to (usually) share the same number of skeletal electron pairs.

Deltahedra

Deltahedral clusters have exclusively triangular faces, n vertices, $2(n-2)$ faces, and $3(n-2)$ edges. Examples are:

n	Structure	Point group
4	Tetrahedron	T_d
5	Trigonal bipyramid	D_{3h}
6	Octahedron	O_h
7	Pentagonal bipyramid	D_{5h}
8	Dodecahedron	D_{2d}
9	Tricapped trigonal prism	D_{3h}
10	Bicapped square antiprism	D_{4d}
11	Octadecahedron	C_{2v}
12	Icosahedron	I_h

There are also a number of less symmetrical deltahedra, including species with capping atoms.

> **The *closo*, *nido* and *arachno* nomenclature**
>
> *closo*—from the Greek κλωβόζ, a cage, taken here to mean a complete deltahedral molecule.
>
> *nido*—from the Latin *nidus*, a nest, taken here to mean a deltahedral structure with a missing vertex.
>
> *arachno*—from the Greek ἀράχνη, a web, taken here to mean a deltahedral structure with two missing vertices.

1.11.2 Three-Connected Main Group Clusters

The hydrocarbons C_nH_n and C_nR_n have just the right number of valence electrons to form the three-connected species illustrated in Figure 1.29 because the CR fragments can form three localized two-center two-electron bonds along the $3n/2$ edges of the polyhedron.[96] For the lower members of the homologous series the strain energies are considerable because of the large deviation from the idealized tetrahedral bond angles. However, for molecules such as dodecahedrane there is little if any strain. The recent synthesis and

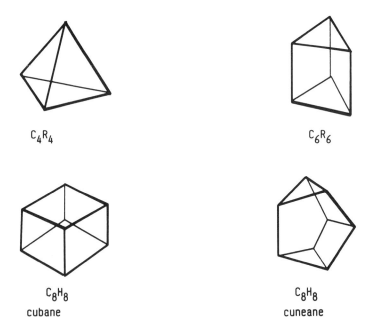

Figure 1.29 Three-connected polyhedral molecules.

Hydrocarbon Polyhedra

The hydrocarbon polyhedra become increasingly stable and inert as the number of carbon atoms increases. For the lower members, replacement of the hydrogen atoms by alkyl groups leads to the isolation of more robust compounds. The high thermal stability of dodecahedrane is particularly noteworthy.

Physical data

Compound			Chemical shifts (δ ppm)		i.r. ν(C—H) (cm^{-1})	C—C (Å)
			^{1}H	^{13}C		
Tetra(t-butyl)tetrahedrane	$C_4Bu^t_4$	Colorless crystals, m.p. 135° (decomp.)	1.18 (Bu^t)	−15 (estimate for C_4H_4)		1.485 (4)
Prismane	C_6H_6	Explosive liquid, decomp. 90°	2.28	30.6	3066	1.500
Hexamethylprismane	C_6Me_6	Violet plate crystals, m.p. 91°				1.585
Cubane	C_8H_8	Rhombohedral crystals, m.p. 130°–131°	4.0	47.3	3000	1.549 (3)
Cuneane	C_8H_8	Liquid, b.p. −1° to 0°	2.7 (2H) 2.3 (2H) 2.1 (4H)		3056	—
Dodecahedrane	$C_{20}H_{20}$	Crystals m.p. >450°	3.38	66.9	2945	1.538 (5) a.v.

Tetrahedral Electron-Precise Molecules

Synthesis of P_4

$$2Ca_3(PO_4)_2 + 6SiO_2 + 10C \xrightarrow{1400°} 6CaSiO_3 + 10CO + P_4$$

White α-P_4 (m.p. 44°, b.p. 281°) converts into hexagonal β form at $-77°$. Insoluble in H_2O, sol. CS_2, PCl_3, $POCl_3$, SO_2, NH_3, C_6H_6. Spontaneously ignites at 35° in air.

	E—E (Å)	D (kJ mol^{-1})
P_4	2.25	209
As_4	2.44	180
Sb_4		142
Si_4^{4-}	2.43	

characterization of dodecahedrane represents a significant tribute to the skill and imagination of organic chemists. For the heavier main group elements, strain energies do not appear to represent a significant influence, because tetrahedral clusters are observed in P_4, As_4, Sb_4, and $Sn_2Bi_2^{2-}$.[97] In principle it should be possible to synthesize analogues of prismane, cubane, and dodecahedrane based on phosphorus, arsenic, and antimony, but this has not been achieved to date.

1.12 CLUSTERS IN MOLECULAR BEAMS

In the previous sections we have described "molecular clusters" which can usually be isolated in sufficient quantities to be studied by X-ray diffraction techniques (if they form single crystals) and in solution using a range of spectroscopic techniques. The study of clusters in the gas phase is a more recent development and can be traced to the pioneering work of Gole, Schumacher, and coworkers. These clusters were originally generated in the gas phase by high-temperature techniques and analyzed using mass spectroscopy. In addition, Knudsen cell methods were used to study the energetics of the vapor equilibrium involving the metal clusters. The study of these molecules with low internal temperatures using modern laser spectroscopy has yielded high-resolution spectroscopic information about their electronic structures. The molecular beam experiments combine a means of producing metal vapors at elevated temperatures with techniques of rarefied gas dynamics to produce an isentropic supersonically expanded "free jet" of "cold" metal clusters which are probed by mass and laser spectroscopy.

Alternatively, metal vaporization in an inert carrier gas stream leads to a "quenching" of the metal vapor. This results in large supersaturation ratios and rapid nucleation and cluster growth. Fluorescence excitation spectra, using tunable dye lasers and multiphoton ionization spectroscopy, have provided particularly useful information concerning cluster stability and electronic structure. For example, it has been established that sodium clusters exhibit distinct "magic numbers" indicating particularly stable nuclearities.[98] Figure 1.30 shows a single continuous mass scan over the range $n = 4$ to 75. Each peak represents the number of clusters of given nuclearity, n, detected during a fixed time interval in a molecular beam of sodium seeded in argon. The peaks corresponding to $n = 8, 20, 40, 58,$ and 92 are particularly large, especially compared with the peaks immediately following. These relatively stable clusters can be understood in terms of a one-electron central potential (electron in a potential well of spherical symmetry) which gives the shell structure shown in the figure. Actually, the same shell structure is seen for a free electron in a spherical well and for a linear combination of atomic orbitals (LCAO) model with coefficients based upon the free-electron

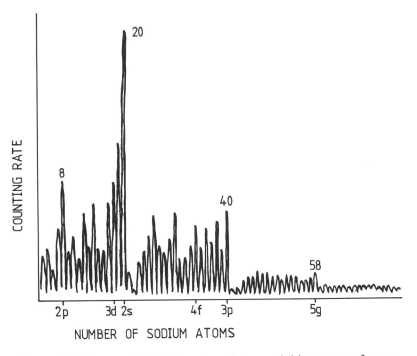

Figure 1.30 Mass spectral data for sodium clusters seeded in a stream of argon with nuclearities in the range 4 to 100. Reproduced with permission from Knight, W. D., et al., *Phys. Rev. Lett.*, **52**, 2141 (1984).

wavefunctions.[99] The order of the shells predicted by a central field calculation for sodium is[98]

$$1s < 2p < 3d < 2s < 3f < 3p < 4g < 4d < 3s < 5h,$$

where $4d$, for example, has two angular nodes and one radial node and is fivefold spatially degenerate in notation analogous to that used for atomic orbitals. When filled in "Aufbau" fashion, this ordering accounts for the observed magic numbers. Similar regularities have been noted in the molecular beams of other alkali metal atoms.

Most recently, supersonic beam techniques have been used to study the reactions of gas-phase clusters with hydrogen, oxygen, and carbon monoxide. These experiments are handicapped by fragmentation phenomena which occur during reactive collisions, electronic transitions, and neutralization and ionization processes. Therefore parent molecules and fragments are no longer easily distinguished.

However, positively and negatively charged clusters may also be generated and studied. The ions are energy-analyzed, mass-separated, and then introduced into an ion drift tube where they are kept at low kinetic energy in a radiofrequency environment. The confined ions are then allowed to react with molecules such as CO. All collision-induced fragmentations or chemical reactions of the confined cluster ions are then analyzed with another mass spectrometer at the exit of the ion drift tube. In this manner clusters of the type $Ni_n(CO)_{12}{}^+$ ($n = 1$ to 13) have been identified and studied.

EXERCISES

1.1 Discuss the differences in structure between the following pairs of compounds:
  (a) $CsMnF_4$ and $CsReCl_4$
  (b) CrO_3 and MoO_3
  (c) $CrCl_2$ and $MoCl_2$
  (d) As_4S_4 and N_4S_4
  (e) $Cr_2Cl_9{}^{3-}$ and $W_2Cl_9{}^{3-}$
  (f) C_2H_4 and $Sn_2(CH(SiMe_3)_2)_4$

1.2 NbO crystallizes in a primitive cubic cell with cell dimensions $a = b = c = 4.192$ Å. The Nb and O atom coordinates are $(0.5, 0.5, 0)$ and $(0, 0, 0.5)$, respectively. Sketch the unit cell and calculate the Nb—O, Nb—Nb and O—O distances. What is the relationship between this structure and those described in this chapter for $[Mo_6Cl_8]^{4+}$ and $[Ta_6Cl_{12}]^{2+}$?

1.3 Discuss the following trends in bond dissociation energies for simple metal-metal dimers in the gas phase:

Cu_2	190	Ag_2	159	Au_2	221
Ni_2	230	Pd_2	105	Pt_2	358
Cr_2	152	Mo_2	404		

(the units are kJ/mol).

1.4 The enthalpy of disruption, ΔH_D, for the process

$$\text{Mo}_2(\text{NMe}_2)_{6(g)} \rightleftharpoons 2\text{Mo}_{(g)} + 6\text{NMe}_{2(g)}$$

is 1929 kJ/mol. A value of 255 kJ/mol has been obtained for $\bar{D}(\text{Mo}—\text{NMe}_2)$ in $\text{Mo}(\text{NMe}_2)_4$, and an estimated value of 190 kJ/mol has been suggested for $\text{Mo}(\text{NMe}_2)_6$. Calculate alternative values of $\bar{D}(\text{Mo}\equiv\text{Mo})$ from these data. Your results should give food for thought regarding estimates of multiple bond strengths.

1.5 Matsumoto, H., et al. (*J. Chem. Soc., Chem. Comm.*, 1083 (1988)) have reported the synthesis of the cage compound Si_8R_8 ($\text{R} = \text{SiMe}_2{}^t\text{Bu}$). It has the following nmr characteristics: ^{29}Si $\delta = -35.03$ and 5.60 (equal intensities) and ^1H $\delta = 0.57$ and 1.28 (with intensity ratio 6:9). Suggest a structure for this molecule and indicate possible isomeric cage structures.

1.6 Von Schnering, H.-G., et al. (*Angew. Chem., Int. Ed.*, **26**, 349 (1987)) have established the following infrared and Raman frequencies for $\text{Si}_4{}^{4-}$ in the solid state: 482 (Raman), 285 (Raman), and 356 (Raman and infrared) all in cm^{-1}. Use standard group theoretical techniques to assign these bands. The authors also report the following force constants: $\text{Si}_4{}^{4-}$ 1.13, P_4 2.07, As_4 1.64, Sb_4 1.20, and Bi_4 0.89 in units of mDyne/Å. Interpret these results in terms of the bonding in these molecules.

1.7 Adatia, T., et al. (*J. Chem. Soc., Chem. Comm.*, 1106 (1988)) have proposed that the cone angles of the phosphines PCy_3 (Cy = cyclohexyl) and $\text{P(CHMe}_2)_3$ influence the structures of copper-ruthenium clusters which they have isolated. Use the Tolman cone angles of these ligands to calculate their cluster cone angles in a tetrahedral and an octahedral cluster with M—M = 2.50 Å. Do you agree with the conclusions of the above authors?

REFERENCES

1. Cotton, F. A., *Q. Rev. Chem. Soc.*, **20**, 389 (1966).
2. Corey, E. R., Dahl, L. F., and Beck, W., *J. Amer. Chem. Soc.*, **85**, 1202 (1963).
3. Chini, P., *J. Organometallic Chem.*, **200**, 37 (1981).
4. Johnson, B. F. G., and Lewis, J., *Adv. Inorg. Chem. Radiochem.*, **24**, 225 (1981).
5. Stone, F. G. A., *Angew. Chem., Int. Ed.*, **23**, 89 (1984).
6. Vahrenkamp, H., *Adv. Organometallic Chem.*, **22**, 169 (1983).
7. Vargas, M. D., and Nicholls, J. N., *Adv. Inorg. Chem. Radiochem.*, **30**, 123 (1987).
8. Mingos, D. M. P., *Acc. Chem. Res.*, **17**, 311 (1984).
9. Green, J. C., Mingos, D. M. P., and Seddon, E. A., *Inorg. Chem.*, **20**, 2595 (1981).
10. Muetterties, E. L., *J. Organometallic Chem.*, **200**, 177 (1980).
11. Cotton, F. A., and Walton, R. A., *Multiple Bonds between Metal Atoms*, Wiley, New York, 1982.
12. Pauling, L., *The Nature of the Chemical Bond*, 3rd ed., Cornell University Press, Ithaca, N.Y., 1980, p. 440.

13. Horwitz, C. P., Holt, E. M., and Shriver, D. F., *Inorg. Chem.*, **23**, 2491 (1984).
    Johnston, V. J., Einstein, F. W. B., and Pomeroy, R. K., *Organometallics*, **7**, 1867 (1988).
14. Woolley, R. G., *Nouv. J. Chim.*, **5**, 219, 227, 441 (1980).
15. Orgel, L. E., *Introduction to Transition Metal Chemistry*, Methuen, London, 1960.
16. Mason, R., and Mingos, D. M. P., *J. Organometallic Chem.*, **50**, 53 (1973).
17. Schugart, K. A., and Fenske, R. F., *J. Amer. Chem. Soc.*, **108**, 5094, 5100 (1989).
18. Kunze, K. L., and Hall, M. B., *J. Amer. Chem. Soc.*, **108**, 5122 (1988).
19. Angerwood, K., Claus, K. H., Goddard, R., and Krüger, C., *Agnew. Chem., Int. Ed.*, **24**, 237 (1985).
20. Connor, J. A., in Johnson, B. F. G. (ed.), *Transition Metal Clusters*, Wiley, New York, 1981, p. 345.
    Connor, J. A., *Topics in Current Chemistry*, **71**, 71 (1977).
21. Pitzer, K. S., *Acc. Chem. Res.*, **12**, 271 (1979).
    Pyykkö, P., and Desclaux, J. P., *Acc. Chem. Res.*, **12**, 276 (1979).
22. Chisholm, M. H., and Rothwell, I. P., *Prog. Inorg. Chem.*, **29**, 1 (1982).
    Poe, A. J., in Moskovits, M. (ed.), *Metal Clusters*, Wiley, New York, 1986, p. 53.
23. Hall, K. P., and Mingos, D. M. P., *Prog. Inorg. Chem.*, **32**, 239 (1983).
24. Mingos, D. M. P., *Inorg. Chem.*, **21**, 466 (1982).
25. Corbett, J. D., *Prog. Inorg. Chem.*, **21**, 129 (1976).
26. Setzer, W. N., and Schleyer, P. von R., *Adv. Organometallic Chem.*, **24**, 353 (1985).
27. Shearer, H. M. M., Snaith, R., Sowerby, J. D., and Wade, K., *J. Chem. Soc., Chem. Comm.*, 1275 (1971).
28. Simon, A., *Struct. Bonding*, **36**, 81 (1979).
29. Simon, A., *Z. Anorg. Allg. Chem.*, **395**, 301 (1973).
30. Gallizot, P., in Moskovits, M. (ed.), *Metal Clusters*, Wiley, New York, 1986, p. 216.
    Minachev, K. L. M., and Isaacov, Y. I., in Rabo, J. A. (ed.), *Zeolite Chemistry and Catalysis*, American Chemical Society, Washington, D.C., 1976, p. 552.
31. Perrin, A., and Sergent, M., *New J. Chem.*, **12**, 337 (1988).
32. Corbett, J. D., *Acc. Chem., Res.*, **14**, 239 (1986).
33. Simon, A., *Angew. Chem., Int. Ed.*, **27**, 159 (1988).
34. Perrin, A., Ihmaine, S., and Sergent, M., *New J. Chem.*, **12**, 321 (1988).
35. Schäffer, H., and von Schnering, H. G., *Angew. Chem., Int. Ed.*, **20**, 833 (1964).
36. Johnston, R. L., and Mingos, D. M. P., *Inorg. Chem.*, **25**, 1661 (1986).
37. Jodden, K., von Schnering, H. G., and Schäffer, H., *Angew. Chem., Int. Ed.*, **14**, 570 (1975).
38. Chisholm, M. H., Errington, R. J., Folting, K., and Huffmann, J. R., *J. Amer. Chem. Soc.*, **104**, 2025 (1982).
39. Stensvad, S., Hellend, B. J., Backich, M. W., Jacobson, R. A., and McCarley, R. E., *J. Amer. Chem. Soc.*, **100**, 6527 (1978).
40. Chevrel, R., Gougeon, P., Potel, M., and Sergent, M. J., *Solid State Chem.*, **57**, 25 (1985).

41. Holm, R. H., *Acc. Chem. Res.*, **10**, 427 (1977).
42. Holm, R. H., and Ibers, J. A., *Science*, **209**, 223 (1980).
43. Christou, G., and Garner, C. D., *J. Chem. Soc., Dalton Trans.*, 2363, 2534 (1980).
44. Holm, R. H., *Chem. Soc. Rev.*, **10**, 455 (1981).
45. Bottomley, F., and Grein, E., *Inorg. Chem.*, **21**, 4170 (1982).
46. Fenske, D., Hachgenei, J., and Ohmer, J., *Angew. Chem., Int. Ed.*, **24**, 706 (1985).
47. Fenske, D., Ohmer, J., and Hachgenei, J., *Angew. Chem., Int. Ed.*, **24**, 993 (1985).
48. McCarley, R. E., *Polyhedron*, **5**, 51 (1986).
49. Wheeler, R. A., and Hoffmann, R., *Angew. Chem., Int. Ed.*, **25**, 822 (1986).
50. Kryuchov, S. V., Kuzina, A. F., and Spitsyn, V. I., *Dokl. Akad. Nauk. SSSR*, **260**, 127 (1982).
51. Chisholm, M. H., Folting, K., Kepert, J. A., Hoffmann, D. M., and Huffman, J. C., *Angew. Chem., Int. Ed.*, **25**, 21, 1014 (1986).
52. Lokken, D. A., and Corbett, J. D., *Inorg. Chem.*, **12**, 556 (1973).
53. Simon, A., Holzer, N., and Mattausch, H., *Z. Anorg. Allg. Chem.*, **456**, 207 (1979).
54. Bullett, D. W., *Inorg. Chem.*, **24**, 3319 (1985).
55. Miller, G. T., Burdett, J. K., Schwartz, C., and Simon, A., *Inorg. Chem.*, **25**, 4437 (1986).
56. Muetterties, E. L., *Bull. Soc. Chim. Belg.*, **84**, 959 (1975).
57. Muetterties, E. L., and Wexler, R. M., *Survey Prog. in Chem.*, **10**, 62 (1983).
58. Ceriotti, A., Demartin, F., Longoni, G., Manaserro, M., Marchionna, M., Pira, G., and Sansoni, M., *Angew. Chem. Int. Ed.*, **24**, 696 (1985).
59. Raghavachari, K., *J. Chem. Phys.*, **84**, 5672 (1986).
60. Kharas, K. C. C., and Dahl, L. F., *Adv. Chem. Phys.*, **70**, 1 (1988).
61. Mingos, D. M. P., and Lin, Z., *J. Organometallic Chem.*, **339**, 367 (1988).
62. Johnson, B. F. G., and Benfield, R. E., in Johnson (ed.), *Transition Metal Clusters*, Wiley, New York, 1980, p. 471.
63. Evans, J., *Adv. Organometallic Chem.*, **16**, 319 (1977).
64. Clarc, B. W., Fairas, M. C., Kepert, D. L., and May, A. S., *Adv. Dynamic Stereochemistry*, **1**, 1 (1985).
65. Wells, A. F., *Structural Inorganic Chemistry*, 5th ed., Oxford University Press, London, 1982, p. 1288.
66. Bradley, J. S., in Moskovits, M. (ed.), *Metal Clusters*, Wiley, New York, 1986, p. 105.
67. Gladfelter, W. L., *Adv. Organometallic Chem.*, **24**, 41 (1985).
68. Ceriotti, A., Longoni, G., Manaserro, M., Perogo, M., and Sansoni, M., *Inorg. Chem.*, **24**, 117 (1985).
    Albano, V. G., Chini, P., Martinengo, S., and Sansoni, S., *J. Chem. Soc., Dalton Trans.*, 463 (1978).
69. Johnson, B. F. G., Lewis, J., Nelson, W. J. H., Perogo, J. N., Raithby, P. R., Rosalis, M. J., Schröder, M. J., and Vargas, M. D., *J. Chem. Soc., Dalton Trans.*, 2447 (1983).

70. Constable, E. C., Johnson, B. F. G., Lewis, J., Pain, G. N., and Taylor, M. J., *J. Chem. Soc., Chem. Comm.*, 754 (1982).
71. Ciani, G., Garlaschelli, K., Sironi, A., and Martinengo, S., *J. Chem. Soc., Chem. Comm.*, 536 (1981).
    Vidal, J. L., Walker, W. E., and Schoening, R. C., *Inorg. Chem.*, **20**, 238 (1981).
72. Vidal, J. L., and Troup, J. M., *J. Organometallic Chem.*, **213**, 351 (1981).
73. Albano, V. G., Ceriotti, A., Chini, P., Martinengo, S., and Anker, W. M., *J. Chem. Soc., Chem. Comm.*, 859 (1975).
74. Halet, J. F., and Mingos, D. M. P., *Organometallics*, **7**, 51 (1988).
75. Schmidt, G., *Struct. and Bonding*, **62**, 51 (1985).
76. Teo, B. K., and Keating, K., *J. Amer. Chem. Soc.*, **106**, 2224 (1984).
77. Gregson, D., Howard, J. A. K., Murray, M., and Spencer, J. L., *J. Chem. Soc., Chem. Comm.*, 716 (1981).
    Frost, P. W., Howard, J. A. K., Spencer, J. L., and Turner, D. G., *J. Chem. Soc., Chem. Comm.*, 1104 (1981).
78. Sherwood, D. E., and Hall, M. B., *Inorg. Chem.*, **21**, 3458 (1982).
79. Churchill, M. R., and Chang, S. W.-Y., *Inorg. Chem.*, **13**, 2413 (1974).
80. Churchill, M. R., Hollander, F. J., and Hutchinson, J. P., *Inorg. Chem.*, **16**, 2697 (1977).
81. Dedieu, A., Albright, T. A., and Hoffmann, R., *J. Amer. Chem. Soc.*, **101**, 3141 (1979).
82. Hoffmann, R., Schilling, B. E. R., Bau, R., Kaesz, H. D., and Mingos, D. M. P., *J. Amer. Chem. Soc.*, **101**, 6088 (1978).
83. Green J. C., Seddon, E. A., and Mingos, D. M. P., *J. Organometallic Chem.*, **185**, C20 (1980).
84. Bau, R., Carroll, W. E., Teller, R. G., and Koetzle, T. F., *Inorg. Chem.*, **23**, 159 (1984).
    Bau, R., Carroll, W. E., Teller, R. G., and Koetzle, T. F., *J. Amer. Chem. Soc.*, **99**, 3872 (1977).
85. Oxton, I. A., Kettle, S. F. A., Jackson, P. F., Johnson, B. F. G., and Lewis, J., *J. Chem Soc., Chem. Comm.*, 687 (1979).
86. Dahl, L. F., and Paquette, M. S., *J. Amer. Chem. Soc.*, **102**, 6621 (1980).
87. Singleton, E., and Oosthuizen, H. E., *Adv. Organometallic Chem.*, **22**, 209 (1983).
88. Mingos, D. M. P., *Transition Metal Chem.*, **3**, 1 (1978).
89. Butler, A. R., Glidewell, C., and Li, M. -H., *Adv. Inorg. Chem.*, **32**, 336 (1988).
90. Mingos, D. M. P., *Nature Phys. Sci.*, **236**, 99 (1972).
91. *Gmelin Handbuch der Anorganischen Chemie*, 8th ed., Springer-Verlag, New York, 1979, Vol. 54, Borverbindungen Teil 20.
92. Scholer, F. R., and Todd, L. J., in Jolly, W. L. (ed.), *Preparative Inorganic Reactions*, Vol. 7, 1971, p. 1.
93. Knoth, W. H., and Muetterties, E. L., *Polyhedral Boranes*, Marcel Dekker, New York, 1968.

94. Wade, K., *Electron Deficient Compounds*, Nelson and Sons, Ltd., London, 1971.
95. Gillespie, R. J., *Chem. Soc. Rev.*, **8**, 315 (1979).
96. Paquette, L. A., *Acc. Chem. Res.*, **4**, 280 (1971).
97. Cisar, A., and Corbett, J. D., *Inorg. Chem.*, **10**, 2482 (1977).
    Corbett, J. D., et al., *J. Amer. Chem. Soc.*, **98**, 7234 (1976).
98. Knight, W. D., Clemenger, K., Saunders, W. A., Chou, M. Y., and Cohen, M. L., *Phys. Rev. Lett.*, **52**, 2141 (1984).
99. Wales, D. J., *Some Theoretical Aspects of Cluster Chemistry*, Ph.D. Thesis, Cambridge University, 1988.

2

Closed-Shell Electronic Requirements for Cluster Compounds

2.1 INTRODUCTION

The survey of cluster compounds in Chapter 1 hinted at the occurrence of relationships between the structures of cluster compounds and the total number of valence electrons present, and we will now develop this relationship more fully. At the beginning of the 1970s, as a result of some inspired pattern recognition and semiempirical molecular orbital calculations, it became apparent that the structures of transition metal and main group clusters could be described in terms of a set of electron counting rules analogous to those developed for simple main group molecules by Sidgwick and Powell and by Nyholm and Gillespie (the valence shell electron pair repulsion theory).[1] The new rules for predicting the electronic closed-shell requirements (i.e., the number of skeletal bonding orbitals) of clusters, evolved from the ideas of Williams,[2] Wade,[3] Mingos,[4] and Rudolph,[5] are now known as the polyhedral skeletal electron pair theory, or PSEPT. This theory relates the number of electron pairs involved in skeletal bonding to the polyhedral geometry. Details of the historical development of these generalizations can be found elsewhere.[6] Here the rules will be summarized and their scope illustrated. The relationships between localized and delocalized descriptions will also be highlighted.

2.2 THE EFFECTIVE ATOMIC NUMBER RULE

Within the valence bond framework the bonding in clusters and ring compounds can sometimes be described in terms of edge-localized two-center two-electron bonds. Through the formation of such element-element bonds, the atoms at the vertices of the ring or polyhedron can achieve an effective inert gas configuration (8 valence electrons for a main group atom and 18 valence electrons for a transition metal atom). In these situations the total number of valence electrons in the cluster, N_e, is given by:

$$N_e = 8n - 2E \quad \text{for main group atoms}$$

and

$$N_e = 18n - 2E \quad \text{for transition metal atoms,}$$

where n is the number of atoms in the cluster skeleton and E is the number of edges of the cluster ring or polyhedron.

Since rings are characterized by n edges, it follows that ring molecules generally have the following electron counts:

$$N_e = 8n - 2n = 6n \quad \text{main group}$$

and

$$N_e = 18n - 2n = 16n \quad \text{transition metal.}$$

Furthermore, three-connected polyhedral molecules have $3n/2$ edges and consequently are characterized by the following electron counts:

$$N_e = 8n - 2 \times \frac{3n}{2} = 5n \quad \text{main group}$$

and

$$N_e = 18n - 2 \times \frac{3n}{2} = 15n \quad \text{transition metal.}$$

These considerations lead to the following three rules of the polyhedral skeletal electron pair theory:

Rule 1

Main group and transition metal ring compounds have a total of 6n and 16n valence electrons, respectively.

Rule 2

Main group and transition metal three-connected cluster compounds have 5n and 15n valence electrons, respectively.

Rule 3

If a transition metal atom occupying a vertex position is replaced by a main group atom, then the characteristic number of valence electrons is reduced by 10.

In main group chemistry, a wide range of homonuclear and heteronuclear ring compounds are known; some examples are illustrated in Figure 2.1.[7] In transition metal chemistry, the known examples are generally limited to three and four metal atoms; some typical examples are illustrated in Figure 2.2.

It is interesting to note that although $Os_3(CO)_{12}$ and $Fe_3(CO)_{12}$ both have triangular metal skeletons and the same number of valence electrons (48), they have rather different arrangements of ligands (see Figure 1.16 in Chapter 1). This occurs because the carbonyl ligand can function as a two-electron donor, irrespective of whether it is behaving as a terminal or an edge-bridging ligand. It is a common feature of metal cluster chemistry that the metal-metal bonding network is supplemented by bridging ligands, and it is important to understand the alternative ligand donor characteristics. Not all ligands donate equal numbers of electrons irrespective of whether they are terminally bound or edge-bridging, as carbon monoxide does. For example,

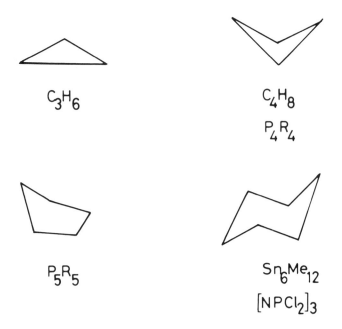

Figure 2.1 Examples of main group ring compounds with $6n$ skeletal electrons.

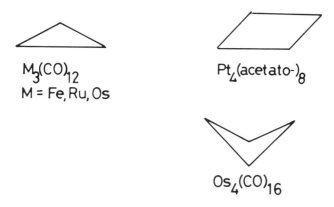

Figure 2.2 Examples of transition metal ring compounds with $16n$ skeletal electrons.

the chloro ligand is a one-electron donor if it is terminally bound, a three-electron donor if it is edge-bridging, and a five-electron donor if it is face-bridging. The donor characteristics of some common ligands found in cluster chemistry are summarized in Table 2.1.

Some examples of three-connected clusters of the main group elements are illustrated in Figure 2.3. Many of the examples shown are derived from organic alicyclic chemistry because the carbyne CR fragment is ideally suited to form three carbon-carbon bonds to adjacent atoms in a three-connected polyhedron. The stabilities of the resulting molecules differ markedly because the symmetry of the skeleton imposes rather different C—C—C bond angles, varying from 60° in tetrahedrane (C_4H_4) to 108° in dodecahedrane ($C_{20}H_{20}$). The latter is close to the idealized sp^3 tetrahedral bond angle, whereas the former results in substantial ring strain. For the heavier main group elements, where the valence s and p orbitals differ significant in energy, sp^3 hybridization is less favorable and the homonuclear element-element bonding can be

TABLE 2.1 DONOR CHARACTERISTICS OF SOME COMMON LIGANDS

Edge-bridging (μ_2)	
One-electron donors	H, CH_3, Ph, $AuPR_3$, SiR_3
Two-electron donors	CO, CS, CNR, CR_2, SO_2, Hg
Three-electron donors	PR_2, SR, OR, NO, Cl, Br, I
Face-bridging (usually μ_3 or μ_4)	
One-electron donors	H, $AuPR_3$
Two-electron donors	CO, CS, $SnCl_2$, Hg
Three-electron donors	NO, CR, P, As, Bi
Four-electron donors	PR, S, O
Five-electron donors	Cl, Br, I, OR

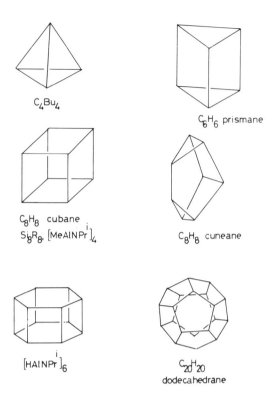

Figure 2.3 Examples of main group three-connected clusters with $5n$ skeletal electrons.

described in terms of p^3 bonding. It follows that in molecules such as P_4 and As_4 the strain energy is a less significant factor.

For transition metal compounds the availability of d orbitals largely eliminates these bond strain considerations, and there are many examples of three-connected transition metal clusters (Figure 2.4).

The tetrahedral and trigonal prismatic clusters are particularly well represented for the later transition metals. One of the clusters in Figure 2.4 contains an interstitial carbon atom, which effectively donates all of its four valence electrons for skeletal bonding. This must not be taken to mean that the carbon has a charge of $+4$, but rather that the carbon $2s$ and $2p$ orbitals overlap strongly with the skeletal molecular orbitals of the metal cage atoms. This generates strongly bonding and antibonding combinations which are delocalized over the carbon atom and the cluster cage. The carbon valence electrons occupy the bonding molecular orbitals and therefore, in a formal sense, donate four electrons to the cage. Hence the interstitial carbon atom acts as an electron donor without producing any additional skeletal bonding orbitals, and is therefore a good remedy for "electron deficiency." Molecular

Figure 2.4 Some three-connected transition metal clusters with $15n$ skeletal electrons. (a) Tetrahedron, 60 valence electrons: $Co_4(CO)_{12}$, $Rh_4(CO)_{12}$. (b) Trigonal prism, 90 electrons: $[Rh_6C(CO)_{15}]^{2-}$; central carbido donates four electrons. (c) Cube, 120 valence electrons: $Ni_8(PPh)_6(CO)_8$; PPh ligands bridge faces and are four-electron donors. (d) "Cuneane," 120 valence electrons: $Co_8S_2(NBu^t)_4(NO)_8$; NBu^t and S act as bridging ligands and both donate four electrons.

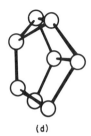

orbital calculations for such clusters indicate that the carbon atom is fairly close to being neutral, emphasizing the validity of the Pauling electroneutrality principle even in these unusual bonding situations. The occurrence of clusters with interstitial atoms is a particularly unique feature of transition metal cluster chemistry, and their donor characteristics are summarized in Table 2.2. Clearly, as the cluster size becomes larger, the size of the interstitial cavity increases and larger atoms can be incorporated within the cluster. For clusters with 12 or more atoms the cavity is sufficiently large for a transition metal atom to be incorporated interstitially; some examples of such clusters are described below.

Rule 3 provides an important link between transition metal and main group cluster compounds, since it makes it possible to predict the electronic requirements of families of structures with varying proportions of transition metal and main group vertex atoms. Two such isostructural series of this type are illustrated in Figure 2.5.

TABLE 2.2 DONOR CHARACTERISTICS OF INTERSTITIAL ATOMS

One-electron donors	H
Two-electron donors	Mg
Three-electron donors	B
Four-electron donors	C, Si, Ge
Five-electron donors	P, As, Bi, Sb
Nine-electron donors	Rh, Co
Ten-electron donors	Pt, Pd
Eleven-electron donors	Au, Ag

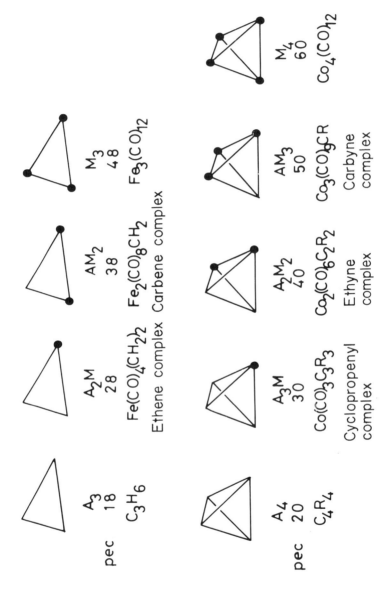

Figure 2.5 Isostructural series of transition metal and main group clusters. Every time a transition metal is replaced by a main group atom, the total valence electron count falls by 10. pec = polyhedral electron count.

These relationships occur because the metal and main group fragments contribute the same numbers of electrons and orbitals for skeletal bonding. Semiempirical molecular orbital calculations by Hoffmann and his coworkers[8] have confirmed that the frontier orbitals of such fragments do indeed have similar nodal characteristics, and they have suggested the term "isolobal" to

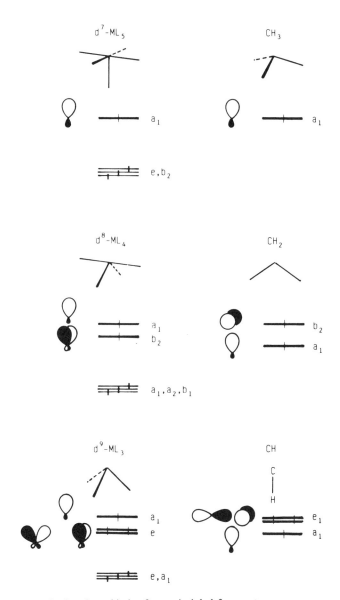

Figure 2.6 The frontier orbitals of some isolobal fragments.

describe this relationship. Figure 2.6 illustrates the frontier orbitals of some isolobal CH_{4-n} and $Mo(CO)_{6-n}$ ($n = 1, 2, 3$) fragments. In general, the bonding capabilities of CH_n fragments are dominated by the requirement that all four valence orbitals should be used for bonding, and if possible correlate with sp^3 hybrids directed toward the vertices of a tetrahedron. For transition metal fragments, d^2sp^3 hybrids pointing toward the vertices of an octahedron form the basis of the isolobal analogy.

For metals with fewer than six electrons in the d manifold, the inert gas rule suggests the utilization of coordination numbers greater than 6. Some possible hybridization patterns are:

d^4 d^3sp^3 7-coordinate, capped octahedron, pentagonal bipyramid, capped trigonal prism

d^2 d^4sp^3 8-coordinate, bicapped trigonal prism, dodecahedron, square antiprism

d^0 d^5sp^3 9-coordinate, tricapped trigonal prism, capped square antiprism

The hybridization schemes relevant to these geometries form the basis of isolobal analogies with CH_{4-n} if up to three hybrids are used for forming cluster molecular orbitals.[9] However, in clusters derived from such fragments, four hybrids may be used for skeletal bonding, and therefore the analogy with CH_{4-n} breaks down. For example, a square-pyramidal ML_5 fragment can have four d^5sp^3 hybrids directed toward the missing vertices of a capped square antiprism (the ligands define the other five vertices). For a d^4 metal ion each of these hybrids will contain a single electron which can be used to form four metal-metal bonds to neighboring metal atoms. In order to form four hybrids of this type, the following atomic orbitals are required:

    one σ-type (s, p_z, d_{z^2})
    two π-type ($p_x, p_y; d_{xz}, d_{yz}$)
    one δ-type ($d_{xy}, d_{x^2-y^2}$)

where the local z axis points radially outward. The σ, π, δ classification refers to the number of nodal planes the atomic orbital contains which also contain the radius vector of the atom from the center of the cluster (see Chapter 3). The local z axis at each vertex is defined to point radially outward from the center of the cluster. We shall make much use of this terminology in subsequent chapters. A hybridization scheme of this type can be used to describe the bonding in halide clusters of the earlier transition metals, as we shall see in Section 2.5.

2.3 LOCALIZED AND DELOCALIZED REPRESENTATIONS

For an Os(CO)$_4$ fragment based on an octahedron, the frontier orbitals can be represented either as a pair of hybrid orbitals (Figure 2.7a) or as a set of a_1 and b_2 molecular orbitals (Figure 2.7b). The two representations are entirely equivalent; the latter merely represent a "symmetrized" representation of the hybrids in Figure 2.7b, which takes into account the C_{2v} symmetry of the fragment.

In a similar manner the localized bonds in rings and three-connected clusters can be employed as the basis set for generating the delocalized cluster bonding orbitals. For example, in Os$_3$(CO)$_{12}$, which belongs to the D_{3h} point group, the localized metal-metal bonds form a basis for the A'_1 and E' irreducible representations. The edges of the triangle form a basis for the following representation of D_{3h}:

D_{3h}	E	$2C_3$	$3C'_2$	σ_h	$2S_3$	$3\sigma_v$
Γ_{edge}	3	0	1	3	0	1

and hence $\Gamma_{\text{edge}} \cong A'_1 \oplus E'$. ($\cong$ and $\oplus$ are used for "equality" and "addition" in representation algebra.) It is apparent from Figure 2.8 that these are

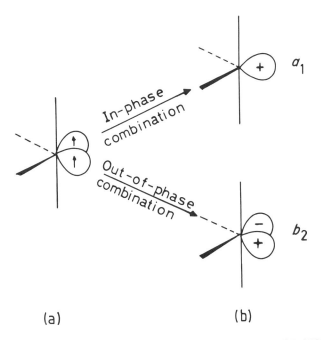

(a)   (b)

Figure 2.7 Two equivalent representations of the frontier orbitals of Fe(CO)$_4$. (a) Hybrid orbitals, (b) symmetry adapted orbitals.

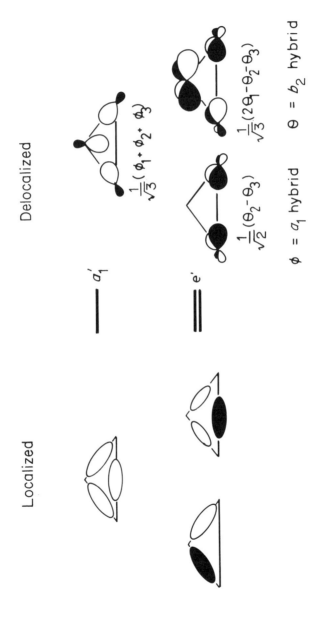

Figure 2.8 Localized and delocalized representations of the skeletal bonding orbitals for $Os_3(CO)_{12}$.

equivalent to the bonding skeletal molecular orbitals derived from the a_1 and b_2 frontier orbitals of the $Os(CO)_4$ fragments.

Ionization of electrons from these skeletal molecular orbitals is clearly discernible in the photoelectron spectrum of $Os_3(CO)_{12}$, illustrated in Figure 2.9,[10-11] because electrons in these orbitals are the least tightly bound. The bands at higher energies are associated with the t_{2g}-type orbitals of the octahedral vertex fragments and the molecular orbitals associated with the ligands. The photoelectron spectrum of $Re_3H_3(CO)_{12}$ is also shown in this figure and shows how the formation of three-center M—H—M bonds shifts the energies of the a_1' and e' molecular orbitals.

Similarly, the edge-localized bonds of a three-connected polyhedron can also form a basis for the bonding skeletal molecular orbitals. For example, in

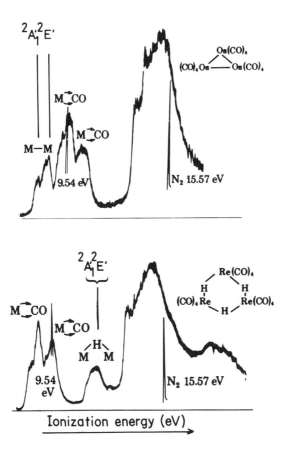

Figure 2.9 The photoelectron spectra of $Os_3(CO)_{12}$ and $Re_3H_3(CO)_{12}$. Note the shift in the metal-metal bonding orbital energies on forming three-center M—H—M bonds. The bands labeled M⇌CO correspond to the "t_{2g}" orbitals of the metal atoms.

prismane, C_6H_6, the nine edges represent the localized two-center two-electron bonds and give rise to the following irreducible representations in the D_{3h} point group: A_1', A_2'', E', A_1', E', and E''. Some of the corresponding molecular orbitals are both face- and edge-bonding, and are more stable than those molecular orbitals which are edge-bonding but face-antibonding (see Figure 2.10).[12-13] In group theoretical terms the former transform in the same way as the faces of the polyhedron. For example, for prismane the five faces transform according to the irreducible representations $2A_1'$, A_2'', and E'. The remaining e' and e'' molecular orbitals are less strongly bonding. The results of a molecular orbital calculation on prismane, shown in Figure 2.10, underline the distinction between the two classes of molecular orbital. The strongly bonding and less strongly bonding edge-based molecular orbitals for other three-connected molecules are summarized in Table 2.3. In each case the molecular orbitals can be derived from group theoretical considerations by using the edges and faces of the polyhedron as bases for the irreducible representations:[14]

D_{3h}	E	$2C_3$	$3C_2'$	σ_h	$2S_3$	$3\sigma_v$
Γ_{edge}	9	0	1	3	0	3
Γ_{face}	5	2	1	3	0	3

so that $\Gamma_{edge} \cong 2A_1' \oplus A_2'' \oplus 2E' \oplus E''$ and $\Gamma_{face} \cong 2A_1' \oplus A_2'' \oplus E'$.

Rules 1 and 2 are applicable to any cluster where the bonding can be described adequately in terms of localized two-center two-electron bonds. The transformation between localized and delocalized representations is only valid, however, if the connectivities of the vertex atoms do not exceed the number of valence orbitals that are available for skeletal bonding. B—H and related

TABLE 2.3 SYMMETRIES OF BONDING MOLECULAR ORBITALS IN THREE-CONNECTED CLUSTERS*

Cluster	Vertices	Strongly bonding	Weakly bonding
Tetrahedron (T_d)	4	A_1, T_2	E
Trigonal prism (D_{3h})	6	$2A_1'$, A_2'', E'	E'', E'
Cube (O_h)	8	A_{1g}, T_{1u}, E_g	T_{2u}, T_{2g}
Pentagonal prism (D_{5h})	10	$2A_1'$, A_2'', E_1', E_2'	E_1', E_2', E_1'', E_2''
Hexagonal prism (D_{6h})	12	$2A_{1g}$, A_{2u}, E_{1u}, E_{2g}, B_{1g}	B_{2g}, E_{1u}, E_{2g}, B_{2u}, E_{1g}, E_{2u}
Truncated tetrahedron (T_d)	12	$2A_1$, $2T_2$	$2E$, T_1, T_2
Dodecahedron (I_h)	20	A_g, T_{1u}, T_{1g}, T_{2u}	G_u, H_g, G_g, H_u

*The strongly bonding orbitals have the same symmetries as are contained in the representation spanned by the faces of the cluster. The symmetries of the weakly bonding orbitals are those contained in the representation spanned by the $3n/2$ edges minus those which have already been counted in the face-generated set.

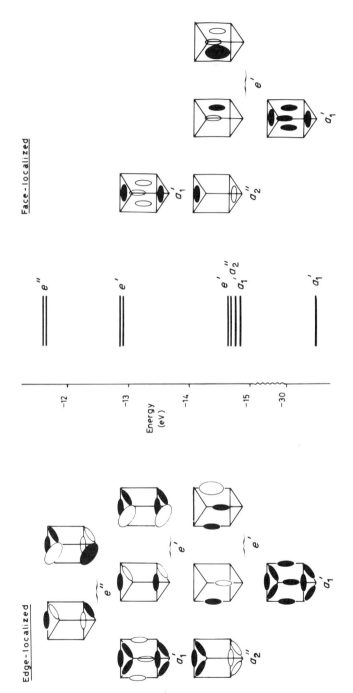

Figure 2.10 The skeletal molecular orbitals of prismane.

main group fragments and conical $M(CO)_3$ fragments based on the octahedron have only three orbitals available for skeletal bonding. Consequently, only three-connected polyhedra can be described in this localized fashion. Higher-connectivity clusters have to be described using delocalized bonding schemes. A powerful theoretical methodology appropriate for analyzing these systems will be discussed in Chapter 4; the relevant closed-shell requirements for these clusters are summarized below.

Rule 4

Four-connected clusters are characterized by a total of $4n + 2$ (main group) or $14n + 2$ (transition metal) valence electrons, respectively, so long as the n vertex atoms lie approximately on a spherical surface.[15]

Some examples of four-connected transition metal clusters are illustrated in Figure 2.11. Although $B_8H_8^{2-}$ has a dodecahedral geometry in the solid state, there is some evidence from nmr studies that it can have a square-antiprismatic four-connected structure or a bicapped trigonal-prismatic structure in solution.[16] The octahedral geometry has been observed for both

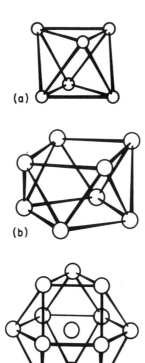

Figure 2.11 Examples of four-connected transition metal clusters. (a) Octahedron, 86 valence electrons: $Rh_6(CO)_{16}$. (b) Square antiprism, 114 valence electrons: $[Co_8C(CO)_{18}]^{2-}$. (c) Anticuboctahedron, 170 valence electrons: $[Rh_{13}(CO)_{24}H_3]^{2-}$. Interstitial rhodium donates nine electrons.

Ru$_6$C(CO)$_{17}$, which has an interstitial carbido ligand, and Rh$_6$(CO)$_{16}$, which has not—both molecules have a total of 86 valence electrons. The square-antiprismatic geometry has been established for [Co$_8$C(CO)$_{18}$]$^{2-}$, and the anticuboctahedral geometry for [Rh$_{13}$H$_3$(CO)$_{24}$]$^{2-}$, which has an interstitial rhodium atom. In the latter example the 4d, 5s, and 5p valence orbitals of the interstitial rhodium are involved in bonding, and this atom is therefore considered to donate formally nine valence electrons to the skeletal molecular orbitals.

In these four-connected clusters the topological connection between the number of bonding molecular orbitals and the numbers of edges and faces of the polyhedron is lost. A four-connected polyhedron with n vertices has $n + 2$ faces and $2n$ edges but $n + 1$ skeletal bonding molecular orbitals.

Rule 5

closo, nido, and *arachno deltahedral molecules based on n main group vertex atoms are characterized by* $4n + 2$, $4n + 4$, *and* $4n + 6$ *valence electrons, respectively. This result is sometimes known as the debor principle. The corresponding isostructural transition metal cluster compounds are characterized by* $14n + 2$, $14n + 4$, *and* $14n + 6$ *valence electrons, respectively.*

closo deltahedral molecules have exclusively triangular faces, and the atoms lie approximately on a single spherical surface. The most complete series of deltahedral main group molecules is B$_n$H$_n{}^{2-}$ ($n = 6$ to 12); these structures are illustrated in Figure 2.12. The corresponding *nido* structures are identical except that one of the vertices is missing. The *arachno* cluster is also structurally related to a *closo* parent geometry, but has two vertices missing.

Bond Lengths in Polyhedral Boranes and Carboranes

The boron-boron bond lengths given in Figure 2.12 were obtained from single-crystal X-ray studies. The bond lengths to capping atoms are generally shorter than those involving the other boron atoms, and in general the bond lengths reflect the connectivities. The higher the average connectivity, the longer the bond length.

Carbon has a smaller covalent radius and therefore B—C distances are about 0.05 Å shorter than comparable B—B distances. They show the same trends with coordination number noted above. In the more stable carborane isomers the carbons occupy the sites of lowest connectivity and adopt positions as far apart from each other as possible, i.e., maximize the number of B—C bonds.

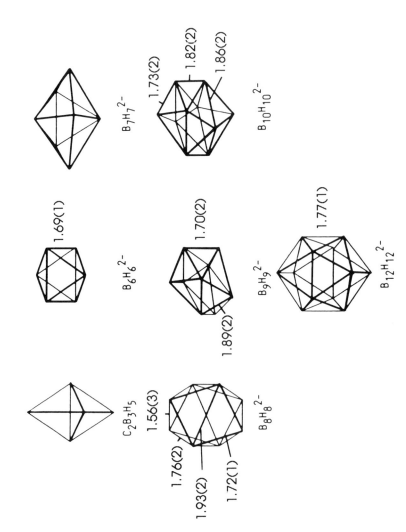

Figure 2.12 The boron skeletons of the $B_nH_n^{2-}$ deltahedra with bond lengths in Å indicated.

Examples of boranes which conform to this generalization were illustrated in Figure 1.28 in Chapter 1. *closo*, *nido*, and *arachno* molecules have the general formulas $B_nH_n^{2-}$, B_nH_{n+4}, and B_nH_{n+6}, respectively, where we assume that all the boron atoms are roughly disposed on the surface of a single sphere. One of the highest-connectivity vertices is generally lost in the notional *closo* to *nido* transformation, and an additional adjacent vertex is lost in the *nido* to *arachno* transformation. The additional hydrogen atoms are bonded around the open faces created in the *closo* to *nido* to *arachno* transformations either as bridging ligands or as B—H terminal bonds.[17-18] The locations of bridging hydrogens in *nido* and *arachno* boranes are indicated in Figure 2.13.

Examples of *closo*, *nido*, and *arachno* transition metal carbonyl clusters are illustrated in Figures 2.14 and 2.15. Note that a number of *arachno* transition metal clusters have structures where non-adjacent vertices (often trans-vertices) are lost.

The *closo*, *nido*, and *arachno* relationships described above are reminiscent of the structural pattern which forms the central feature of the valence shell electron pair repulsion theory for main group molecules.[1] *closo*, *nido*, and *arachno* molecules share the same deltahedral template and also have the same number of skeletal electron pairs. This can be more easily appreciated if reference is made to the hypothetical deprotonated molecules $B_nH_n^{4-}$ (*nido*) and $B_nH_n^{6-}$ (*arachno*). The series of molecules $B_nH_n^{2-}$, $B_{n-1}H_{n-1}^{4-}$, and $B_{n-2}H_{n-2}^{6-}$ all have $n+1$ skeletal electron pairs if two-center two-electron bonds are assigned to the B—H terminal bonds. Therefore, the spherical nature of these clusters leads to the occurrence of $n+1$ bonding skeletal molecular orbitals even when the atoms do not define a complete spherical shell. The quantum mechanical basis of this pattern is discussed in Chapter 5. Examples of isostructural *closo*, *nido*, and *arachno* cluster compounds containing varying combinations of transition metal and main group atoms are illustrated in Figure 2.16. The molecules adopt the same skeletal geometries because the number of bonding skeletal molecular orbitals remains constant throughout each series.

In addition to the borane and hydrocarbon clusters described above there is an extensive series of "naked" clusters of the post-transition metals. Examples of these clusters are given in Table 2.4.[19] These clusters, with one exception, conform to the rules developed above, indicating that in these molecules there are lone-pair orbitals which point out from the center of the cluster and therefore emulate the B—H or C—H bonds in $B_nH_n^{2-}$ and C_nH_n. The ion Bi_9^{5+} does not conform to the generalizations, since it has a *closo*-tricapped trigonal prismatic geometry rather than the *nido* structure anticipated for a cluster with $4n+4$ valence electrons. The long intertriangular Bi—Bi bond lengths in Bi_9^{5+} suggest that this exception to the rules arises because there is a weakly antibonding skeletal molecular orbital which is occupied by the additional electron pair.[20] A similar distortion has been noted for Te_6^{4+}, which has a related elongated trigonal prismatic structure.

Sec. 2.3 Localized and Delocalized Representations

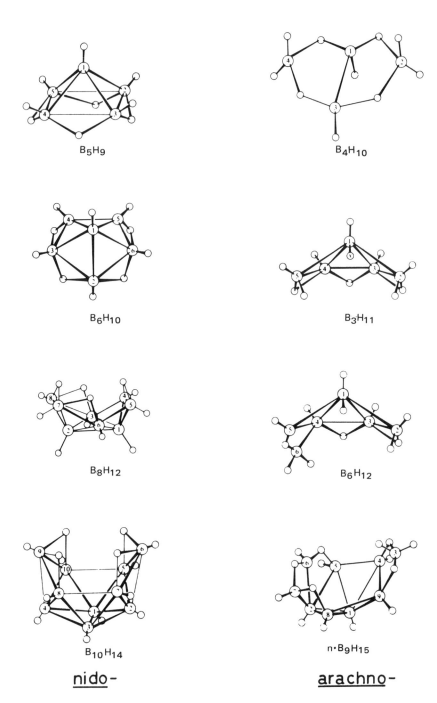

Figure 2.13 Structural relationships in *nido* and *arachno* boranes, and the locations of the bridging protons.

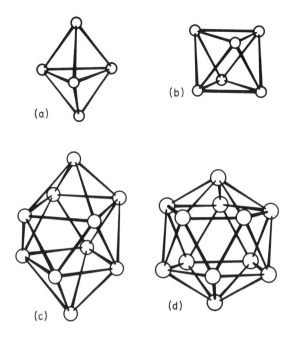

Figure 2.14 Some examples of *closo* metal carbonyl clusters with $14n + 2$ valence electrons. (a) Trigonal bipyramid, 72 electrons: $[Os_5(CO)_{15}]^{2-}$. (b) Octahedron, 86 electrons: $Rh_6(CO)_{16}$. (c) Bicapped square antiprism, 142 valence elctrons: $[Rh_{10}S(CO)_{22}]^{2-}$. (d) Icosahedron, 170 valence electrons: $[Rh_{12}Sb(CO)_{27}]^{3-}$.

TABLE 2.4 EXAMPLES OF "NAKED" METAL CLUSTERS

Cluster	Geometry	Number of valence electrons
Cationic clusters		
Bi_5^{3+}	Trigonal bipyramid	*closo*-$4n + 2$ (22)
Se_4^{2+}	Square planar	*arachno*-$4n + 6$ (22)
Bi_8^{2+}	Square antiprism	*arachno*-$4n + 6$ (38)
Bi_9^{5+}	Tricapped trigonal prism	*closo*-$4n + 4$* (40)
Anionic and neutral clusters		
P_4, As_4, Sb_4	Tetrahedral	Three-connected-$5n$ (20)
Bi_4^{2-}	Square planar	*arachno*-$4n + 6$ (22)
Sn_5^{2-}, Pb_5^{2-}	Trigonal bipyramid	*closo*-$4n + 2$ (22)
Ge_9^{2-}, $TlSn_8^{3-}$	Tricapped trigonal prism	*closo*-$4n + 2$ (38)
Sn_9^{4-}, Pb_9^{4-}	Capped square antiprism	*nido*-$4n + 4$ (40)
$TlSn_9^{3-}$	Bicapped square antiprism	*closo*-$4n + 2$ (42)

*The exception; see Chapter 5 for discussion.

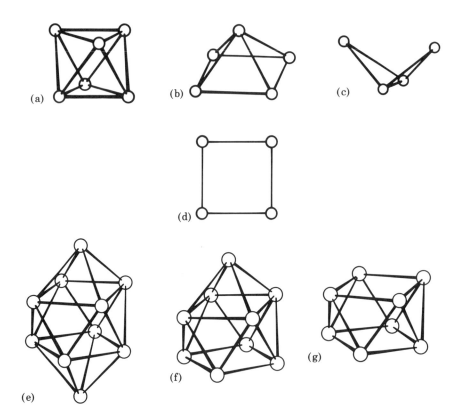

Figure 2.15 Examples of related (a, e) *closo*, (b, f) *nido*, and (c, d, g) *arachno* transition metal carbonyl clusters with $14n+2$, $14n+4$, and $14n+6$ valence electrons, respectively. (n is the number of vertices.) (a) Octahedron, 86 valence electrons: $Rh_6(CO)_{16}$. (b) Square pyramid, 74 valence electrons: $Os_5C(CO)_{15}$. (c) Butterfly, 62 valence electrons: $[Os_4N(CO)_{12}]^-$. (d) Square, 62 electrons: $Fe_4(CO)_{11}(PPh)_2$. (e) Bicapped square antiprism, 142 valence electrons: $[Rh_{10}S(CO)_{22}]^{2-}$, $14n+2$. (f) Capped square antiprism, 130 valence electrons: $[Ni_9C(CO)_{17}]^{2-}$, $14n+4$. (g) Square antiprism, 118 valence electrons: $[Ni_8C(CO)_{16}]^{2-}$, $14n+6$.

The presence of S, P, and related "bare" atoms in borane and hydrocarbon clusters follows the same principles, and it is assumed that each of these atoms has a nonbonding lone pair of electrons. Similarly, other E—H fragments (where E is a main group atom) can replace B—H and C—H in these molecules without causing any major breakdowns in the generalizations. The carboranes $C_2B_nH_{n+2}$ represent a particularly extensive series of *closo* molecules which are isostructural and isoelectronic with $B_nH_n^{2-}$. There are also many examples of *nido* and *arachno* carborane molecules with $4n+4$ and $4n+6$ valence electrons, respectively.

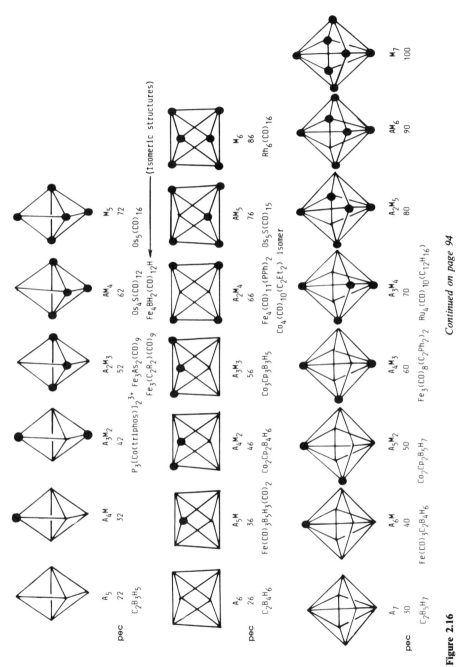

Figure 2.16

Continued on page 94

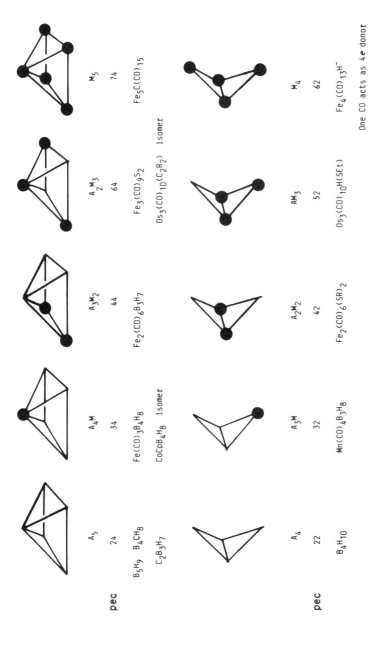

Figure 2.16 Series of isostructural clusters with different combinations of main group and transition metal atoms.

Rule 6

The total electron count in a condensed cluster is equal to the sum of the electron counts for the parent polyhedra, A and B, minus the electron count characteristic of the atom, pair of atoms, or face of atoms common to both polyhedra. These characteristic electron counts are 18 for a common metal vertex, 34 for a common pair of atoms, 48 for a common triangular face, and 62 for a common square face. This is usually called the "condensation principle."[21]

For main group condensed polyhedra, the corresponding characteristic electron counts are: 4 for a common vertex, 12 for a common edge, and 18 for a common triangular face. In main group chemistry the vast majority of clusters have an approximately spherical geometry and the condensation principle rarely has to be used. Nevertheless, there are some examples, as illustrated in Table 2.5. In $Si(C_2B_9H_{11})_2$ there are two icosahedral $SiC_2B_9H_{11}$ polyhedra sharing a common silicon atom. The main group metallocenes $M(\eta\text{-}C_5H_5)_2$ (M = Mg, etc.) can be considered as two *nido* polyhedra sharing a common vertex.[22] There are in addition a number of *nido* and *arachno* boranes which condense through a common edge; some examples are given in Table 2.5 and illustrated in Figure 2.17. $B_{20}H_{16}(NCCH_3)_2$ provides a rare example of two borane polyhedra sharing a common triangular face; its structure is also illustrated in this figure.

Figure 2.18 shows examples of transition metal carbonyl condensed clusters. It is noteworthy that when a triangle is condensed with a *closo* skeleton, an edge-bridged structure results and there is an increment of 14 in the electron count. Therefore, series of edge-bridged structures can be generated in this fashion and it is an easy matter to build up, for example, the raft

TABLE 2.5 EXAMPLES OF CONDENSED MAIN GROUP CLUSTERS

Compound	Valence electrons	Component polyhedra	Condensation rule
Vertex sharing			
$Si(C_2B_9H_{11})_2$	96	Two icosahedra	$A + B - 4$
$Mg(C_5H_5)_2$	52	Two pentagonal pyramids (*nido*-)	$A + B - 4$
Edge sharing			
$B_{12}H_{16}$	52	$B_6H_{10} + B_8H_{12}$	$A + B - 12$
$B_{13}H_{19}$	58	$B_6H_{10} + B_9H_{15}$	$A + B - 12$
$B_{14}H_{18}$	60	$B_6H_{10} + B_{10}H_{14}$	$A + B - 12$
$B_{16}H_{20}$	68	$B_8H_{12} + B_{10}H_{14}$	$A + B - 12$
$B_{18}H_{22}$	76	$2 \times B_{10}H_{14}$	$A + B - 12$
Triangular face-sharing			
$B_{20}H_{16}(NCCH_3)_2$	80	$B_{12}H_{12}^{2-} + nido\text{-}B_{11}H_{11}^{4-}$	$A + B - 18$

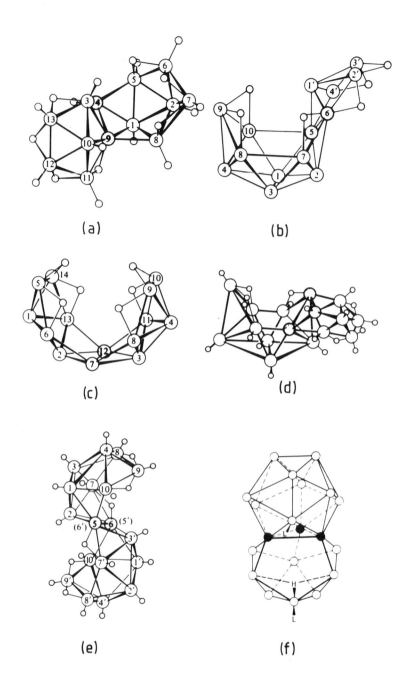

Figure 2.17 Some main group clusters with condensed structures: (a) $B_{13}H_{19}$, (b) proposed structure of $B_{14}H_{18}$, (c) $B_{14}H_{20}$, (d) $B_{16}H_{20}$, (e) n-$B_{18}H_{22}$ (centrosymmetric), (f) $B_{20}H_{16}(NCMe)_2$.

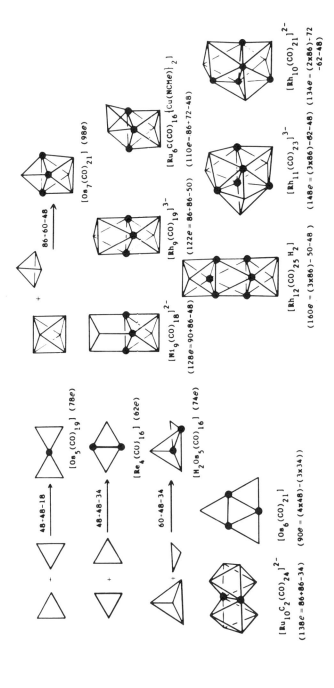

Figure 2.18 Some condensed transition metal clusters.

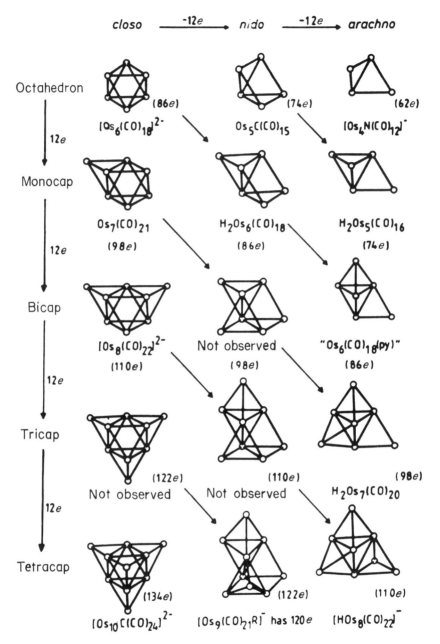

Figure 2.19. Structural relationships in condensed transition metal clusters based on the capping principle.

clusters illustrated in Figure 2.18. Similarly, the condensation of a tetrahedron and a *closo* deltahedron produces a capped triangular face and an increment in the electron count of 12. This process has been described as the "capping principle." If the capping principle is combined with the *closo*, *nido*, and *arachno* relationships developed in Rule 5, then it is possible to generate the panoply of structures illustrated in Figure 2.19. Across the top are *closo*, *nido*, and *arachno* molecules based on the octahedron, and successive capping of these structures leads to the columns of the figure. Since practically every entry represents a structurally determined metal carbonyl cluster, the power of the rules for rationalizing and predicting new compounds is apparent.

The theoretical basis of the capping principle will be presented in Chapter 7. This analysis will also provide a basis for explaining apparent exceptions to the rules and for explaining differences between main group and transition metal condensed clusters. As stated above, it is, for example, not immediately apparent why the characteristic electron count for condensed transition metal clusters conforms to the inert gas rule, i.e., 18 for a common vertex and 34 for a common edge, but this is not the case for main group clusters. The difference is due to the availability of a larger number of valence orbitals in the transition metal clusters; a detailed explanation is given in Chapter 7.

2.4 FACE-LOCALIZED BONDING SCHEMES

The discussion above has been limited to element-element bonds localized on the edges of a cluster. There are, however, alternative localization procedures. For example, on each triangular face of a polyhedron a three-center two-electron bond could be formed. For a tetrahedron, four such bonds are possible and could accommodate a total of eight valence electrons. The bonding in tetrahedral B_4Cl_4 could therefore be analyzed in this manner. These three-center two-electron bonds could also form the basis for a delocalized treatment, and in the specific case of B_4Cl_4 the resultant equivalent orbitals would transform as $A_1 \oplus T_2$.[23]

If the same methodology is applied to the octahedral cluster $B_6H_6^{2-}$, the localized three-center two-electron bonds generate orbitals with the following symmetries: A_{1g}, T_{1u}, T_{2g}, and A_{2u}. Interestingly, the a_{1g}, t_{1u}, and t_{2g} orbitals match the bonding molecular orbitals generated from the *s* and *p* valence orbitals of the B—H fragments (see Figure 2.20), but the a_{2u} combination finds no match, because it is doubly noded at each vertex atom. Indeed, such a combination could only be generated if the vertex atoms had doubly noded *d* orbitals of the right energy for bonding. This underlines the very important and general point that a localized description is valid only when the number of edges radiating from each vertex is equal to or less than the number of valence orbitals available for bonding.

2.5 OCTAHEDRAL HALIDE CLUSTERS

We noted in Section 2.2 that a square pyramidal ML_5 fragment has four hybrid orbitals available for skeletal bonding and therefore, localized bonding descriptions are valid for four-connected clusters derived from these fragments. For example, in an octahedral cluster, localized bonds can be formed either along the edges, in which case they are two-center two-electron bonds, or in the faces, in which case they are three-center two-electron bonds. The former scheme produces a total of 24 valence electrons, and the latter, 16 valence electrons.[24-25]

The 12 edge-localized bonds can be transformed into an equivalent set of a_{1g}, t_{1u}, t_{2g}, e_g, and t_{2u} molecular orbitals. The a_{1g} combination can be generated from σ-type atomic orbitals, the t_{1u} and t_{2g} combinations can be generated from π-type atomic orbitals, and the e_g and t_{2u} combinations can be generated only from δ-type atomic orbitals. In $Mo_6Cl_8L_6$ and related clusters the chloride ligands are located over the faces of the octahedron and, together with the terminal ligands L, define a square pyramid around each molybdenum atom. These metal-ligand bonds involve the s, p, and d_{xy} orbitals of molybdenum, leaving the d_{z^2}, d_{xz}, d_{yz}, and $d_{x^2-y^2}$ orbitals for forming the 12 edge-delocalized molecular orbitals. These schemes are discussed in more detail in Chapter 8.

The eight molecular orbitals which are formed from linear combinations of face-localized bonds transform as $A_{1g} \oplus T_{1u} \oplus T_{2g} \oplus A_{2u}$ and can also be generated from d_{z^2}, d_{xz}, d_{yz}, and d_{xy} atomic orbitals (see Figure 2.20) at the vertex atoms. The d_{xy} orbital is doubly noded but, in contrast to $d_{x^2-y^2}$, points toward the faces of the octahedron rather than the edges. In $Ta_6Cl_{12}L_6$, the chloride ligands are located on the edges and when taken with L complete

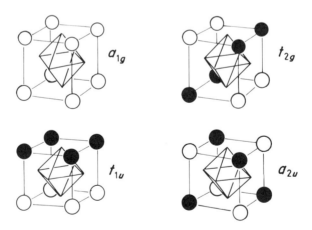

Figure 2.20 Symmetry-adapted linear combinations of the three-center two-electron face-bonding orbitals of an octahedron.

a square pyramid. The metal-metal bonding occurs through the face-delocalized molecular orbitals. Hence, the two molecules have complementary bonding patterns whereby all the faces and edges of the octahedron are involved either in metal-metal or metal-chlorine bonding.

The electrons in these clusters (which are illustrated in Figure 2.21) can be audited as follows:

$Mo_6Cl_8L_6^{4+}$
Mo—L	12e	6L	12e
Mo—Cl	48e (face-bridging)	8Cl	40e
Mo—Mo	24e (edge-localized)	Mo_6^{4+}	32e
Totals	84e		84e

$Ta_6Cl_{12}L_6^{2+}$
Ta—L	12e	6L	12e
Ta—Cl	48e (edge-bridging)	12Cl	36e
Ta—Ta	16e (face-localized)	Ta_6^{2+}	28e
Totals	76e		76e

These closed-shell requirements contrast with those noted above for octahedral metal carbonyl clusters, i.e., 86 valence electrons, because the metal atoms in the latter do not utilize four orbitals for skeletal bonding. In a cluster molecule like $Co_6(CO)_{14}^{4-}$, the arrangement of ligands around the cobalt atom is very similar to that in $Mo_6Cl_8L_6^{4+}$, suggesting that the CoL_5 fragment could also use four orbitals for skeletal bonding. However, for cobalt with a d^9 configuration, the doubly noded $d_{x^2-y^2}$ orbital has an electron pair associated with it and therefore cannot contribute to hybrids localized along the edges of the polyhedron. Therefore, only three orbitals per vertex atom are utilized in skeletal bonding, and the spectrum of delocalized molecular orbitals resembles that found in $B_6H_6^{2-}$.

The localized approach developed above for d^4 ML_5 fragments can also be utilized for clusters with multiple metal-metal bonds. For example, two

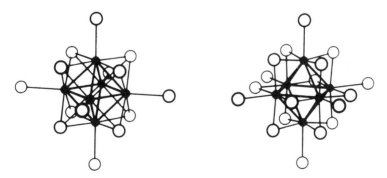

Figure 2.21 The structures of $[Mo_6Cl_8L_6]^{4+}$ (*left*) and $[Ta_6Cl_{12}L_6]^{2+}$ (*right*); the metal atoms are shaded.

such fragments can come together to form a quadruple metal-metal bond. Such compounds have been studied in great detail by Cotton and his coworkers, and this large and important area of cluster chemistry has been described in a recent book and review.[26] These localized hybrids can also be used to form rectangular clusters with alternating triple and single bonds such as $Mo_4Cl_8(PEt_3)_4$.[27] Triangular clusters with three double bonds can also be generated by using the four hybrid orbitals, for example, $Re_3Cl_9L_3$.[28] Finally, these hybrids can be used to form a trigonal prismatic structure with double bonds connecting the triangular faces and single bonds within the faces, for example, $Tc_6Cl_{12}^{2-}$. Examples of cluster compounds with these alternative bonding modes are illustrated in Figure 2.22.[29]

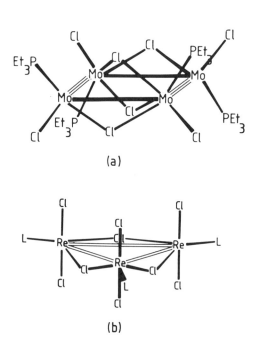

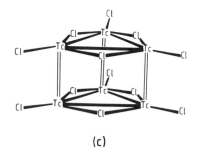

Figure 2.22 Clusters with alternative multiple bonding modes. In each example the metal atoms form four metal-metal bonds: (a) $Mo_4Cl_8(PEt_3)_4$, (b) $Re_3Cl_9L_3$, and (c) $[Tc_6Cl_{12}]^{2-}$.

Since one inaccessible antibonding skeletal molecular orbital results for each edge bond, and two for each face bond, these localized descriptions may be summarized as:

$$N_e = 18n - 2e - 4f \quad \text{for transition metal clusters}$$

and

$$N_e = 8n - 2e - 4f \quad \text{for main group clusters,}$$

where e is the number of edge-localized two-center two-electron bonds and f is the number of face-localized three-center two-electron bonds. For example:

$Mo_6Cl_8L_6^{2+}$	$e = 12, f = 0$	$N_e = 18 \times 6 - 2 \times 12 = 84$
$Nb_6Cl_{12}L_6^{4+}$	$e = 0, f = 8$	$N_e = 18 \times 6 - 4 \times 8 = 76$
B_4Cl_4	$e = 0, f = 4$	$N_e = 8 \times 4 - 4 \times 4 = 16$
C_4R_4	$e = 6, f = 0$	$N_e = 8 \times 4 - 6 \times 2 = 20$

It cannot be emphasized too strongly, however, that such a localized description is only valid when the number of orbitals used by each vertex atom is equal to or exceeds the number of edges or faces emanating from that vertex.

2.6 THE STYX METHOD

Lipscomb's "styx" scheme provides a localized description of the bonding in boranes.[18] It involves the five structural elements illustrated in Figure 2.23, each of which is labeled by a lower-case letter as indicated. Three-center two-electron bonds involving a bridging hydrogen atom are denoted by the

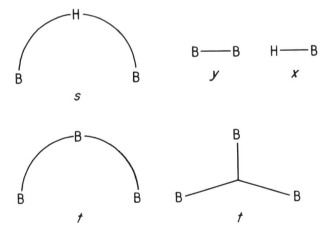

Figure 2.23 Structural elements of the styx scheme.

letter s, while similar arrangements with a central boron atom are denoted by t. There are two possibilities for the latter structural fragment: one with three hybrids directed toward the center of the boron triangle (termed a "closed" three-center bond) and the other involving a p orbital and two inwardly directed hybrids (termed an "open" three-center bond). Two-center two-electron B—B and B—H bonds are denoted by y and x, respectively. It is assumed that each vertex has at least one terminally bound hydrogen, and these are not counted in the x B—H bonds.

For $(BH)_n H_q$, equations of balance for the three-center bonds, hydrogen atoms, and electrons can be derived. If each boron atom is terminally bound to at least one hydrogen atom, then the remaining hydrogen atoms must come from the s bridging hydrogens and the x additional terminally bound hydrogens. Hence

$$q = s + x. \tag{2.1}$$

The number of skeletal electrons donated by the n BH units and the q additional hydrogen atoms must equal the number of electrons involved in all the possible bonds so that

$$2n + q = 2(s + t + y + x). \tag{2.2}$$

Finally, if each boron atom utilizes four valence orbitals for bonding, but supplies only three electrons, then the number of three-center two-electron bonds must equal the number of boron atoms:

$$n = s + t. \tag{2.3}$$

The above results may be combined to produce the key equation $y = (s - x)/2$, and nonnegative integer solution sets are then sought subject to the additional constraint $q/2 \leq s \leq q$, which is required for consistency.[30]

The advantages of this method are its simplicity, its ability to rationalize and predict borane structures systematically, and its success in simple considerations of structure and reactivity. The main difficulty is that the equations of balance do not generally have unique solutions, and a large number of resonance structures need to be considered for symmetrical boranes. The latter are better described within a delocalized framework.

The localized styx formalism can be related to the idea that *closo*, *nido*, and *arachno* boranes are characterized by $2n + 1$, $2n + 2$, and $2n + 3$ electron pairs, respectively, in the following manner:

$$n + (s + t + y + x) = \begin{cases} 2n + 1 & \text{for } closo \text{ clusters} \\ 2n + 2 & \text{for } nido \text{ clusters} \\ 2n + 3 & \text{for } arachno \text{ clusters} \end{cases}$$

These arise because $(s + t + y + x)$ is equal to the number of bonding orbitals of the cluster except for the one terminal B—H bond per vertex which is not counted in this sum. Examples of these relationships are summarized below (see Figure 2.13).

Borane	n	s	t	y	x	$(s+t+y+x)$	Structure
B_5H_9	5	4	1	2	0	7	*nido*
B_6H_{10}	6	4	2	2	0	8	*nido*
$B_{10}H_{14}$	10	4	6	2	0	12	*nido*
B_4H_{10}	4	4	0	1	2	7	*arachno*
B_5H_{11}	5	3	2	0	3	8	*arachno*

For the *closo*-borane anions, $B_nH_n^{2-}$, $s = x = 0$ and consequently $y = 0$. Hence the bonding is described in terms of $n + 1$ three-center two-electron bonds. Since these do not correspond to the number of faces of the deltahedron $(2n - 2)$, resonance between the alternative permutations of these bonding arrangements must be invoked.

2.7 ELECTRON-RICH CAGE MOLECULES

From the rules developed above it is apparent that the maximum number of valence electrons associated with a three-dimensional main group cluster is $5n$, i.e., the number associated with a three-connected structure. There exist in addition a range of molecules with electron counts between $5n$ and $6n$ with structures which are neither complete three-dimensional cages nor ring compounds. Three series of cage molecules of this type are illustrated in Figure 2.24.[31] In the first series there is only one "butterfly" structure between P_4 (tetrahedral, with 20 valence electrons) and P_4Ph_4 (ring with 24 valence electrons). For the eight-vertex series there are examples of compounds with 44 and 46 valence electrons between C_8H_8 and S_8. These intermediate structures are described as "electron-rich" cage molecules, and their structures are related to the symmetrical structures at the ends of the series.[4] The great majority of these compounds can be described in terms of localized bonding schemes: the addition of each successive electron pair results in the breaking of an element-element bond and the formation of lone pairs on what are now separated atoms (Figure 2.25).

The above process corresponds to the addition of an electron pair to an antibonding cage orbital and the selective opening up of the structure to relieve the antibonding interaction. The result is the formation of a pair of molecular orbitals localized predominantly on the separated atoms and directed away from the other cage atoms. This pattern is clearly discernible in the series of related structures in the figure: the bonds are broken in a selective fashion and the intermediate cage structures begin to open out and resemble more closely the ring compounds at the end of the series.

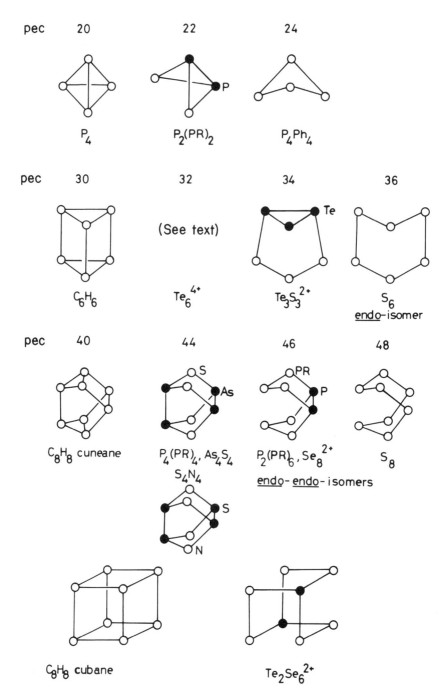

Figure 2.24 Molecules with structures which are intermediate between a cage and a ring.

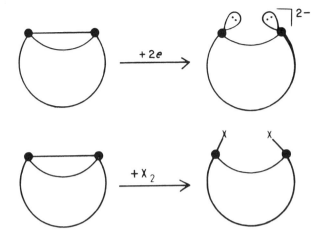

Figure 2.25 Notional bond breaking by addition of an electron pair to a localized bond to give two lone pairs. Oxidative addition also leads to a bond cleavage.

The delocalized view of the bonding also suggests an alternative way of reducing the antibonding character of the newly occupied molecular orbital. If several bonds in the structure are simultaneously lengthened, the net effect could be energetically more favorable than the selective lengthening of one bond. For example, Te_6^{4+} has a total of 32 valence electrons, and a localized description of the bonding would suggest that it should adopt a structure based on a trigonal prism (30 valence electrons), with one edge broken. In fact it retains a trigonal prismatic structure with the Te—Te distance between the triangular faces (3.13 Å) substantially longer than those within the triangles (2.67 Å).[32] Therefore, in molecular orbital terms this would be associated with a HOMO which is antibonding between the two triangles. An equivalent localized description would be based on resonance between the canonical structures shown in Figure 2.26 leading to a formal bond order of $\frac{2}{3}$ for the intertriangular bonds. Interestingly, the 34-electron molecules $Te_3S_3^{2+}$ and $Te_2Se_4^{2+}$ have a boat structure (Figure 2.27) which can be described satisfactorily in localized terms.[33]

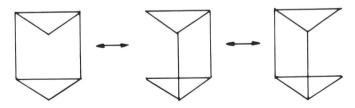

Figure 2.26 Resonance structures for the localized description of the bonding in Te_6^{4+}.

Sec. 2.7   Electron-rich Cage Molecules

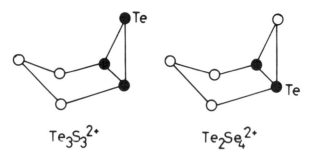

Figure 2.27 The boat structure of the 34-electron species $Te_3S_3^{2+}$ and $Te_2Se_4^{2+}$.

Similar structural patterns are discernible in "electron-rich" transition metal cage molecules. For example, $Co_3(\eta\text{-}C_5H_5)_3S(CO)$ has a regular triangular skeleton and metal-metal bond lengths (2.45 Å) commensurate with the presence of three localized metal-metal bonds (i.e., it has a total of 48 valence electrons). The related 50-electron molecules $Co_3(\eta\text{-}C_5H_5)_3S_2$ and $Fe_3(CO)_9S_2$ illustrate the localized-delocalized duality most effectively, since the former has a symmetrical D_{3h} structure, with all the Co—Co bond lengths equal, but lengthened by about 0.24 Å; whereas the latter has an open C_{2v} structure with two of the Fe—Fe distances (2.59 Å) indicating the presence of bonds and the third (3.37 Å) suggesting a nonbonded contact.[34-36]

$Fe_4(CO)_4(\eta\text{-}C_5H_5)_4$ and $Mo_4S_4(\eta\text{-}C_5H_5)_4$ have conventional tetrahedral geometries with 60 valence electrons and metal-metal bond lengths consistent with the presence of single bonds. $Co_4S_4(\eta\text{-}C_5H_5)_4$ has an additional 12 valence electrons, which is sufficient to populate all the metal-metal antibonding molecular orbitals, and consequently the formal cobalt-cobalt bond orders are zero. The bond length data are consistent with this interpretation, since the Co-Co lengths lie between 3.236 and 3.343 Å. $Fe_4S_4(\eta\text{-}C_5H_5)_4$, with 68 valence electrons, can be described within a localized framework, since it has four long and two short bonds. However, $Fe_4S_4(\eta\text{-}C_5H_5)_4^{2+}$, which has 66 valence electrons, does not have the anticipated C_{3v} structure with three short and three long bonds. Instead it has a D_{2d} structure with two nonbonded Fe—Fe contacts and four lengthened but equivalent bonds. In order to describe this structure in localized terms, resonance would again have to be invoked, leading to a formal bond order of $\frac{3}{4}$ for the shorter bonds.

2.8 METAL CARBONYL CLUSTERS IN MOLECULAR BEAMS

In some recent molecular beam experiments Fayet, McGlinchey, and Wöste have mass-selected individual Ni_x^+ ($x = 1$ to 20) cluster ions and studied their reactions with carbon monoxide.[37] A range of cluster ions are formed with varying numbers of carbonyl ligands. The following primary reactions have been identified:

$$\text{Ni}_n^+ + k\text{CO} \rightarrow \text{Ni}_n(\text{CO})_k^+,$$

$$\text{Ni}_n^+ + l\text{CO} \rightarrow \text{Ni}_n\text{C}(\text{CO})_l^+,$$

$$\text{Ni}_n^+ + m\text{CO} \rightarrow \text{Ni}_{n-1}(\text{CO})_m^+.$$

The limiting number of carbonyls in these reactions for a given number of metal atoms (Table 2.6) can be related to the number of accessible orbitals of the neutral cluster $\text{Ni}_n(\text{CO})_k$. For example, $\text{Ni}_4(\text{CO})_{10}$ has 60 valence electrons and is predicted to have a tetrahedral geometry. Similarly, $\text{Ni}_5(\text{CO})_{12}$, $\text{Ni}_6(\text{CO})_{13}$, and $\text{Ni}_7(\text{CO})_{15}$ should have octahedral, square-pyramidal, and pentagonal bipyramidal geometries, respectively. The next five members of the series involve successive increments in the number of valence electrons by 12. This strongly suggests that the structures involved are based upon successive cappings of the pentagonal bipyramid. The last member of the series, $\text{Ni}_{13}(\text{CO})_{20}$, has 170 valence electrons, which suggests that it is an icosahedron with an interstitial nickel atom (or perhaps a centered cuboctahedron). If the metal skeletons are not changed by ligation, then this sequence of structures will reflect the growth sequence in the molecular beam (Figure 2.28 (right hand side)). From our understanding of the bonding in transition metal carbonyls of this type, this sequence seems more plausible than that shown on the left of the figure. A similar sequence may occur in the formation of inert gas atom clusters, but this is the first experimental evidence that such a pattern may arise for transition metals.

The second series of nickel cluster ions has an interstitial carbon atom in each cluster, and consequently the largest number of carbonyls observed is

TABLE 2.6 REACTIONS OF NICKEL ATOM CLUSTERS WITH CARBON MONOXIDE IN MOLECULAR BEAMS

	Number of valence electrons		Predicted structure
$\text{Ni}_4(\text{CO})_{10}$	60	$15n$	Tetrahedral
$\text{Ni}_5(\text{CO})_{12}$	74	$14n + 4$	Square pyramid
$\text{Ni}_6(\text{CO})_{13}$	86	$14n + 2$	Octahedral
$\text{Ni}_7(\text{CO})_{15}$	100	$14n + 2$	Pentagonal bipyramid
$\text{Ni}_8(\text{CO})_{16}$	112	$14n$	Capped pentagonal bipyramid
$\text{Ni}_9(\text{CO})_{17}$	124	$14n - 2$	Bicapped pentagonal bipyramid
$\text{Ni}_{10}(\text{CO})_{18}$	136	$14n - 4$	Tricapped pentagonal bipyramid
$\text{Ni}_{11}(\text{CO})_{19}$	148	$14n - 6$	Tetracapped pentagonal bipyramid
$\text{Ni}_{12}(\text{CO})_{20}$	160	$14n - 8$	Pentacapped pentagonal bipyramid
$\text{Ni}_{13}(\text{CO})_{20}$	170	$14(n-1) + 2$	Icosahedron

Adapted from Fayet, McGlinchey, and Wöste.[37]

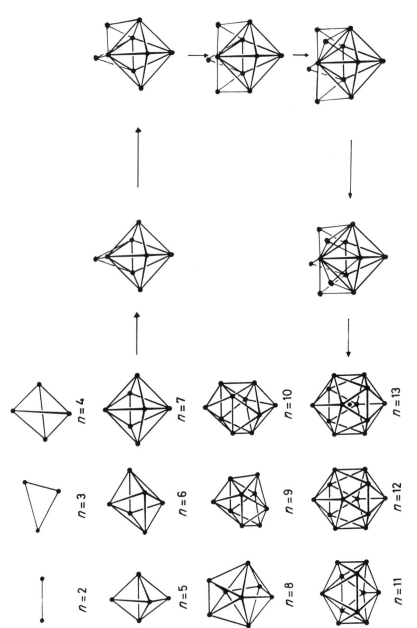

Figure 2.28 Candidate metal cluster skeletons for the $Ni_n(CO)_k$ series. The capped structures suggested by the mass spectrometric data provide an alternative pathway to the centered icosahedron.

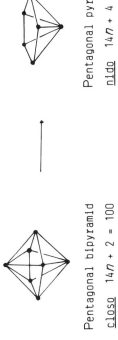
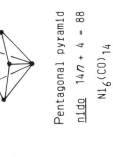

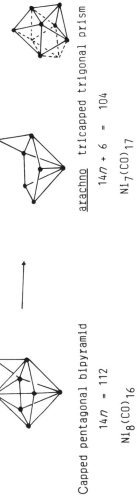

Pentagonal bipyramid
<u>closo</u> $14n + 2 = 100$

$Ni_7(CO)_{15}$

Pentagonal pyramid
<u>nido</u> $14n + 4 = 88$

$Ni_6(CO)_{14}$

Capped pentagonal bipyramid
$14n = 112$

$Ni_8(CO)_{16}$

<u>arachno</u> tricapped trigonal prism
$14n + 6 = 104$

$Ni_7(CO)_{17}$

Tricapped trigonal prism

Figure 2.29 Candidate *nido* and *arachno* structures for the metal cluster skeletons of the $Ni_{n-1}(CO)_m$ series.

TABLE 2.7 MOLECULAR IONS WHICH RESULT FROM FRAGMENTATION OF THEIR PRECURSORS

	Number of valence electrons		Predicted structure	
$Ni_5(CO)_{11}$	72	$14n + 2$	Trigonal bipyramid	
$Ni_6(CO)_{14}$	88	$14n + 4$	Pentagonal pyramid	
$Ni_7(CO)_{17}$	104	$14n + 6$		*arachno*-tricapped trigonal prism
$Ni_8(CO)_{18}$	116	$14n + 4$	Capped	*arachno*-tricapped trigonal prism
$Ni_9(CO)_{19}$	128	$14n + 2$	Bicapped	*arachno*-tricapped trigonal prism
$Ni_{10}(CO)_{20}$	140	$14n$	Tricapped	*arachno*-tricapped trigonal prism
$Ni_{11}(CO)_{21}$	152	$14n - 2$	Tetracapped	*arachno*-tricapped trigonal prism
$Ni_{12}(CO)_{22}$	164	$14n - 4$	Pentacapped	*arachno*-tricapped trigonal prism

2 fewer than for the above series. This is consistent with our view of an interstitial carbon atom as a four-electron donor (see Table 2.2).

The cluster ions of the final series are derived by fragmentation and have the general formula $Ni_{n-1}(CO)_m^+$. The total number of valence electrons associated with these clusters is summarized in Table 2.7. $Ni_6(CO)_{14}$ has the correct electron count for a pentagonal pyramid, which could result from the loss of an apical atom from a pentagonal bipyramid. There are then a series of six clusters which show an incremental electron count of 12, which again suggests successive capping of a "parent" structure. In this case the neutral parent cluster with seven metal atoms has 104 valence electrons, suggesting that it has an *arachno* geometry. The loss of an apical atom from a capped pentagonal bipyramid (Figure 2.29) generates an *arachno* tricapped trigonal prism. The remaining structures could arise from capping the remaining five faces of the putative pentagonal bipyramid. Although the aesthetic appeal of the related series of structures which we have proposed to account for the experimental observations is obvious, we should emphasize that this is not a unique solution (see Exercise 2.12).

2.9 SUMMARY

This chapter has indicated that the structures of a wide range of clusters can be interpreted using a set of simple rules. The success of these rules is clearly not fortuitous and represents a challenge to valence theories. It has been apparent that many of the generalizations are not amenable to interpretation using localized bonding descriptions, and it is necessary to seek an understanding of the problem using delocalized models. If successful, such an interpretation should not only account for the rules described above, but should also be sufficiently flexible to indicate situations where the rules may no longer be valid. In addition, a completely satisfactory theoretical model should be able to interpret not only the structures of these molecules, but also their rearrangements and their spectroscopic properties.

EXERCISES

2.1 Use localized bonding descriptions to derive the symmetries of the unoccupied skeletal molecular orbitals of $Os_3(CO)_{12}$ and prismane (C_6H_6) giving the TSH notation in each case. (See Chapter 3.)

2.2 Why do the valence shell s and p orbitals generally become further apart in energy on descending a group in the periodic table? What effect does this have on the energetics of hybrid formation?

2.3 Deduce the symmetries of the strongly bonding and weakly bonding skeletal orbitals (i.e., the C—C σ bonds) in dodecahedrane, $C_{20}H_{20}$.

2.4 When do you expect a localized description of the bonding in a cluster to be appropriate?

2.5 Sketch qualitative energy-level diagrams for the B—H—B bridge bonds in diborane, B_2H_6, using the following hybridization schemes: (a) sp^2 hybridized boron atoms (with two of the hybrids directed toward the terminal hydrogen atoms), and (b) sp^3 hybridized boron atoms (with the four hybrids directed toward two bridging and two terminal hydrogen atoms). (You should obtain the same answer in each case!) Does the experimental evidence suggest that diborane is literally electron-deficient?

2.6 Using the styx method, predict the structure of B_5H_{11} and sketch a bonding scheme in terms of the styx structural elements.

2.7 Use Rule 5 (the *debor* pattern) to deduce the structures of the following borohydrides and carboranes: B_5H_{11}, $B_{11}H_{11}^{2-}$, B_5H_9, $C_2B_9H_{11}^{2-}$, and B_6H_{10}. Where would you expect the bridging and extra terminal hydrogen atoms to be positioned in these clusters?

2.8 Use Rule 3 to account for the following series of isostructural clusters: $B_6H_6^{2-}$, $Co_4(CO)_{10}C_2Et_2$, $C_2B_3H_5(Fe(CO)_3)$, and $Rh_6(CO)_{16}$ (octahedral); $C_2B_5H_7$, $C_2B_3H_5(Fe(CO)_3)_2$, $Fe_3(CO)_8(Ph_2C_2)_2$, and $C_2B_3H_5Co_2(C_5H_5)_2$ (pentagonal bipyramidal); and As_4, $As_3Co(CO)_3$, $As_2Co_2(CO)_6$, $AsCo_3(CO)_9$, and $Co_4(CO)_{12}$ (tetrahedral).

2.9 Would you expect the following clusters to have *closo*, *nido*, or *arachno* geometries? (a) $C_2B_3H_5$, (b) $Fe_3(CO)_9(Ph_2C_2)$ (consider the alkyne carbons as vertex atoms), (c) Pb_5^{2-}, (d) $Fe_3(CO)_9As_2$ (consider As as a vertex atom), (e) $Fe_3(CO)_9S_2$ (consider S as a vertex atom), (f) $C_2B_3H_5Co(C_5H_5)$, and (g) $C_2B_3H_{10}Co_2(C_5H_5)_2$.

2.10 Using the condensation rule (Rule 6), rationalize the structures of the following: (a) $Os_6(CO)_{18}$ (capped trigonal bipyramid), (b) $Co_3(\eta\text{-}C_5H_5)_3B_3H_3$ (capped octahedron), and (c) $Re_4(CO)_{16}^{2-}$ (two triangles sharing an edge).

2.11 $Fe_4S_3(NO)_7$ has a distorted tetrahedral C_{3v} metal geometry with three Fe—Fe bond lengths at around 2.71 Å and the other three at around 3.51 Å. Show how this geometry can be described using localized bonding schemes.

2.12 In Section 2.8 we described some possible structures for nickel cluster carbonyls of the type $Ni_x(CO)_y^+$. These ions are all odd electron species, and all the arguments are based upon bonding schemes for the corresponding neutral clusters $Ni_x(CO)_y$, assuming that the odd electron resides in the HOMO of this species. This seems a reasonable comparison, but there are other possibilities. Suppose instead that we relate the structures not to the neutral species but to the ions $Ni_x(CO)_y^{2+}$, which require 1 less accessible orbital. Show that the series of fragmented clusters could then be partly based on capped trigonal prisms. This assumption is tantamount to suggesting that the odd electron in the $Ni_x(CO)_y^+$ ion resides in an orbital like the LUMO of the dipositive ion. What factors might favor such structures?

2.13 Fehlner, T. P., et al. (*Angew. Chem., Int. Ed.*, **27**, 424 (1988)) have described the synthesis and characterization of $Co_3(C_5H_5)_3(PPh)(BPh)$. What structure would you predict for this molecule if the phosphorus and boron atoms are considered as vertex atoms? If they are considered as bridging groups, what is the skeletal electron count of the cluster?

2.14 Baudler, M., et al. (*Angew. Chem., Int. Ed.*, **27**, 1059 (1988)) have shown that the reaction of white phosphorus with sodium naphthelide in dioxymethane gives $NaHP_4$, which is stable only below $-78°C$. The ion shows ^{31}P resonances at $\delta = 71$, -329, and -355 ppm with intensity ratios 1:1:2 and $J(P-H)$ coupling constants of 16.5, 129.2, and 3.2 Hz, respectively. Suggest a structure for this ion and indicate whether the proposed structure is consistent with the generalizations developed in this chapter.

2.15 Haubold, W., et al. (*Angew. Chem. Int. Ed.*, **27**, 925 (1988)) have isolated $P_2B_4Cl_4$ from the reaction of B_2Cl_2 and PCl_3 and demonstrated that it has a polyhedral structure with average distances $\bar{d}(P-P) = 2.222(3)$ Å and $\bar{d}(B-B) = 1.73(1)$ Å. Use the rules presented in this chapter to suggest a structure for this molecule, and use the following nmr data to deduce its isomeric form:

^{11}B $\delta = 2.5$ and 22.1 ppm, intensity ratio 1:1

^{31}P $\delta = -187$ ppm

2.16 Vahrenkamp, H. et al. (*Organometallics*, **7**, 1946 (1988)) have described the chemical and redox properties of $Fe_4(CO)_{11}(PR)_2$ and $Fe_4(CO)_{12}(PR)_2$, which both have a square plane of metal atoms with μ_4-PR groups above and below the metal plane. Consider the structures first as Fe_4P_2 octahedra and decide which of the compounds conforms to the rules developed in this chapter. Now consider the compounds as square-planar Fe_4 clusters. Are they electronically related to cyclobutadiene? For a more detailed theoretical analysis, see Halet.[38]

2.17 Gillespie, R. J., et al. (*J. Chem. Soc., Chem. Comm.*, 902 (1988)) have reported that the structure of $[Te_4S_4]^{2+}$ is very similar to that of As_4S_4, P_4S_4, and S_4N_4. Why is this surprising? Can one use the ideas developed for Te_6^{4+} to rationalize this exception to the localized bond-cleavage model for electron-rich clusters?

2.18 In Section 2.5 we noted that d^4 metal fragments could either dimerize to form a quadruple bond or condense to give rectangular clusters with alternating single and double bonds, triangular clusters with double bonds, or octahedral clusters with single bonds. Explore the corresponding possibilities for d^3 metal fragments. The following examples may help to confirm your conclusions:

$Mo_2(OCH_2Bu^t)_6$ Chisholm, M. H., et al., *Inorg. Chem.*, **18**, 2266 (1979)

$W_4(OPr^i)_{12}$ Chisholm, M. H., et al., *J. Amer. Chem. Soc.*, **109**, 7750 (1987)
Chisholm, M. H., et al., *Angew. Chem., Int. Ed.*, **25**, 1014 (1986)

$Mo_4S_4Br_4$ Perrin, et al., *Seances Acad. Sci. Ser. C*, **281**, 23 (1975)

REFERENCES

1. Sidgwick, N. V., and Powell, H. E., *Proc. Roy. Soc. A*, **176**, 153 (1940).
   Nyholm, R. S., and Gillespie, R. J., *Quart. Rev. Chem. Soc.*, **11**, 339 (1957).
2. Williams, R. E., *Inorg. Chem.*, **10**, 210 (1971).
   Williams, R. E., *Prog. Inorg. Chem. Radiochem.* **18**, 67 (1976).
3. Wade, K., *J. Chem. Soc., Chem. Comm.*, 792 (1971).
   Wade, K., *Adv. Inorg. Chem. Radiochem.* **18**, 1 (1976).
4. Mingos, D. M. P., *Nature Phys. Sci.*, **236**, 99 (1972).
   Mingos, D. M. P., *Acc. Chem. Res.*, **17**, 311 (1984).
5. Rudolph, R. W., and Pretzer, W. R., *Inorg. Chem.*, **11**, 1974 (1972).
   Rudolph, R. W., *Acc. Chem. Res.*, **9**, 446 (1976).
6. Mingos, D. M. P., and Johnston, R. L. *Struct. Bond.*, **68**, 29 (1987).
7. Aylett, B. J., *Organometallic Compounds*, Vol. 1, Part II, Chapman & Hall, London, 1979.
8. Elian, M., Chen, M. M. L., Mingos, D. M. P., and Hoffmann, R., *Inorg. Chem.* **15**, 1148 (1976).
   Hoffmann, R., *Angew. Chem., Int. Ed.*, **21**, 711 (1982).
9. Mingos, D. M. P., and Zhenyang, L., *Struct. Bond.*, **72**, 73 (1989).
10. Green, J. C., Mingos, D. M. P. and Seddon, E. L., *Inorg. Chem.* **20**, 2595 (1981).
11. Sherwood, D. E., and Hall, M. B., *Inorg. Chem.*, **21**, 3458 (1982).
12. Johnston, R. L., and Mingos, D. M. P., *J. Organometallic Chem.*, **280**, 407 (1985).
13. Schulman, J. M., Fischer, C. R., and Venanzi, T. J., *J. Amer. Chem. Soc.*, **100**, 2949 (1978).
14. Mingos, D. M. P., and Zhenyang, L., *New J. Chem.*, **12**, 787 (1988).
15. Johnson, R. L., and Mingos, D. M. P., *J. Organometallic Chem.*, **280**, 419 (1970).
16. Muetterties, E. L., Wiersema, R. J., and Hawthorne, M. F., *J. Amer. Chem. Soc.*, **95**, 7520 (1973).
17. *Gmelin Handbuch der Anorganischen Chemie*, 8th ed., Boron Hydrogen Compounds Part 3. Springer-Verlag, New York, 1979, Vol. 54, Borverbindungen Teil 20.

18. Lipscomb, W. N., *Boron Hydrides*, Benjamin, New York, 1963.
19. Corbett, J. D., *Prog. Inorg. Chem.*, **21**, 129 (1976).
    Corbett, J. D., *Chem. Rev.*, **85**, 383 (1985).
20. O'Neill, M. E., and Wade, K., *Inorg. Chem.*, **21**, 461 (1982).
21. Mingos, D. M. P., *J. Chem. Soc., Chem. Comm.*, 706 (1983).
22. Eisenstein, O., Canadill, E., and Thanh, B. T., *Nouv. J. Chimie*, **10**, 421 (1986).
23. Kettle, S. F. A., and Tomlinson, V., *J. Chem. Soc. A*, 2002, 2007 (1969).
24. Cotton, F. A., and Haas, T. E., *Inorg. Chem.*, **3**, 10 (1964).
25. Kettle, S. F. A., *Theo. Chim. Acta*, **3**, 211 (1965).
26. Cotton, F. A., and Walton, R. A., *Multiple Bonds between Metal Atoms*, Wiley, New York, 1982.
27. McCarley, R. E., *Philos. Trans. R. Soc. Lond. A*, **308**, 141 (1982).
28. Perrin, A., and Sergent, M., *New J. Chem.*, **12**, 337 (1988).
29. Wheeler, R. A., and Hoffmann, R., *J. Amer. Chem. Soc.*, **108**, 6605 (1986).
30. See, for example, Purcell, K. F., and Kotz, J. F., *Inorganic Chemistry*, Holt-Saunders, Philadelphia, 1977.
31. Gillespie, R. J., *Chem. Soc. Rev.*, **8**, 315 (1979).
32. Burns, R. C., Gillespie, R. J., Luk, W. C., and Slim, D. R., *Inorg. Chem.*, **18**, 3086 (1979).
33. Schrobilgen, G. J., Burns, R. C., and Granger, P., *J. Chem. Soc., Chem. Comm.*, 951 (1978).
34. Wei, C.-H, and Dahl, L. F., *J. Amer. Chem. Soc.*, **88**, 1821 (1988).
35. Strouse, C. E., and Dahl, L. F., *Discuss. Farad. Soc.*, **47**, 93 (1969).
    Strouse, C. E., and Dahl, L. F., *J. Amer. Chem. Soc.*, **93**, 6032 (1971).
36. Stevenson, D. L., Wei, C.-H., and Dahl, L. F., *J. Amer. Chem. Soc.*, **93**, 6027 (1971).
37. Fayet, P., McGlinchey, M. J., and Wöste, L. M., *J. Amer. Chem. Soc.*, **109**, 1733 (1987).
38. Halet, J.-F., *Nouv. J. Chimie*, **11**, 315 (1987).

3

Introduction to Tensor Surface Harmonic Theory

3.1 INTRODUCTION

In this chapter we will describe the basic theory and application of Stone's tensor surface harmonic (TSH) theory of the bonding in cluster compounds. Tensor surface harmonic theory was presented in its original form by Stone in 1980,[1] and the same author discussed applications to the bonding in transition metal clusters in 1981;[2] since then there have been a number of further developments.[3-7]

At this point a word of reassurance may be in order for readers with a less mathematical background. Please do not be tempted to skip through these sections simply because they contain the majority of the mathematics, for you will not then truly appreciate the beauty of some of the results which are to follow in later chapters. Some knowledge of group theory and the use of character tables is assumed, but we will proceed at a gentle pace wherever unfamiliar concepts may arise.

3.2 SOME HELPFUL ANALOGIES

3.2.1 The Linear Combination of Atomic Orbitals Method

Before we attempt to study three-dimensional cluster skeletons, it will be instructive to discuss some well-known results from simple Hückel theory[8] which will serve to introduce the linear combination of atomic orbitals–

molecular orbital (LCAO-MO) method. In the LCAO approach the solutions of the electrostatic Schrödinger equation

$$\hat{\mathcal{H}}\psi = E\psi, \qquad (3.1)$$

(where $\hat{\mathcal{H}}$ is the Hamiltonian operator and E the energy) are approximated by a linear combination of N atomic orbitals, χ_r, so that

$$\psi_{\text{LCAO}} = \sum_{s=1}^{N} c_s \chi_s. \qquad (3.2)$$

In the Rayleigh-Ritz method the expansion coefficients are treated as variable parameters and the variation principle is applied. This tells us that the best representation of the wavefunction in this form is obtained when the energy, evaluated with the trial function, is as low as possible. The expectation value of the energy for the expansion of Equation (3.2) is

$$E = \frac{\langle \psi | \hat{\mathcal{H}} | \psi \rangle}{\langle \psi | \psi \rangle} = \frac{\sum_r \sum_s c_r c_s \langle \chi_r | \hat{\mathcal{H}} | \chi_s \rangle}{\sum_r \sum_s c_r c_s \langle \chi_r | \chi_s \rangle}, \qquad (3.3)$$

where the coefficients are chosen to be real. We now set all the first-order partial derivatives of this expectation value with respect to the expansion coefficients equal to zero to find a turning point. This turning point is a minimum, and leads us to the so-called secular equations as follows:

$$\frac{\partial}{\partial c_t}\left[\frac{\langle \psi | \hat{\mathcal{H}} | \psi \rangle}{\langle \psi | \psi \rangle} \sum_r \sum_s c_r c_s \langle \chi_r | \chi_s \rangle - \sum_r \sum_s c_r c_s \langle \chi_r | \hat{\mathcal{H}} | \chi_s \rangle\right] = 0$$

$$\Rightarrow \underbrace{\frac{\partial}{\partial c_t}\frac{\langle \psi | \hat{\mathcal{H}} | \psi \rangle}{\langle \psi | \psi \rangle}}_{\text{(zero at a turning point)}} + 2\frac{\langle \psi | \hat{\mathcal{H}} | \psi \rangle}{\langle \psi | \psi \rangle}\sum_s c_s c_t \langle \chi_s | \chi_t \rangle - 2\sum_s c_s c_t \langle \chi_s | \hat{\mathcal{H}} | \chi_t \rangle = 0$$

$$\Rightarrow \sum_s c_s (\langle \chi_s | \hat{\mathcal{H}} | \chi_t \rangle - E \langle \chi_s | \chi_t \rangle) = 0 \qquad \text{for all } t. \qquad (3.4)$$

This may be written more compactly in matrix notation as

$$\det |\mathcal{H}_{st} - ES_{st}| = 0 \qquad (3.5)$$

where det denotes the determinant and the matrix elements $\mathcal{H}_{rt}$ and S_{rt} correspond to the integrals in the previous equation. Solving this $N \times N$ dimension determinantal problem gives N linearly independent molecular orbitals each with an associated energy eigenvalue.

3.2.2 The Hückel Approach

In the Hückel method for the π electron system of a conjugated polyene we consider linear combinations of the p^π orbitals, one per atom. A simple parameterization is adopted for the matrix elements of the Hamiltonian operator and the overlap matrix:

$$\mathcal{H}_{ts} = \begin{cases} \alpha & \text{if } t = s \\ \beta & \text{if } t \text{ and } s \text{ are on adjacent atoms} \\ 0 & \text{otherwise} \end{cases}$$

and (3.6)

$$S_{ts} = \begin{cases} 1 & \text{if } t = s \\ 0 & \text{otherwise.} \end{cases}$$

For a linear polyene, $C_n H_{n+2}$, the secular equations may be written

$$(\alpha - \epsilon)c_1 + \beta c_2 = 0,$$

$$\beta c_{t-1} + (\alpha - \epsilon)c_t + \beta c_{t+1} = 0 \quad \text{for } 1 < t < n-1,$$

and

$$\beta c_{n-1} + (\alpha - \epsilon)c_n = 0, \tag{3.7}$$

where ϵ is the energy and c_t is the LCAO coefficient of the p^π orbital at atom t. Using an appropriate trial solution and applying boundary conditions at the end-points of the molecule, we find that the orbitals and energy levels are

$$\psi_m = \sqrt{\frac{2}{n+1}} \sum_t p(t) \sin \frac{\pi t m}{n+1}$$

with (3.8)

$$\epsilon_m = \alpha + 2\beta \cos \frac{\pi m}{n+1},$$

where $p(t)$ is the p orbital at atom t and $1 \leq m \leq n$. These solutions are illustrated for some simple polyenes in Figure 3.1.

For a free electron in a one-dimensional box of length L, the Schrödinger equation is

$$-\frac{\hbar^2}{2m_e} \frac{d^2\psi}{dx^2} + V\psi = E\psi, \tag{3.9}$$

where the potential energy, V, is zero for $0 < x < L$ inside the box and infinitely large outside it. $\hbar$ is Planck's constant divided by 2π, and m_e is the

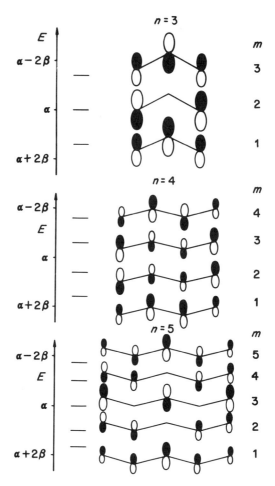

Figure 3.1 The Hückel π orbitals and energy levels of some linear conjugated polyenes.

mass of the electron. This equation is readily solved, and on applying boundary conditions at the ends of the box we find that

$$\psi_m = \sqrt{\frac{2}{L}} \sin \frac{m\pi x}{L}. \quad (3.10)$$

If we compare this result with the Hückel molecular orbitals for the linear polyene, above, we see that the coefficients in the sum can be interpreted as the values of the particle in-a-box wavefunction calculated at the appropriate atom sites. This merely requires us to identify $t/(n+1)$ in the Hückel coefficient at atom t with x/L in the corresponding particle in-a-box wavefunction. This correlation is illustrated for C_3H_5 in Figure 3.2. The atom positions

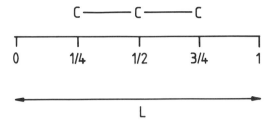

Figure 3.2 The atom positions for C_3H_5.

are at $t/(n + 1) = \tfrac{1}{4}, \tfrac{1}{2}$, and $\tfrac{3}{4}$, so that the appropriate box length, L, is actually four bond lengths ($n + 1$ bond lengths in general) with walls one bond length from either end of the molecule. This analysis suggests that the LCAO wavefunctions for a polyene can be derived from a free-electron model. This is a very important relationship, which will be developed further below for cyclic polyenes and cluster cages.

There is a similar correspondence between the Hückel wavefunctions for a cyclic polyene, $C_n H_n$, and the wavefunctions for a free electron constrained to move on a ring. In this case the secular equations are

$$(\alpha - \epsilon)c_1 + \beta c_2 + \beta c_n = 0,$$
$$\beta c_{t-1} + (\alpha - \epsilon)c_t + \beta c_{t+1} = 0 \quad \text{for } 1 < t < n, \quad (3.11)$$
$$\beta c_1 + \beta c_{n-1} + (\alpha - \epsilon)c_n = 0,$$

with unnormalized wavefunctions

$$\psi_m = \sum_t p(t) \sin \frac{2mt\pi}{n} \quad \text{or} \quad \sum_t p(t) \cos \frac{2mt\pi}{n} \quad (3.12)$$

and
$$\epsilon_m = \alpha + 2\beta \cos \frac{2m\pi}{n}.$$

There are n linearly independent wavefunctions which may be constructed for integral m starting from $m = 0$. The free electron constrained to move on a ring of radius R has Schrödinger equation

$$-\frac{\hbar^2}{2m_e R^2} \frac{d^2\psi}{d\phi^2} = E\psi, \quad (3.13)$$

where ϕ is the polar angle and ranges between zero and 2π. $f_{mc}(\phi) = \sqrt{1/\pi} \cos m\phi$ and $f_{ms}(\phi) = \sqrt{1/\pi} \sin m\phi$ are real solutions of this equation for integer m. The Hückel linear combinations for a cyclic hydrocarbon can be obtained from the free-electron wavefunctions using the following expansion:

$$\psi_{m\mu} = \sum_t f_{m\mu}(\phi_t) p(t), \quad (3.14)$$

where ϕ_t is the angular coordinate of carbon atom t and $\mu = c$ or s for cosine or sine combinations. For example, for benzene, $\phi_1 = 0$, $\phi_2 = \pi/3$, $\phi_3 = 2\pi/3$, $\phi_4 = \pi$, $\phi_5 = 4\pi/3$, and $\phi_6 = 5\pi/3$. Hence, since $f_{0c} = \sqrt{1/\pi}$, we find that

$$\psi_{0c} \propto p(1) + p(2) + p(3) + p(4) + p(5) + p(6). \tag{3.15}$$

Similarly,

$$\psi_{1s} \propto p(1) \sin 0 + p(2) \sin \frac{\pi}{3} + p(3) \sin \frac{2\pi}{3} + p(4) \sin \pi + p(5) \sin \frac{4\pi}{3} + p(6) \sin \frac{5\pi}{3}$$

$$\propto \frac{\sqrt{3}}{2} p(2) + \frac{\sqrt{3}}{2} p(3) - \frac{\sqrt{3}}{2} p(5) - \frac{\sqrt{3}}{2} p(6). \tag{3.16}$$

When we include normalization within the Hückel approximation, we simply obtain

$$\psi_{0c} = \frac{1}{\sqrt{6}} [p(1) + p(2) + p(3) + p(4) + p(5) + p(6)]$$

and $\quad \psi_{1s} = \frac{1}{2} [p(2) + p(3) - p(4) - p(5)].$
$$\tag{3.17}$$

Hence the Hückel linear combinations may be obtained by using the free electron on-a-ring wavefunctions evaluated at the atom positions as coefficients, so that ϕ is identified with $2\pi t/n$ at atom t. Substitution in Equation (3.12) gives the following forms for the wavefunctions and energy levels of the benzene π system, as illustrated in Figure 3.3 (normalization within the Hückel approximation is included by taking the square root of the reciprocal of the sum of squared coefficients for each function):

m	ψ_m	ϵ_m
0	$\sqrt{\frac{1}{6}}[p(1) + p(2) + p(3) + p(4) + p(5) + p(6)]$	$\alpha + 2\beta$
1	$\left\{ \begin{array}{l} \sqrt{\frac{1}{3}}[p(1) - \frac{1}{2}p(2) - p(3) - \frac{1}{2}p(4) + \frac{1}{2}p(5) + p(6)] \\ \frac{1}{2}[p(1) + p(2) - p(4) - p(5)] \end{array} \right\}$	$\alpha + \beta$
2	$\left\{ \begin{array}{l} \sqrt{\frac{1}{3}}[-\frac{1}{2}p(1) - \frac{1}{2}p(2) + p(3) - \frac{1}{2}p(4) - \frac{1}{2}p(5) + p(6)] \\ \frac{1}{2}[p(1) - p(2) + p(4) - p(5)] \end{array} \right\}$	$\alpha - \beta$
3	$\sqrt{\frac{1}{6}}[-p(1) + p(2) - p(3) + p(4) - p(5) + p(6)]$	$\alpha - 2\beta$

In summary, the Hückel linear combinations of p^π orbitals may be constructed using the wavefunctions of a corresponding free-electron problem evaluated at the atom positions as expansion coefficients. We should also note at this point that the energy of the free-electron wavefunction used in a given expansion, which increases as m^2 for the particle on a ring, is not a particularly accurate guide to the energy of the corresponding Hückel molecular orbital. However, the general trend is correct.

We now enter the realms of TSH theory by considering how similar

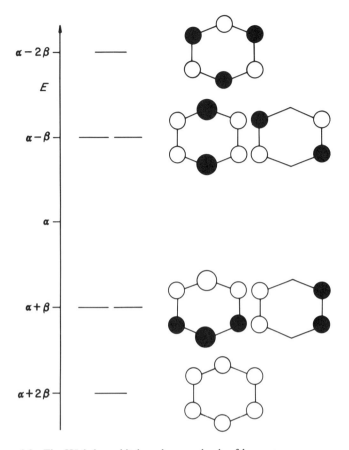

Figure 3.3 The Hückel π orbitals and energy levels of benzene.

results are of use for a cluster compound in which the cluster skeletal atoms lie approximately on the surface of a sphere. By analogy with the foregoing analyses, we might hope to produce the linear combinations which solve the Hückel problem from the wavefunctions for a free electron constrained to move on the surface of a sphere, namely, the spherical harmonics. In fact, this is a considerably harder problem than those considered above, but the underlying descent in symmetry from the sphere to an approximately spherical cluster with a finite number of vertices is indeed the basis of TSH theory.

3.3 INTRODUCTION TO SPHERICAL HARMONICS

The spherical harmonics are probably the most important and ubiquitous of the special functions encountered in theoretical chemistry. Readers who have persevered this far probably already know a great deal about their properties, even if they do not yet realize it, from their knowledge of the form of atomic orbitals.

3.3.1 The One-Electron Atom

The Schrödinger equation for a one-electron atom of atomic number Z is

$$-\frac{\hbar^2}{2m_e}\nabla^2\psi - \frac{Ze^2}{4\pi\varepsilon_0 r}\psi = E\psi, \quad (3.18)$$

where e is the electronic charge, ε_0 is the permittivity of free space, r is the electron-nucleus distance, and

$$\nabla^2 = \frac{1}{r^2}\frac{\partial}{\partial r}r^2\frac{\partial}{\partial r} + \frac{1}{r^2\sin\theta}\frac{\partial}{\partial\theta}\sin\theta\frac{\partial}{\partial\theta} + \frac{1}{r^2\sin^2\theta}\frac{\partial^2}{\partial\phi^2}. \quad (3.19)$$

We may rearrange Equation (3.18) so that the left-hand side is independent of θ and ϕ and the right-hand side is independent of r:

$$\left[\frac{\partial}{\partial r}r^2\frac{\partial}{\partial r} + \frac{2m_e r^2}{\hbar^2}\left(E + \frac{Ze^2}{4\pi\varepsilon_0 r}\right)\right]\psi = -\left(\frac{1}{\sin\theta}\frac{\partial}{\partial\theta}\sin\theta\frac{\partial}{\partial\theta} + \frac{1}{\sin^2\theta}\frac{\partial^2}{\partial\phi^2}\right)\psi. \quad (3.20)$$

The solutions to this equation take the familiar form $\phi_{NLM} = R_{N,L}(r)Y_{L,M}(\theta, \phi)$, where $Y_{L,M}(\theta, \phi)$ is a spherical harmonic. This follows because the separation in the above equation implies that the solutions can be written as a simple product of functions depending upon r alone and upon θ and ϕ. Furthermore, both sides must be equal to a constant, independent of r, θ, and ϕ, so that the part of the solution which depends upon θ and ϕ is simply a spherical harmonic. Because the radial part of the solution is invariant to rotations and reflections which leave the origin unchanged, it follows that all the angular variation is contained in $Y_{L,M}$. The same is true for the solutions of Equation (3.18) with any potential energy term which does not depend upon orientation (i.e., θ and ϕ); the Coulomb potential appropriate here is one such example.

L and M are the quantum numbers for the spherical harmonic $Y_{L,M}(\theta, \phi)$. L may be any nonnegative integer, and M is any integer satisfying $-L \leq M \leq L$. For $M > 0$, these functions are partly imaginary, and we will usually find it more convenient to use the real linear combinations

$$Y_{L,Mc} = \sqrt{\frac{1}{2}}[(-1)^M Y_{L,M} + Y_{L,-M}]$$

$$Y_{L,Ms} = \frac{1}{i}\sqrt{\frac{1}{2}}[(-1)^M Y_{L,M} - Y_{L,-M}], \quad (3.21)$$

where $i = \sqrt{-1}$. The c and s subscripts are used because the above combinations are proportional to $\cos M\phi$ and $\sin M\phi$, respectively, and the functions with $L = 0$, 1, and 2 are given in Table 3.1.

TABLE 3.1 THE FUNCTIONS Y_{L,M_c} AND Y_{L,M_s} FOR $L = 0, 1,$ AND 2. r IS THE RADIAL DISTANCE

	Spherical polar form		Cartesian form	
L, M	$Y_{L,M_c}(\theta, \phi)$	$Y_{L,M_s}(\theta, \phi)$	$Y_{L,M_c}(x, y, z)$	$Y_{L,M_s}(x, y, z)$
0, 0	$\sqrt{\dfrac{1}{4\pi}}$		$\sqrt{\dfrac{1}{4\pi}}$	
1, 0	$\sqrt{\dfrac{3}{4\pi}} \cos\theta$		$\sqrt{\dfrac{3}{4\pi}} \dfrac{z}{r}$	
1, 1	$\sqrt{\dfrac{3}{4\pi}} \sin\theta \cos\phi$	$\sqrt{\dfrac{3}{4\pi}} \sin\theta \sin\phi$	$\sqrt{\dfrac{3}{4\pi}} \dfrac{x}{r}$	$\sqrt{\dfrac{3}{4\pi}} \dfrac{y}{r}$
2, 0	$\sqrt{\dfrac{5}{4\pi}} \dfrac{1}{2} (3\cos^2\theta - 1)$		$\sqrt{\dfrac{5}{4\pi}} \dfrac{1}{2} \left(\dfrac{3z^2}{r^2} - 1 \right)$	
2, 1	$\sqrt{\dfrac{5}{4\pi}} \dfrac{3}{4} \sin\theta \cos\theta \cos\phi$	$\sqrt{\dfrac{5}{4\pi}} \dfrac{3}{4} \sin\theta \cos\theta \sin\phi$	$\sqrt{\dfrac{5}{4\pi}} \dfrac{3}{4} \dfrac{xz}{r^2}$	$\sqrt{\dfrac{5}{4\pi}} \dfrac{3}{4} \dfrac{yz}{r^2}$
2, 2	$\sqrt{\dfrac{5}{4\pi}} \dfrac{3}{16} \sin^2\theta \cos 2\phi$	$\sqrt{\dfrac{5}{4\pi}} \dfrac{3}{16} \sin^2\theta \sin 2\phi$	$\sqrt{\dfrac{5}{4\pi}} \dfrac{3}{16} \dfrac{x^2 - y^2}{r^2}$	$\sqrt{\dfrac{5}{4\pi}} \dfrac{3}{16} \dfrac{2xy}{r^2}$

Of course, when we wish to visualize atomic orbitals, we normally think about real functions, such as p_x and p_y, which correspond to the combinations $Y_{L,Mc}$ and $Y_{L,Ms}$ already discussed. For example, the angular parts of d_{z^2}, d_{xz}, d_{yz}, $d_{x^2-y^2}$, and d_{xy} atomic orbitals arise in Table 3.1 as $Y_{2,0}$, $Y_{2,1c}$, $Y_{2,1s}$, $Y_{2,2c}$, $Y_{2,2s}$. Hence the spherical harmonics describe the angular dependence of simple atomic orbitals. The latter may therefore be used as a mnemonic for many of the properties of the spherical harmonics (parity under inversion, for example).

3.3.2 Motion of a Free Electron on a Sphere

The Schrödinger equation for a free electron constrained to move on the surface of a sphere is

$$-\frac{\hbar^2}{2m_e}\nabla^2\psi = E\psi, \qquad (3.22)$$

where
$$\nabla^2 = \frac{1}{\sin\theta}\frac{\partial}{\partial\theta}\sin\theta\frac{\partial}{\partial\theta} + \frac{1}{\sin^2\theta}\frac{\partial^2}{\partial\phi^2} \qquad (3.23)$$

in spherical polar coordinates and the radial terms have been dropped. The solutions of this equation are simply the spherical harmonics, and the energy eigenvalues are $E_L = \hbar^2 L(L+1)/2m_e R^2$, where R is the radius of the sphere.

3.4 σ CLUSTER ORBITALS

In TSH theory the basis atomic orbitals out of which combinations are formed are classified according to how many nodal planes containing the radius vector they possess. They are designated σ, π, and δ for 0, 1, and 2 such nodes, respectively, and a linear combination of σ-type basis functions, for example, is called a σ cluster orbital. Clearly, an s atomic orbital has σ symmetry in this context, and so have radially pointing p_z and d_{z^2} orbitals, where the local z axis at each vertex is chosen to point radially outwards. Examples are illustrated in Figure 3.4.

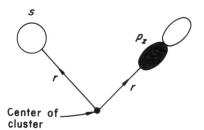

Figure 3.4 s and p_z σ-type basis atomic orbitals.

The general (unnormalized) form of a σ cluster orbital derived from the wavefunctions for a free electron on-a-sphere (see 3.3.2) is then taken to be

$$L^\sigma_\mu = \sum_t Y_{L,\mu}(\theta_t, \phi_t)\sigma(t), \tag{3.24}$$

where $\mu = Mc$ or Ms, $\sigma(t)$ is the σ orbital at vertex t, and θ_t and ϕ_t are the angular coordinates of this vertex in the global coordinate system. Alternatives to the L^σ_μ notation are to give the Cartesian form of the spherical harmonic in the subscript or simply the irreducible representation spanned by the orbitals in question. Hence for the octahedron the following forms are equivalent: $P^\sigma_{0,1c,1s}$, $P^\sigma_{z,x,y}$, and $P^\sigma(T_{1u})$. The appropriate Cartesian forms can easily be found from Table 3.1. These σ cluster orbitals transform in the same manner as the spherical harmonic expansion functions themselves under the operations of the molecular point group.

Let us now illustrate all the above points for an octahedral cluster such as Li_6 or $BeLi_6$ and $MgNa_6$ (with interstitial atoms), for which extensive *ab initio* calculations have been performed.[9] The same considerations would obviously apply to octahedral CLi_6 with an interstitial carbon atom.[10] The coordinate system is defined in Figure 3.5. The appropriate point group is clearly O_h, and the s and p orbitals of the carbon atom span A_{1g} and T_{1u}, respectively (note that these central p orbitals transform like x, y, and z, so that these results merely require consultation with the right-hand column of the character table for O_h). The representation spanned by the six lithium $2s$ orbitals, Γ_s, is found by inspection to be

O_h (m3m)	E	$8C_3$	$6C_2$	$6C_4$	$3C_2$ ($=C_4^2$)	i	$6S_4$	$8S_6$	$3\sigma_h$	$6\sigma_d$
Γ_s	6	0	0	2	2	0	0	0	4	2

since the character of Γ_s for any operation of the point group is simply the number of orbitals which remain unshifted by that operation. Reducing this, we find that $\Gamma_s = A_{1g} \oplus T_{1u} \oplus E_g$ and the required cluster orbitals are S^σ_0

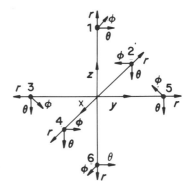

Figure 3.5 The labeling and axis system used for octahedral molecules.

(spanning A_{1g}), P_x^σ, P_y^σ, and P_z^σ (spanning T_{1u} and abbreviated $P_{0,1c,1s}^\sigma(T_{1u})$), and $D_{x^2-y^2}^\sigma$ and $D_{z^2}^\sigma$ (spanning E_g and abbreviated $D_{0,2c}^\sigma(E_g)$). These combinations are illustrated in Figure 3.6, and we see that the energy is expected to increase with the number of nodes. Since the number of angular nodes in the expansion function $Y_{L,Mc}$ or $Y_{L,Ms}$ is simply L, we see that the cluster orbital energies should increase with L. This result is confirmed by more detailed considerations[6] and by numerous calculations performed using cluster orbital basis sets.

In general the required cluster orbitals are found by matching the symmetries contained in Γ_s by the irreducible representations spanned by the spherical harmonics starting from $L = 0$, $M = 0$. This produces a set of cluster orbitals which span the same representation as the original basis set. Usually they will also be linearly independent, especially if the molecule has high symmetry. If it turns out that the set is not linearly independent, then one may remove the offending function with the largest values of L and M and replace it with another independent function (based on higher M and/or L) which must transform in the same way. For example, $D_{xz,yz,xy}^\sigma$ and $P_{x,y,z}^\sigma$ are not

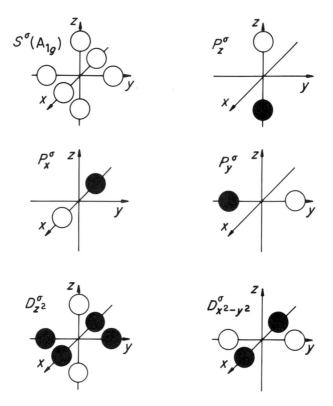

Figure 3.6 The σ cluster orbitals of an octahedron.

linearly independent for a tetrahedron, and we could choose either set (they are in fact equivalent). By convention the smallest values of L and M are used.

Using the appropriate formulas for the spherical harmonics (Table 3.1), the linear combinations for these σ cluster orbitals can be rapidly derived. The angular coordinates, (θ, ϕ), of the octahedron in degrees are $(0, 0)$, $(90, 180)$, $(90, 270)$, $(90, 0)$, $(90, 90)$, and $(180, 0)$ (in order corresponding to Figure 3.5), and hence we obtain (after dividing by the square root of the sum of squared coefficients for normalization):

Orbital	LCAO form	Energy
$S_0^\sigma(A_{1g})$	$\sqrt{\tfrac{1}{6}}[\sigma(1) + \sigma(2) + \sigma(3) + \sigma(4) + \sigma(5) + \sigma(6)]$	$\alpha + 4\beta$
$P_{0,1c,1s}^\sigma(T_{1u})$	$\begin{cases}\sqrt{\tfrac{1}{2}}[\sigma(1) - \sigma(6)] \\ \sqrt{\tfrac{1}{2}}[\sigma(4) - \sigma(2)] \\ \sqrt{\tfrac{1}{2}}[\sigma(3) - \sigma(5)]\end{cases}$	α
$D_{0,2c}^\sigma(E_g)$	$\begin{cases}\tfrac{1}{2}[\sigma(2) + \sigma(4) - \sigma(3) - \sigma(5)] \\ \sqrt{1/12}[2\sigma(1) + 2\sigma(6) - \sigma(2) - \sigma(3) - \sigma(4) - \sigma(5)]\end{cases}$	$\alpha - 2\beta$

For example, to form P_z^σ, simply use $Y_{1,1c} = \sqrt{3/4\pi} \cos\theta$ to obtain

$$P_z^\sigma \propto \sigma(1) - \sigma(6). \tag{3.25}$$

The normalization within the Hückel approximation is then clearly $[1^2 + (-1)^2]^{-1/2} = \sqrt{\tfrac{1}{2}}$, and

$$\left\langle \sqrt{\tfrac{1}{2}}[\sigma(1) - \sigma(2)] \middle| \hat{\mathscr{H}} \middle| \sqrt{\tfrac{1}{2}}[\sigma(1) - \sigma(2)] \right\rangle = \frac{2\alpha}{2} = \alpha, \tag{3.26}$$

again within the Hückel approximation.

These linear combinations are identical to those found using conventional group theoretical techniques and will be familiar to those readers who have derived the a_{1g}, e_g, and t_{1u} symmetry-adapted linear combinations of the ligand donor σ orbitals for an octahedral transition metal complex. This connection between the TSH theory methodology and the group theoretical approach is explained in more detail below.

The Hückel energies reflect the number of atoms with nonzero coefficients and the relative signs of these coefficients. For example, for S^σ, the energy of $\alpha + 4\beta$ can be deduced simply from the fact that all the orbitals have the same coefficient and four nearest neighbors. Using the above results, we may now construct the energy-level diagram of Figure 3.7 for the σ cluster orbitals of an octahedron. *Ab initio* calculations show that the clusters BeLi$_6$ and MgNa$_6$, with eight valence electrons each and interstitial beryllium and magnesium atoms, have the greatest atomization energy per atom. In both these clusters the S^σ and P^σ shells must be precisely filled—the effect of the interstitial atom is simply to generate bonding and antibonding combinations of s/S^σ and p/P^σ,

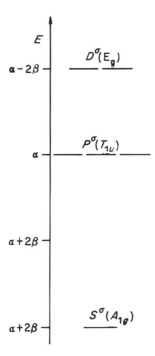

Figure 3.7 A qualitative energy-level diagram for Li_6.

as we shall see in Chapter 4. We would also expect the isoelectronic CLi_6^{2+} molecule, with an interstitial carbon atom, to be more stable than the neutral molecule.

The above example is particularly clear-cut because the atoms of the cluster shell have only one simple σ-type valence orbital. The procedures for dealing with π and δ orbitals will be discussed later. Before we do this, however, we should explain how the method deals with clusters where each vertex has more than one σ-type valence orbital, since the vast majority of interesting systems are actually of this type. For example, in a simple borohydride cluster, such as $B_6H_6^{2-}$, each boron atom has two σ valence orbitals: one $2s$ and one radial $2p_z$. The representation spanned by the $2s$ orbitals is obviously the same as that spanned by the $2p_z$ orbitals, so we simply form cluster orbitals using the same sets of coefficients for both types of basis functions. Hence there will be $S_0^\sigma(s)$ and $S_0^\sigma(p_z)$ molecular orbitals with the same coefficients for basis functions at the same vertex. In fact, to reduce mixing between cluster orbitals, it is more appropriate to combine the s and p_z orbitals to give inwardly and outwardly directed sp hybrid orbitals. We may then form cluster orbitals using the same expansion coefficients again; hence there is an S_0^σ cluster orbital for both the inwardly and the outwardly pointing hybrids.

The bonding in some apparently more complex clusters can also be understood by considering only the σ cluster orbitals, as discussed in Chapter 4. In these molecules, exemplified by a number of gold clusters, the valence-shell

p orbitals lie much higher in energy than the s orbitals, and the tangential p orbitals do not contribute any accessible skeletal orbitals.

3.4.1 Transformation Properties of Spherical Harmonics

In much of the following work we will have to determine the transformation properties of spherical harmonics in particular point groups. The rotation group of the spherical one-electron atom considered above is the *full rotation group*, consisting of all the possible rotations which leave the origin unchanged. All such rotations are symmetry elements because the energy of the atom does not depend upon its orientation in space, and each set of spherical harmonics $Y_{L,M}$ with $-L \leq M \leq L$ is a basis for a $(2L+1)$-dimensional irreducible representation of the group.

There are several ways to work out the representations spanned by any given set of spherical harmonics in a lower symmetry point group, and we will now go through the most useful of these. Here, and in subsequent sections, the reader will find a good set of character tables invaluable.[11] The first way of finding the representation spanned by the $2L+1$ spherical harmonics with fixed L and $-L \leq M \leq L$ is to simply work out the character for each class of operation in the point group. If $\chi^L(\hat{\mathcal{R}})$ represents the character for operation $\hat{\mathcal{R}}$, then the following formulas may be applied:

$$\chi^L(E) = 2L + 1$$

$$\chi^L(C_n) = \frac{\sin(L + \tfrac{1}{2})\, 2\pi/n}{\sin(\pi/n)}$$

$$\chi^L(i) = (-1)^L(2L + 1) \quad (3.27)$$

$$\chi^L(\sigma) = 1$$

$$\chi^L(S_n) = (-1)^L \frac{\cos(L + \tfrac{1}{2})\, 2\pi/n}{\cos(\pi/n)},$$

where E is the identity, C_n is a rotation through $2\pi/n$, i is the inversion operation, σ is a reflection, and S_n is a rotation through $2\pi/n$ followed by reflection in a plane perpendicular to the rotation axis. The representation so formed is then reduced using the standard formula

$$n_\alpha = \sum_{\hat{\mathcal{R}}} \chi^L(\hat{\mathcal{R}})\chi^\alpha(\hat{\mathcal{R}}), \quad (3.28)$$

where n_α is the number of times irreducible representation α of the lower symmetry point groups occurs with the spherical harmonic representation of character χ^L.

The second method simply involves looking up the effect of reduction in symmetry using descent in symmetry tables such as Table 3.2. However, where

TABLE 3.2 SYMMETRIES OF THE SPHERICAL HARMONICS IN SOME FINITE POINT GROUPS

	O_h	T_d	C_{3v}	D_3
S^σ	A_{1g}	A_1	A_1	A_1
P^σ, P^π	T_{1u}	T_2	$A_1 + E$	$A_2 + E$
D^σ, D^π	$E_g + T_{2g}$	$E + T_2$	$A_1 + 2E$	$A_1 + 2E$
F^σ, F^π	$A_{2u} + T_{1u} + T_{2u}$	$A_1 + T_1 + T_2$	$2A_1 + A_2 + 2E$	$A_1 + 2A_2 + 2E$
G^σ, G^π	$A_{1g} + E_g + T_{1g} + T_{2g}$	$A_1 + E + T_1 + T_2$	$2A_1 + A_2 + 3E$	$2A_1 + A_2 + 3E$
$\bar{\Gamma}^\sigma_u$	A_{1u}	A_2	A_2	A_1
$\bar{P}^\pi$	T_{1g}	T_1	$A_2 + E$	$A_2 + E$
$\bar{D}^\pi$	$E_u + T_{2u}$	$E + T_1$	$A_2 + 2E$	$A_1 + 2E$
$\bar{F}^\pi$	$A_{2g} + T_{1g} + T_{2g}$	$A_2 + T_1 + T_2$	$A_1 + 2A_2 + 2E$	$A_1 + 2A_2 + 2E$
$\bar{G}^\pi$	$A_{1u} + E_u + T_{1u} + T_{2u}$	$A_2 + E + T_1 + T_2$	$A_1 + 2A_2 + 3E$	$2A_1 + A_2 + 3E$

such a tabulation is not available (which is often the case for larger values of L), the method of the previous paragraph may be employed.

Finally, we can make use of the transformation properties of x, y, and z and functions thereof which are usually given in the final columns of character tables. Generally, such information is provided for all the real combinations of spherical harmonics given in Table 3.1, and some tables go to higher L as well. These data are often the most valuable of all because they tell us how individual $Y_{L,Mc}$ and $Y_{L,Ms}$ transform. For example, in O_h, S^σ transforms like $x^2 + y^2 + z^2$, i.e., A_{1g}; P^σ transforms like x, y, and z, i.e., T_{1u}; and D_0^σ and D_{2c}^σ transform like z^2 and $x^2 - y^2$, i.e., like E_g. Direct products may also be taken to give the combinations appropriate for higher values of L, and this may also be helpful when used in conjunction with the results obtained by these two methods.

3.5 π CLUSTER ORBITALS

3.5.1 How Do π Orbitals Transform?

Notice that a simple-minded approach which just involves counting spherical harmonic nodes would indeed give a reasonable guide to the energies of the σ cluster orbitals in the previous section. However, for molecular orbitals which are linear combinations of π-type valence orbitals, this could not be expected to work, because of the intrinsic nodes in the functions themselves. Stone first solved this problem when he noted that a set of tangential p orbitals, for instance, transform among themselves under rotations like a set of unit vectors. This is a very important point, so we shall take some trouble to explain it clearly.

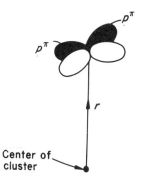

Figure 3.8 Tangential p^π orbitals.

Consider the two tangential p orbitals at any cluster vertex; these functions are orthogonal and lie entirely in the plane which makes a tangent to the (hypothetical) sphere described by the cluster skeleton at the given vertex, as illustrated in Figure 3.8. Now consider any cluster molecular orbital and suppose that the coefficients of these two p orbitals are a and b, so that the term $ap^a + bp^b$ appears in the expansion. Now imagine rotating the two p orbitals about the axis which passes through the origin and the vertex in question to give $p^{a'}$ and $p^{b'}$. The molecular orbital remains unchanged if the term $ap^a + bp^b$ is replaced by $a'p^{a'} + b'p^{b'}$ where a' and b' are new coefficients whose values can always be determined. The p orbital coefficients are like components of a vector: when the p orbitals are rotated, the components of that vector change accordingly, but the molecular orbital does not change. Explicitly, we can denote unit vectors in the direction of the two tangential p orbitals by **a** and **b**. Then we can write

$$ap^a + bp^b = \mathbf{V} \cdot \mathbf{a} p^a + \mathbf{V} \cdot \mathbf{b} p^b \qquad (3.29)$$

for some vector **V**. But

$$\mathbf{V} \cdot \mathbf{a} p^a + \mathbf{V} \cdot \mathbf{b} p^b = \mathbf{V} \cdot \mathbf{a'} p^{a'} + \mathbf{V} \cdot \mathbf{b'} p^{b'} \qquad (3.30)$$

then defines the coefficients with respect to the rotated basis, so that the p orbitals can be interpreted as unit vectors specifying an orientation. Furthermore, it is clear that we can choose the orientation arbitrarily, and hence to give the clearest picture of the bonding.

3.5.2 Choosing Coefficients for π Orbitals

The problem remains of how to choose the coefficients in a linear combination of π orbitals so as to produce useful approximate molecular orbitals. The previous paragraphs showed that such a choice is equivalent to the selection of a tangential vector at each vertex, and that the coefficients are then the components of this vector along the tangential directions defined by the p orbitals. Hence we use the gradient vector of a spherical harmonic evaluated at each vertex point to form the coefficients in this case. This vector is denoted by $\nabla Y_{L,\mu}$ and is given by

$$\nabla Y = \left(\frac{\partial Y}{\partial x}, \frac{\partial Y}{\partial y}, \frac{\partial Y}{\partial z}\right) \quad \text{in Cartesian components}$$

or (3.31)

$$\nabla Y = \frac{\partial Y}{\partial r}\hat{\mathbf{e}}_r + \frac{1}{r}\frac{\partial Y}{\partial \theta}\hat{\mathbf{e}}_\theta + \frac{1}{r\sin\theta}\frac{\partial Y}{\partial \phi}\hat{\mathbf{e}}_\phi = \mathbf{V} \quad \text{in spherical polar coordinates}$$

where $\hat{\mathbf{e}}_r$, $\hat{\mathbf{e}}_\theta$, and $\hat{\mathbf{e}}_\phi$ are unit vectors in the direction of r, θ, and ϕ increasing, respectively. Since spherical harmonics have no dependence upon r (when written in terms of θ and ϕ), the $\hat{\mathbf{e}}_r$ component is always zero, so that the gradient vector is always tangential. Furthermore, **V** transforms under rotations in the same way as the parent spherical harmonic, and satisfies the free electron on-a-sphere problem. Hence we form π cluster orbitals by evaluating the gradient vector at each cluster vertex and taking its dot product with the unit vectors in the direction of the tangential p orbitals. The gradient vector, $\mathbf{V} = \nabla Y$, is a vector surface harmonic, or a tensor surface harmonic of rank 1.

The vector function $\bar{\mathbf{V}} = \mathbf{r} \wedge \nabla Y = \mathbf{r} \wedge \mathbf{V}$ is perpendicular to **V**, because $\mathbf{r} \wedge \mathbf{V} \cdot \mathbf{V} = 0$, and it is also tangential. ($\wedge$ represents the vector cross product operation.) Again it transforms like the parent spherical harmonic, Y, under rotations and is a second kind of vector surface harmonic. Since there are only two independent tangential directions at any point, these are the only types of vector which we need to consider, and the nonzero components are given by

$$\mathbf{V} = \frac{\partial Y}{\partial \theta}\hat{\mathbf{e}}_\theta + \frac{1}{\sin\theta}\frac{\partial Y}{\partial \phi}\hat{\mathbf{e}}_\phi$$

and (3.32)

$$\bar{\mathbf{V}} = -\frac{1}{\sin\theta}\frac{\partial Y}{\partial \phi}\hat{\mathbf{e}}_\theta + \frac{\partial Y}{\partial \theta}\hat{\mathbf{e}}_\phi.$$

Under the inversion operation, $\mathbf{V}_{L,\mu}$ has parity $(-1)^L$, the same as the parent spherical harmonic, while $\bar{\mathbf{V}}_{L,\mu}$ has the opposite parity $(-1)^{L+1}$. Hence the two types of vector are known as even and odd, respectively, as are the cluster orbitals formed from them. From the previous equation we also note that the θ component of $\bar{\mathbf{V}}$, written $\bar{V}^\theta$, is equal to the negative ϕ component of **V**, while $\bar{V}^\phi = V^\theta$. Odd π cluster orbitals may therefore be obtained from their even partners by 90° rotations of the π orbitals and associated coefficients about the radius vectors of their respective atoms. Note that the θ and ϕ components would be appropriate to p orbitals oriented in the θ and ϕ increasing directions, and the linear combinations may then be written

$$\psi^\pi_{L,\mu} = \sum_i [V^\theta_{L,\mu}(i)p^\theta(i) + V^\phi_{L,\mu}(i)p^\phi(i)]$$

(3.33)

$$\bar{\psi}^\pi_{L,\mu} = \sum_i [\bar{V}^\theta_{L,\mu}(i)p^\theta(i) + \bar{V}^\phi_{L,\mu}(i)p^\phi(i)].$$

TABLE 3.3 VECTOR SPHERICAL HARMONICS IN SPHERICAL POLAR FORM*

L, M	$V^\theta_{L,Mc}$	$V^\theta_{L,Ms}$	$V^\phi_{L,Mc}$	$V^\phi_{L,Ms}$
1, 0	$\sqrt{\dfrac{3}{4\pi}}\sin\theta$			
1, 1	$\sqrt{\dfrac{3}{4\pi}}\cos\theta\cos\phi$	$\sqrt{\dfrac{3}{4\pi}}\cos\theta\sin\phi$	$-\sqrt{\dfrac{3}{4\pi}}\sin\phi$	$\sqrt{\dfrac{3}{4\pi}}\cos\phi$
2, 0	$-\sqrt{\dfrac{5}{4\pi}}\,3\cos\theta\sin\theta$			
2, 1	$\sqrt{\dfrac{15}{16\pi}}\cos 2\theta\cos\phi$	$\sqrt{\dfrac{15}{16\pi}}\cos 2\theta\sin\phi$	$-\sqrt{\dfrac{15}{16\pi}}\cos\theta\sin\phi$	$\sqrt{\dfrac{15}{16\pi}}\cos\theta\cos\phi$
2, 2	$\sqrt{\dfrac{15}{64\pi}}\sin 2\theta\cos 2\phi$	$\sqrt{\dfrac{15}{64\pi}}\sin 2\theta\sin 2\phi$	$-\sqrt{\dfrac{15}{16\pi}}\sin\theta\sin\phi$	$\sqrt{\dfrac{15}{16\pi}}\sin\theta\cos 2\phi$

*Since $\bar{V}^\theta = -V^\phi$ and $\bar{V}^\phi = V^\theta$, the odd vector spherical harmonic components can also be found from this table.

These components are tabulated in Table 3.3 for $L = 1$ and 2; for $L = 0$, the derivatives are all zero, so that S^π cluster orbitals do not exist.

A simple way to sketch σ and π cluster orbitals involves drawing an atomic orbital based upon the same spherical harmonic, $Y_{L,\mu}(\theta, \phi)$, at the center of a sphere. The corresponding σ cluster orbital expansion coefficients match this function in phase and magnitude: there are nodes matching the nodal surfaces of $Y_{L,\mu}(\theta, \phi)$ and extrema matching the turning points of $Y_{L,\mu}(\theta, \phi)$. For the π cluster orbitals, on the other hand, the gradient vector may be represented by an arrow on the surface of the sphere, whose magnitude is largest at the nodes of $Y_{L,\mu}(\theta, \phi)$ and zero at the extrema. (The rate of change of $Y_{L,\mu}(\theta, \phi)$ is greatest at the nodes, and zero, by definition, at the extrema.) The corresponding odd vector spherical harmonics can be obtained by rotating each vector by 90°, all in the same sense. Some examples are given in Figure 3.9.

To form a basis of cluster π orbitals, we start by forming and reducing the representation spanned by the π basis orbitals. Even for large, complicated clusters it is easy to do this by inspection. For example, if the cluster has n vertices, then the number of π orbitals of any given type is $2n$, and the character under the identity operation is obviously $2n$. The character under the inversion operation is zero, because all the atoms move. The character under all reflections is also zero, because at any vertex which does not move

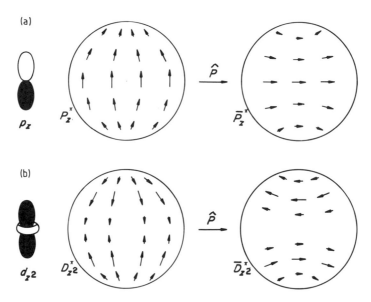

Figure 3.9 Sketching the gradient vector of $Y_{L,\mu}(\theta, \phi)$ using a central atomic orbital: (a) P_z^π and $\bar{P}_z^\pi$ from p_z, (b) $D_{z^2}^\pi$ and $\bar{D}_{z^2}^\pi$ from d_{z^2}.

under the operation we can choose one p orbital to lie in the plane and the other to lie perpendicular to it. The character under rotations is only nonzero when one or two vertices lie on the axis in question. For a rotation axis of order n, one may find the character simply by looking up how x and y transform in the point group C_n. In fact, there is a simple relationship between the representations spanned by the σ and π cluster orbitals defined by

$$\Gamma_{\pi/\bar{\pi}} \cong \Gamma_\sigma \otimes \Gamma_{x,y,z} - \Gamma_\sigma, \qquad (3.34)$$

where $\otimes$ denotes the direct product. This follows because the character of Γ_σ is simply the number of cluster vertices which are invariant under any given point group operation, while the character of $\Gamma_{x,y,z}$ defines the transformation properties of three p orbitals under the same operation.[12] Since the p_z orbital is actually of the σ type, we need to subtract Γ_σ from the direct product to obtain Γ_π.

Starting from $L = 1$, $M = 0$, we construct L_μ^π and $\bar{L}_\mu^\pi$ cluster orbitals until we have matched the irreducible representations spanned by the basis π functions in number and type. This will usually give a linearly independent set of functions, but if it does not then one can substitute even and odd π cluster orbitals with higher values of L and/or M for functions of the same symmetry until linear independence is achieved. L_μ^π cluster orbitals transform in the same way as the parent spherical harmonic $Y_{L,\mu}$ under all the operations of the point group. $\bar{L}_\mu^\pi$ orbitals transform in the same way under rotations, but with a sign change under reflections, inversions, and rotation-reflection operations. Two examples are given below.

For transition metal clusters, there are d orbitals of π symmetry as well as p orbitals. In local axes these are the d_{xz} and d_{yz} functions as illustrated in Figure 3.10. We can form $L_\mu^\pi(d)$ and $\bar{L}_\mu^\pi(d)$ cluster orbitals from these in precisely the same way as we would for cluster orbitals composed of p^π orbitals. The p^π and d^π orbitals could also be hybridized and expansions formed from the resulting hybrids, as for s^σ and p^σ orbitals in the previous Section 3.4.

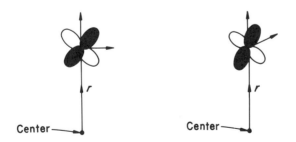

Figure 3.10 d orbitals of π-type symmetry.

Sec. 3.5 π Cluster Orbitals

3.5.3 Example: An Equilateral Triangular Cluster

As a first example consider the equilateral triangular cluster illustrated in Figure 3.11, with D_{3h} point group symmetry. It is a simple matter to show that $\Gamma_\sigma \cong A_1' \oplus E'$ and $\Gamma_{\pi/\bar{\pi}} \cong A_2'' \oplus E' \oplus A_2' \oplus E''$ using the rules discussed above to form the required representations. We can also derive $\Gamma_{\pi/\bar{\pi}}$ from Equation (3.34): $\Gamma_{\pi/\bar{\pi}} \cong \Gamma_\sigma \otimes \Gamma_{x,y,z} - \Gamma_\sigma \cong (A_1' \oplus E') \otimes (A_2'' \oplus E') - (A_1' \oplus E')$, etc. The σ cluster orbitals are therefore $S_0^\sigma(A_1')$ and $P_{1c,1s}^\sigma(E') \equiv P_{x,y}^\sigma$, as shown in Figure 3.12.

The π cluster orbitals are more interesting. Because the p^θ orbitals are perpendicular to the horizontal mirror plane, they must form bases for irreducible representations which are antisymmetric with respect to reflection in this plane. In contrast, the p^ϕ orbitals form bases for irreducible representations which are symmetric with respect to reflection in the plane of the triangle. Hence, for this cluster, the π cluster orbitals are linear combinations of p^θ or p^ϕ orbitals, but not both.

The angular coordinates of the vertices in degrees are (90, 0), (90, 120), and (90, 240), so using Table 3.2, we obtain the following linear combinations, which are illustrated in Figure 3.13 (the P_0^π and $\bar{P}_0^\pi$ combinations can be related to the sketches of Figure 3.9):

Orbital	LCAO form	Energy
$P_0^\pi(A_2'')$	$\dfrac{1}{\sqrt{3}}[p^\theta(1) + p^\theta(2) + p^\theta(3)]$	$\alpha + 2\beta$
$\bar{P}_0^\pi(A_2')$	$\dfrac{1}{\sqrt{3}}[p^\phi(1) + p^\phi(2) + p^\phi(3)]$	$\alpha - 2\beta$
$P_{1c,1s}^\pi(E')$	$\left\{\begin{array}{l}\frac{1}{\sqrt{6}}[2p\phi(1) - p^\phi(2) - p^\phi(3)] \\ \frac{1}{\sqrt{2}}[p^\phi(2) - p^\phi(3)]\end{array}\right\}$	$\alpha + \beta$
$\bar{P}_{1c,1s}^\pi(E'')$	$\left\{\begin{array}{l}\frac{1}{\sqrt{6}}[2p^\theta(1) - p^\theta(2) - p^\theta(3)] \\ \frac{1}{\sqrt{2}}[p^\theta(2) - p^\theta(3)]\end{array}\right\}$	$\alpha - \beta$

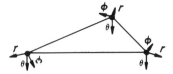

Figure 3.11 The coordinate system for a triangular cluster.

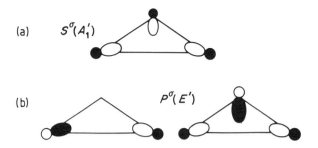

Figure 3.12 σ cluster orbitals for the triangular cluster: (a) $S_0^\sigma(A_1')$; (b) the two $P_{1c,1s}^\sigma(E')$ orbitals.

The $P_{x,y}^\pi \equiv P_{1c,1s}^\pi$ orbitals, for example, are constructed as follows:

$$P_x^\pi \propto -\sqrt{\frac{3}{4\pi}}[p^\phi(1)\sin 0° + p^\phi(2)\sin 120° + p^\phi(3)\sin 240°]$$
$$\propto p^\phi(2) - p^\phi(3), \qquad (3.35)$$

while $\quad P_y^\pi \propto -\sqrt{\frac{3}{4\pi}}[p^\phi(1)\cos 0° + p^\phi(2)\cos 120° + p^\phi(3)\cos 240°]$

$$\propto 2p^\phi(1) - p^\phi(2) - p^\phi(3).$$

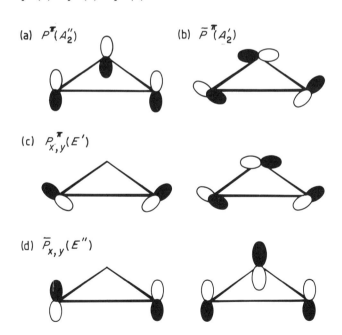

Figure 3.13 π cluster orbitals for the triangular cluster: (a) $P_0^\pi(A_2'')$; (b) $\bar{P}_0^\pi(A_2')$; (c) the two $P_{1c,1s}^\pi(E')$ orbitals; (d) the two $\bar{P}_{1c,1s}^\pi(E'')$ orbitals.

Sec. 3.5 $\quad \pi$ Cluster Orbitals

The energies of all orbitals given above were obtained by resolving the matrix elements of $\hat{\mathcal{H}}$, such as $\langle p^\phi(2)|\hat{\mathcal{H}}|p^\phi(3)\rangle$, into σ- and π-type overlaps, and then assuming that $\beta^\sigma = \beta^\pi$. These energies are given merely for illustration, and it is important to emphasize that TSH theory is not synonymous with Hückel theory. In fact, the results of any (minimal-basis) calculation can be expressed in a cluster orbital basis, and interpretations of the bonding based upon such descriptions represent one of the most important uses of TSH theory.

Note that the odd π orbitals can be obtained from their even partners by rotating each basis function through 90° about the appropriate radius vector, as discussed above. In this case such a transformation interchanges p^θ and p^ϕ orbitals. Hence, for $C_3H_3^+$, the occupied skeletal orbitals (i.e., excluding C—H bonds) are $S_0^\sigma(A_1')$, $P_{1c,1s}^{\sigma/\pi}(E')$, and $P_0^\pi(A_2'')$. The latter orbital is out of the plane of the triangle and is composed only of p^θ orbitals which are antisymmetric under σ_h (note that all such orbitals have $''$ superscripts).

3.5.4 Example: $B_6H_6^{2-}$

Our second example is the octahedral borohydride cluster $B_6H_6^{2-}$ illustrated in Figure 3.14. It is appropriate in this case to form inwardly and outwardly directed sp hybrids from the valence s and p_z orbitals, so that each boron atom is bonded to a radial hydrogen atom by a two-center two-electron bond. These orbitals do not mix greatly with the remaining three orbitals per boron vertex, and we shall not consider them further in this section. The representations spanned by a set of σ orbitals and a set of π orbitals are found to be

O_h (m3m)	E	$8C_3$	$6C_2$	$6C_4$	$3C_2$ ($=C_4^2$)	i	$6S_4$	$8S_6$	$3\sigma_h$	$3\sigma_d$
Γ_σ	6	0	0	2	2	0	0	0	4	2
Γ_π	12	0	0	0	-4	0	0	0	0	0

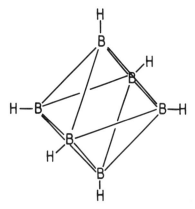

Figure 3.14 The structure of $B_6H_6^{2-}$.

so that $\Gamma_\pi \cong T_{1g} \oplus T_{2g} \oplus T_{1u} \oplus T_{2u}$ while $\Gamma_\sigma \cong A_{1g} \oplus E_g \oplus T_{1u}$ as for Li_6. The σ cluster orbitals are therefore $S_0^\sigma(A_{1g})$, $P_{x,y,z}^\sigma(T_{1u})$, and $D_{0,2c}^\sigma(E_g)$, as above, while the π orbitals are $P_{x,y,z}^\pi(T_{1u})$, $\bar{P}_{x,y,z}^\pi(T_{1g})$, $D_{1c,1s,2s}^\pi(T_{2g})$, and $\bar{D}_{1c,1s,2s}^\pi(T_{2u})$. The π cluster orbitals are illustrated in Figure 3.15, and qualitative energy-level diagrams are given in Figure 3.16. The σ and π cluster orbital energies were evaluated parametrically within the Hückel approximation to obtain the latter diagrams.[6]

The σ cluster orbitals naturally have the same form as those of Li_6 discussed in Section 3.4. The L^π orbitals may be constructed from the table of vector spherical harmonics as follows, using the same angular coordinates as for Li_6:

Orbital	LCAO form	Energy
$P_{0,1c,1s}^\pi(T_{1u})$	$\begin{cases} \frac{1}{2}[p^\theta(2) + p^\theta(3) + p^\theta(4) + p^\theta(5)] \\ \frac{1}{2}[p^\phi(1) - p^\phi(6) - p^\phi(3) + p^\phi(5)] \\ \frac{1}{2}[p^\theta(1) + p^\theta(6) - p^\phi(2) + p^\phi(4)] \end{cases}$	$\alpha + 2\beta$
$\bar{P}_{0,1c,1s}^\pi(T_{1g})$	$\begin{cases} \frac{1}{2}[p^\phi(2) + p^\phi(3) + p^\phi(4) + p^\phi(5)] \\ \frac{1}{2}[-p^\theta(1) + p^\theta(6) + p^\theta(3) - p^\theta(5)] \\ \frac{1}{2}[p^\phi(1) + p^\phi(6) + p^\theta(2) - p^\theta(4)] \end{cases}$	$\alpha - 2\beta$
$D_{1c,1s,2s}^\pi(T_{2g})$	$\begin{cases} \frac{1}{2}[p^\phi(2) + p^\phi(4) - p^\phi(5) - p^\phi(3)] \\ \frac{1}{2}[p^\phi(1) + p^\phi(6) - p^\theta(2) + p^\theta(4)] \\ \frac{1}{2}[p^\theta(1) - p^\theta(6) - p^\theta(5) + p^\theta(3)] \end{cases}$	$\alpha + 2\beta$
$\bar{D}_{1c,1s,2s}^\pi(T_{2u})$	$\begin{cases} \frac{1}{2}[-p^\theta(2) - p^\theta(4) + p^\theta(5) + p^\theta(3)] \\ \frac{1}{2}[-p^\theta(1) - p^\theta(6) - p^\phi(2) + p^\phi(4)] \\ \frac{1}{2}[p^\phi(1) - p^\phi(6) - p^\phi(5) + p^\phi(3)] \end{cases}$	$\alpha - 2\beta$

The local θ and ϕ directions at the north and south poles are chosen as indicated in Figure 3.5. Note that the odd orbitals can be obtained from their even partners by setting $p^\phi \to -p^\theta$ and $p^\theta \to p^\phi$. The orbitals can be sketched in the manner indicated in 3.5.2 using a central atomic orbital with the angular variation of the appropriate parent spherical harmonic. The normalizations and energies were evaluated in the Hückel approximation, assuming that the resonance integrals for σ-type and π-type overlap are both equal to β, as in Section 3.4.

As a specific example, consider the construction of the $P_z^\pi \equiv P_0^\pi$ orbital using $V_{1,0}^\theta = \sqrt{3/4\pi} \sin\theta$. Since $\sin\theta = 1$ for the vertices in the xy plane, and zero for the other two, we obtain

$$P_z^\pi \propto p^\theta(2) + p^\theta(3) + p^\theta(4) + p^\theta(5). \tag{3.36}$$

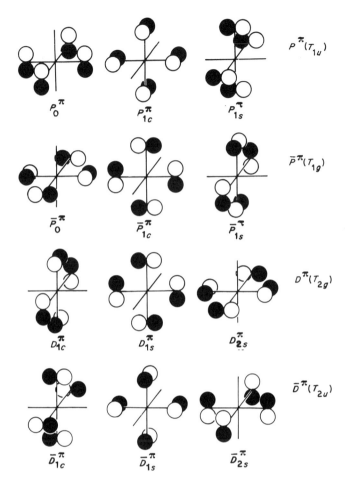

Figure 3.15 The $B_6H_6^{2-}$ L^π and $\bar{L}^\pi$ orbitals.

The normalization in the Hückel approximation is $1/\sqrt{4} = \frac{1}{2}$, and the orbital energy is

$$\tfrac{1}{4}\langle p^\theta(2) + p^\theta(3) + p^\theta(4) + p^\theta(5)|\mathcal{H}|p^\theta(2) + p^\theta(3) + p^\theta(4) + p^\theta(5)\rangle$$
$$= \tfrac{1}{4} \times (4\alpha + 8\beta)$$
$$= \alpha + 2\beta. \tag{3.37}$$

Finally, note that all the above π orbitals contain nodes, even those which are bonding. For example, the $P^\pi(T_{1u})$ orbitals all have a nodal plane containing four atoms, but this does not lead to an antibonding orbital because the node is *intrinsic*. The presence of such intrinsic nodes in the p

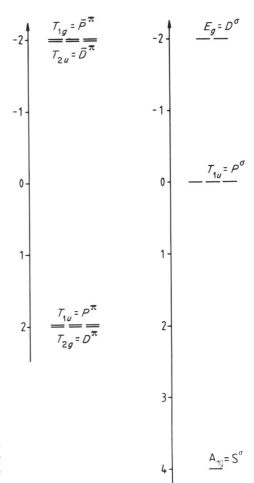

Figure 3.16 Qualitative energy-level diagram for the π (*right*) and σ (*left*) cluster orbitals of $B_6H_6^{2-}$. The energy units are parameterized.

orbital basis functions is precisely the reason why tensor surface harmonics must be used as expansion functions.

When we consider the effects of mixing between the $P^\sigma(T_{1u})$ and $P^\pi(T_{1u})$ cluster orbitals, it is clear that the following seven strongly bonding functions result: $S_0^\sigma(A_{1g})$, $P_{0,1c,1s}^{\sigma/\pi}(T_{1u})$, and $D_{1c,1s,2s}^\pi(T_{2g})$. Hence, the 14 electrons left after forming two-center two-electron bonds to the radial hydrogen atoms are precisely accommodated in the available skeletal bonding orbitals, as can be seen from the energy-level diagram of Figure 3.17. The complementary nature of P^σ and P^π ensures that there is a significant degree of overlap between these orbitals.

In both our examples there is a mirror relationship between the energies

Sec. 3.5 π Cluster Orbitals

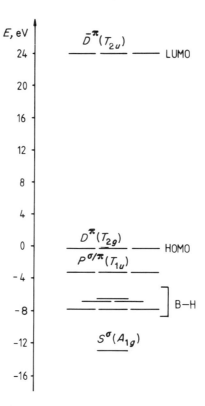

Figure 3.17 Energy-level diagram for $B_6H_6^{2-}$ calculated using the Fenske-Hall method.[13]

of the even π orbitals and their odd partners, relative to $\alpha = 0$. This illustrates the TSH pairing principle, discussed further in Section 3.6, which states that even cluster orbitals and their odd partners generally have opposite bonding characteristics. Furthermore, in this *closo* cluster all the even π orbitals plus the S_0^σ orbital correlate with accessible orbitals. This is actually a general result for *closo* deltahedral clusters, and explains the usual $n + 1$ skeletal electron pairs found in these systems (n is the number of vertices). The same pattern emerges for molecules with four-connected cluster skeletons, in accord with the observation that these systems usually have $n + 1$ skeletal electron pairs too.

The σ and π cluster orbitals which arise for a variety of cluster geometries are given in Figure 3.18. There are some interesting complementary relationships[14] which occur because the vector surface harmonics have zero modulus when the scalar harmonics have maxima and minima.

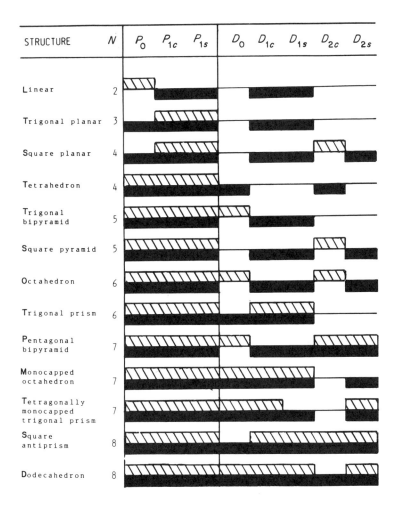

Figure 3.18 P^σ, D^σ, P^π, and D^π functions arising for some common cluster geometries. L^σ orbitals are represented by a shaded box and L^π orbitals by a solid box.

3.6 THE TSH PAIRING PRINCIPLE

We have just seen an illustration of the TSH theory pairing principle in the energy-level spectrum of $B_6H_6^{2-}$. In this section we will discuss the implications of the theorem in more detail. First, however, it may be helpful to draw an analogy with the Coulson-Rushbrooke pairing theorem for alternant hydrocarbons.[15]

3.6.1 Another Analogy

A conjugated hydrocarbon polyene system is described as "alternant" if the atoms can be partitioned into two sets (denoted "starred" and "unstarred") so that starred atoms are connected only to unstarred atoms and vice versa. By convention, the larger set of atoms is the starred set. Such a partition is always possible so long as the molecule contains no odd-membered rings, so that all linear polyenes are alternants, for example. When the π electron systems of such molecules are treated using Hückel theory, some very interesting general principles emerge, many of which are based on the Coulson-Rushbrooke pairing theorem.

First we note that the secular equations may be divided into two sets for starred (s) and unstarred (u) atoms:

$$(\alpha - \epsilon)c_s + \sum_u c_u \beta = 0$$

and (3.38)

$$(\alpha - \epsilon)c_u + \sum_s c_s \beta = 0,$$

where the sums are over the nearest neighbor unstarred and starred atoms, respectively. Hence if we have found sets of coefficients $\{C_u\}$ and $\{C_s\}$ which satisfy the above equations with $\epsilon = \alpha - \lambda$, it follows that the coefficients $\{-C_u\}$ and $\{C_s\}$ satisfy the secular equations with $\epsilon = \alpha + \lambda$ because

$$\lambda C_s + \sum_u C_u \beta = 0 \quad \Rightarrow \quad (-\lambda)C_s + \sum_u (-C_u)\beta = 0, \quad (3.39)$$

and similarly

$$\lambda C_u + \sum_s C_s \beta = 0 \quad \Rightarrow \quad (-\lambda)(-C_u) + \sum_s C_s \beta = 0. \quad (3.40)$$

Hence the molecular orbitals of an alternant hydrocarbon occur in pairs $\alpha \pm \lambda\beta$, and the coefficients for one orbital of the pair can be obtained from the other by reversing the signs at all the unstarred (or starred) atoms. This can be verified for benzene, which we treated in Section 3.2.

Such knowledge of the eigenvalue spectrum of an alternant hydrocarbon can be very useful, and the TSH pairing principle is of equal importance for discussing the electron count and properties of a cluster, as we shall show.

3.6.2 Some Simple Applications of the TSH Pairing Principle

We have already met the first application of the pairing principle in Section 3.5. There we mentioned that Stone first analyzed the general $n + 1$ skeletal electron pair count of n-vertex deltahedral boranes and carboranes in terms of occupation of the S^σ orbital and of the n even $L^\pi(p)$ orbitals. When σ/π mixing is taken into account, we find that the symmetries of the occupied orbitals are not changed and that this interpretation of the bonding is still useful. n-vertex transition metal carbonyl clusters generally have $7n + 1$ valence electron pairs in total, and this may again be rationalized quite simply, as we will show in Chapter 5.

Again we should emphasize that the success of this approach for defining the occupied orbitals of a cluster lies largely in its descent-in-symmetry character.[16] When mixing occurs between the cluster orbitals, the composition of the bonding levels may change somewhat, but the symmetries of these accessible levels generally remain the same. Where deviations from the general rules do occur, there may be some particularly interesting lessons to be learned.

We noted above the analogy of the TSH pairing principle to the pairing theorem for alternant hydrocarbons. The reader may have wondered in both cases what would happen if there were an odd number of orbitals in the manifold, and perhaps even guessed that nonbonding levels would result, as is indeed the case. We shall not consider the hydrocarbon analogy any further here, since the results which may be obtained by this approach can be deduced more rapidly by a technique due to Longuet-Higgins.[17]

For the more complicated three-dimensional case, we will demonstrate the general requirements for a symmetry-forced deviation from the usual $n + 1$ skeletal electron pairs of a borane or carborane cluster, or the $7n + 1$ electron pairs of a transition metal carbonyl cluster. Consider, for illustration, a cluster belonging to an axial C_{nv} point group which has a single atom on the principal rotation axis. If the axis is of order 3 or more, then the p^π orbitals of the unique vertex span an E-type irreducible representation. We will prove in Chapter 6 that the total number of e-type orbitals spanned by the tangential p^π orbitals must be odd in this case. The pairing principle then indicates that there will be a nonbonding doubly degenerate e-type pair of orbitals around the frontier region, and that n or $n + 2$ skeletal electron pairs are required for a closed shell. An example is B_4Cl_4, which has four skeletal electron pairs rather than $n + 1 = 5$, while the P_4 tetrahedron has six skeletal electron pairs corresponding to a localized framework of two-center two-electron bonds. (Here we make use of the fact that C_{3v} is a subgroup of T_d.) Deltahedral clusters with systematic symmetry-forced deviations from $n + 1$ skeletal electron pairs must have $3p + 1$ vertices, where p is an integer. This follows simply from the geometric constraints that all the faces must be triangular and that there is a single atom on the principal rotation axis. This axis must be of order 3, and all the atoms, except for the one which lies on the axis, therefore appear in sets of 3.

3.7 δ ORBITALS

Consideration of δ orbitals has been postponed until this point because some more complicated mathematics is involved for rather less reward than in the previous sections. In transition metal cluster carbonyls, we have already noted that the $6n$ "external" orbitals generally all correlate with accessible levels, and since this set includes the d^δ-type functions, the reader may wonder why any special considerations are necessary at all. In actual fact, some molecules do not conform to such a simple generalization, and the TSH treatment of the δ orbitals may then be important.

δ-type basis atomic orbitals have two nodal planes containing the radius vector to the vertex they are centered upon; in local axes these are the d_{xy} and $d_{x^2-y^2}$ orbitals illustrated in Figure 3.19. The question we must now address is what the appropriate expansion coefficients are for these basis functions. By extrapolation from the σ and π cluster orbitals, it will probably not surprise the reader to learn that second-rank tensor surface harmonics are involved. To see how this arises, we first note that a set of five d orbitals transform among themselves under rotations about their common origin like the components of a traceless, symmetric Cartesian tensor of rank 2.[18] This basically arises because the d orbitals can also be described as transforming like the five independent components of $Y_{2,\mu}$. A general 3×3 matrix, A_{ij}, has nine independent components, but if we add the condition that the matrix is to be symmetric there are then only six. If we further specify that the sum of the diagonal components (the trace) must vanish, then we are left with only five independent components, and a correspondence may be developed with the five components of $Y_{2,\mu}$.

Appropriate components of the 3×3 matrix are given by $\nabla \nabla Y_{L,\mu}$ for even δ cluster orbitals, and by $\mathbf{r} \wedge \nabla \nabla Y_{L,\mu}$ (or $\nabla \mathbf{r} \wedge \nabla Y_{L,\mu}$) for odd δ cluster

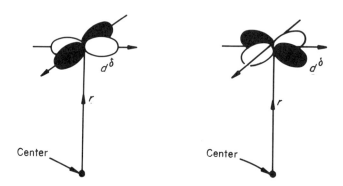

Figure 3.19 d orbitals of δ-type symmetry.

orbitals. For example, in Cartesian coordinates the even tensor components required, T_{ij}, are given by

$$T_{ij} = \frac{\partial}{\partial x_i} \frac{\partial}{\partial x_j} Y_{L,\mu}. \tag{3.41}$$

The five independent components of interest may be identified in spherical polar coordinates as

$$T_{rr} = \frac{L(L+1)}{3} Y_{L,\mu}$$

$$T_{r\phi} = -\frac{1}{\sin\theta} \frac{\partial Y_{L,\mu}}{\partial \phi} = -V^\phi_{L,\mu}$$

$$T_{r\theta} = -\frac{\partial Y_{L,\mu}}{\partial \theta} = -V^\theta_{L,\mu} \tag{3.42}$$

$$T_{\theta\phi} = \frac{\partial}{\partial \theta} \frac{1}{\sin\theta} \frac{\partial Y_{L,\mu}}{\partial \phi}$$

$$T_{\theta^2 - \phi^2} = -2\left(\frac{\partial^2}{\partial \theta^2} + \frac{L(L+1)}{3}\right) Y_{L,\mu}.$$

Hence we see that the appropriate expansion coefficients for the d^σ orbitals are the spherical harmonics, $Y_{L,\mu}$, and for the d^π orbitals are the vector spherical harmonics, as stated in previous sections. (Note that the direction of ϕ increasing is along the negative local x axis at any vertex, while the direction of θ increasing is along the positive local y axis. The r increasing direction is always along the positive local z axis.) δ cluster orbitals may therefore be written as

$$\psi^\delta_{L,\mu} = \sum_i [T^{(x^2-y^2)}_{L,\mu}(i) d^{(x^2-y^2)}(i) + 2 T^{xy}_{L,\mu}(i) d^{xy}(i)]$$

$$\bar{\psi}^\delta_{L,\mu} = \sum_i [\bar{T}^{(x^2-y^2)}_{L,\mu}(i) d^{(x^2-y^2)}(i) + 2 \bar{T}^{xy}_{L,\mu}(i) d^{xy}(i)], \tag{3.43}$$

where $\bar{T}^{(x^2-y^2)}_{L,\mu}(i) = -2T^{xy}_{L,\mu}(i)$ and $2\bar{T}^{xy}_{L,\mu}(i) = T^{(x^2-y^2)}_{L,\mu}(i)$. (The factor of 2 arises on converting from spherical polars to Cartesian form.) Odd δ cluster orbitals may be obtained from their even partners by rotation of the atomic δ functions through 45° about the radius vectors of their respective atoms. Note also that the spherical harmonics with $L = 0$ and $L = 1$ all vanish when differentiated twice in the above manner. Hence, when we build up a basis of δ cluster orbitals, we start from D^δ and $\bar{D}^\delta$.

Again it is straightforward to form representations of δ orbitals—the short-cuts are similar to those for π-type basis functions discussed in Section

3.5. If there are $2n$ basis δ functions, then the character of the representation they span is $2n$ for the identity operation, and is otherwise nonzero only for rotations where there are cluster vertices lying on the rotation axis. In the latter case, the character can be deduced by looking up the characters of the IRs (irreducible representations) spanned by xy and $x^2 - y^2$ in an axial point group of the same order as the rotation axis in question. If two vertices lie on the axis, then the result must be doubled. An alternative approach is to use the formula[12]

$$\Gamma_{\delta/\bar{\delta}} \cong \Gamma_{\pi/\bar{\pi}} \otimes \Gamma_{x,y,z} - \Gamma_{\pi/\bar{\pi}} - \Gamma_{\mathscr{P}} \otimes \Gamma_\sigma - \Gamma_\sigma, \tag{3.44}$$

where $\Gamma_{\mathscr{P}}$ is the irreducible representation which transforms in the same way as the TSH parity operator, $\hat{\mathscr{P}}$. As we shall see in Chapter 5, $\Gamma_{\mathscr{P}}$ is simply $+1$ under proper rotations and -1 under reflections, rotation-reflections, and the inversion operation.

The development of TSH cluster orbitals given in this chapter may be generalized to linear combinations of basis functions which have any number of nodal planes containing the radius vector, but no such analyses have yet proved necessary.

3.8 SUMMARY

In this chapter we have reviewed the basic principles which lie behind the TSH method and discussed some simple applications. The fundamental reason why the method is so useful is that it provides reasonable first approximations to the cluster wavefunctions, and that subsequent mixing between these orbitals is sufficiently small to have no effect on the number of bonding orbitals deduced from the behavior of the diagonal matrix elements of the Hamiltonian alone.[16] Many of the principles discussed above will be reinforced by their use in later chapters, and we have provided some exercises for practice.

The inspiration for this approach is found in the relation between the Hückel wavefunctions and those of the appropriate free-electron systems for simple conjugated polyenes. Complications arise because in three-dimensional cluster skeletons there will also be mixing between the σ cluster orbitals and π and δ orbitals transforming according to the same irreducible representation of the molecular point group. In low symmetry, mixing within the σ and π sets of cluster orbitals also occurs. However, the TSH cluster orbitals are generally very useful, because such mixings are relatively small, or at least unimportant for the purposes of electron counting.

To fully exploit these desirable properties it is clearly necessary for the cluster orbitals to transform according to specific irreducible representations of the molecular point group. For molecules belonging to axial point groups it is easy to show that the $Y_{L,Mc}$ and $Y_{L,Ms}$ functions always transform in this way,[19] because (for given L) only spherical harmonics with the same value of M are mixed by the operations of an axial point group. In point groups of higher symmetry, where there is more than one rotational axis of order 3 or more, it is necessary in general to choose linear combinations of the $Y_{L,Mc}$ and $Y_{L,Ms}$ with the same L but different M to span definite irreducible representations.[6] However, this need not trouble us any further here.

APPENDIX: PROOF OF THE TSH PAIRING PRINCIPLE

We have noted in Section 3.5 that the orientations of the two tangential p orbitals at a cluster vertex can be chosen arbitrarily, so long as the LCAO coefficients are changed concomitantly as if they were vector components. Now consider the diagonal Hamiltonian matrix element for some π cluster orbital L^π. It is sufficient to consider the terms involving the p^π orbitals of two particular vertices as illustrated in Figure 3.20—the orientations are chosen to make the proof as simple as possible. The nonzero matrix elements between these functions (relative to $\alpha = 0$) are

$$\langle L^\pi | \hat{\mathscr{H}} | L^\pi \rangle = \cdots + \beta^\|(1,2) V^\|(1) V^\|(2) - \beta^\perp(1,2) V^\perp(1) V^\perp(2) + \cdots \quad (3.45)$$

where $\beta^\|(1,2)$ and $\beta^\perp(1,2)$ are the resonance integrals for parallel and perpendicular p orbital overlap and $V^\|(i)$ and $V^\perp(i)$ are the coefficients at vertex i in the parallel and perpendicular directions. We now resolve $\beta^\|$ and $\beta^\perp$ into the resonance integrals for σ- and π-type overlap, denoted by β^σ and β^π, respectively. Clearly $\beta^\| = \beta^\pi$, and if the vertices lie on a sphere and subtend an angle ω, then

$$\begin{aligned}\beta^\perp &= \beta^\sigma \cos^2 \frac{\omega}{2} + \beta^\pi \sin^2 \frac{\omega}{2} \\ &\approx \beta^\sigma \end{aligned} \quad (3.46)$$

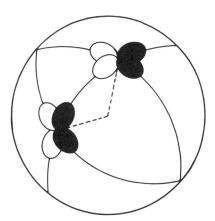

Figure 3.20 p^π orbitals overlapping parallel and perpendicular to their common great circle.

if $\beta^\sigma \approx \beta^\pi$. Now, for the corresponding $\bar{L}^\pi$ orbital, the coefficients of the perpendicular orbitals are $V^\parallel(i)$, and of the parallel orbitals are $-V^\perp(i)$, as we found in Section 3.5. Hence the corresponding terms in the diagonal Hamiltonian matrix element of the odd cluster orbital are precisely the negative of those of the even cluster orbital, relative to $\alpha = 0$, if $\beta^\sigma = \beta^\pi$. The latter assumption is quite reasonable, at least for borohydrides. The other assumption made was that the vertices all lie on the surface of a sphere, but we did not need to make the Hückel approximations—any pair of vertices could have been chosen.

In practice, the pairing principle for odd and even cluster orbitals does not hold exactly, and mixing with other cluster orbitals also spoils the mirror relationship. However, the pattern is generally sufficiently good for us to produce very powerful predictions and analyses of bonding patterns, as we will see in the following chapters. The above proof of the pairing principle was for p^π orbitals and cannot be carried over precisely to cluster orbitals composed of d^π functions, because such orbitals have δ components of overlap. However, the mirror relationship for these orbitals again appears to hold sufficiently well for most transition metal clusters to be a useful tool, as many studies have shown.[20]

EXERCISES

More mathematical problems are marked with an asterisk.

3.1* By substituting the trial solution $c_r = Ax^r$, where A is a constant to be determined and x is a variable, and using the boundary condition $c_0 = c_n$, show that Equation (3.12) gives the solutions to the secular problem for the cyclic polyene set up in Equation (3.11).

3.2* By substituting the trial solution $c_r = Ax^r + Bx^{-r}$, where A and B are constants to be determined and x is a variable, and using the boundary conditions $c_0 = c_{n+1} = 0$, show that Equation (3.8) gives the solutions to the secular problem for the linear polyene as set up in Equation (3.7).

3.3 Using the formulas in Equation (3.27), find and reduce the representation spanned by the seven spherical harmonics with $L = 3$ and $-3 \leq M \leq 3$ in the point group O_h. Check your answer using descent-in-symmetry tables.

3.4 Show that for the bipyramidal molecule $B_5H_5^{2-}$, $\Gamma_\sigma \cong S^\sigma \oplus P^\sigma_{x,y,z} \oplus D^\sigma_{z^2}$ and $\Gamma_{\pi/\bar{\pi}} \cong P^\pi_{x,y,z} \oplus \bar{P}^\pi_{x,y,z} \oplus D^\pi_{xz,yz} \oplus \bar{D}^\pi_{xz,yz}$. Sketch the skeletal energy-level diagram including point group symmetry and TSH labels.

3.5 Using the relation $\Gamma_{\pi/\bar{\pi}} \cong \Gamma_\sigma \otimes \Gamma_{x,y,z} - \Gamma_\sigma$, find Γ_π for the p^π orbitals of an octahedron given that $\Gamma_\sigma \cong A_{1g} \oplus T_{1u} \oplus E_g$. The reader may care to check the TSH theory analysis of $B_6H_6^{2-}$ given in Section 3.5, or of $B_5H_5^{2-}$ in Exercise 3.4, using his or her favorite computational package.

3.6 Using the results of Section 3.2, construct the Hückel π orbitals for the pentadienyl anion $CH_2CHCHCHCH_2^-$ and for cyclooctatetraene. Show how the Coulson-Rushbrooke pairing theorem implies that there must be a nonbonding orbital in the π system of the former species.

3.7 Consider a square planar cluster where the p^θ orbitals are parallel to the principal axis and the p^ϕ orbitals lie in the horizontal mirror plane. Find the TSH π cluster orbitals which are generated by the p^π orbitals in this molecule, and construct these functions. Check your answers by using projection operators to generate linear combinations of the required symmetries. Sketch the p^π energy-level diagram assuming that $\beta^\sigma = \beta^\pi$, giving point group symmetry and TSH labels. What important TSH theory principle does this diagram illustrate?

3.8 Consider B_7H_7 with a capped octahedral C_{3v} skeleton of boron atoms rather than the usual pentagonal bipyramidal structure which is found for $B_7H_7^{2-}$. Work out which TSH σ and π cluster orbitals are generated, and hence predict the symmetries of the occupied orbitals and explain why this cluster has $2n$ instead of the usual $2n + 1$ valence electron pairs. (n is the number of cluster vertices.)

3.9 Use the method described in Section 3.5 to sketch the even and odd vector spherical harmonics for $P^\pi_{1c,1s}$, $\bar{P}^\pi_{1c,1s}$, $D^\pi_{1c,1s,2c,2s}$, and $\bar{D}^\pi_{1c,1s,2c,2s}$, which are illustrated in references 1 and 2. Either by inspection or using the relation between $\Gamma_{\delta/\bar{\delta}}$, $\Gamma_{\pi/\bar{\pi}}$, and Γ_σ, determine which TSH cluster orbitals are required for an octahedron.

REFERENCES

1. Stone, A. J., *Molec. Phys.*, **41**, 1339 (1980).
2. Stone, A. J., *Inorg. Chem.*, **20**, 563 (1981).
3. Stone, A. J., and Alderton, M., *Inorg. Chem.*, **21**, 2297 (1982).
4. Stone, A. J., *Polyhedron*, **3**, 1299 (1984).
5. Fowler, P. W., *Polyhedron*, **4**, 2051 (1985).
6. Stone, A. J., and Wales, D. J., *Molec. Phys.*, **61**, 747 (1987).
7. Wales, D. J., and Stone, A. J., *Inorg. Chem.*, **26**, 3845 (1987).
8. Murrell, J. N., Kettle, S. F. A., and Tedder, J. M., *Valence Theory*, Wiley, New York, 1965.
9. Koutecký, J., and Fantucci, P., *Z. Physik D*, **3**, 147 (1986).
10. Schleyer, P. von R., Würthwein, E.-U., Kaufman, E., and Clark, T., *Amer. Chem. Soc.*, **105**, 5930 (1983).
11. For example, Atkins, P. W., Child, M. S., and Phillips, C. S. G., *Tables for Group Theory*, Oxford University Press, London, 1970.
12. Redmond, D. B., Quinn, C. M., and McKiernan, J. G. R., *J. Chem. Soc., Faraday Trans. II*, **79**, 1791 (1983).
13. Hall, M. B., and Fenske, R. F., *Inorg. Chem.*, **11**, 768 (1972).
14. Mingos, D. M. P., and Hawes, J. C., *Structure and Bonding*, **63**, 1 (1985).
15. Coulson, C. A., and Rushbrooke, S., *Proc. Camb. Phil. Soc.*, **36**, 193 (1940).
16. Wales, D. J., *Molec. Phys.*, **67**, 303, 1989.
17. Longuet-Higgins, H. C., *J. Chem. Phys.*, **18**, 265, 275, 283 (1950).
18. Weissbluth, M., *Atoms and Molecules*, Academic Press, New York, 1978, p. 174.
19. Griffith, J. S., *The Theory of Transition Metal Ions*, Cambridge University Press, London, 1971.
20. Ceulemans, A., and Fowler, P. W., *Inorg. Chim. Acta*, **105**, 75 (1985).

4

Clusters Where Radial Bonding Predominates

4.1 INTRODUCTION

In Chapter 3, the tensor surface harmonic theory was developed in rather general terms; in the remainder of the book we shall fill in some of the details by means of a number of examples. The generality of the previous chapter arose in part because it was necessary to consider the possibility that σ-, π-, and δ-type interactions might all be important. There are, however, specific classes of cluster molecules where the bonding patterns are considerably simplified because either the δ or π (or both) interactions are sufficiently small that they can be ignored without introducing large errors into the ultimate conclusions. In particular, when the σ interactions can be considered in isolation, the bonding pattern takes on a very simple form. This simplified bonding analysis is appropriate for alkali metal and gold cluster compounds. In addition, it has proved to have predictive value; for example, the structure and stoichiometry of the icosahedral gold cluster ion $[Au_{13}Cl_2(PMe_2Ph)_{10}]^{3+}$ was predicted several years before it was actually synthesized and characterized by single-crystal X-ray crystallographic studies.[1] In addition, the ideas developed here may also be used to rationalize the stabilities of alkali metal clusters generated in molecular beams.

4.2 THEORETICAL CONSIDERATIONS

The radial or σ cluster orbitals of a cluster in which the vertices lie roughly on the surface of a sphere are given by the following spherical harmonic expansion (in an unnormalized form):

$$L^\sigma_{L,\mu} = \sum_t Y_{L,\mu}(\theta_t, \phi_t)\sigma(t), \qquad (4.1)$$

where $\mu = Mc$ or Ms, $\sigma(t)$ is the σ orbital at vertex t, and θ_t and ϕ_t are the angular coordinates of this vertex. Using the appropriate formulas for the spherical harmonics given in the previous chapter, it is a relatively easy matter to identify the relevant spherical harmonic σ cluster orbital expansions.

In Figure 4.1, the radial cluster molecular orbitals for some deltahedral clusters are summarized in the form of a histogram to illustrate the manner in which the spherical harmonic functions are used in a sequential fashion.[2] The following points are illustrated in the figure and have a general significance:

(1) For a cluster with n atoms, n linear combinations are generated.
(2) If the spherical harmonic function $Y_{L,\mu}$ has nodes coincident with all the atom positions of the cluster, then it is not possible to generate a linear combination based upon it, because the expansion represented by Equation (4.1) would be null. For example, for the pentagonal bipyramid the D^σ_{1c} and D^σ_{1s} orbitals vanish in the $z = 0$ plane and along the z axis, and consequently the coefficients are zero for all the atom positions in this cluster. The parent spherical harmonics for these cluster orbitals are those found in the atomic orbitals d_{xz} and d_{yz}, and we can use this fact to sketch the D^σ functions as described in Chapter 3 (see Figure 4.2). These spherical harmonics produce only zero terms in the expansion formula given above for the D^σ_{1c} and D^σ_{1s} orbitals of any bipyramid. Note the corresponding spaces in the histogram in Figure 4.1. Not all the spaces correspond to null functions, however; more often they result from the necessity to match the symmetry species contained in the representation spanned by all the σ basis functions. Furthermore, the choice of cluster orbitals within an incomplete L^σ set may not be unique.
(3) For deltahedral clusters, the L^σ orbitals are generated in a sequential manner such that a new shell is not started until the one with a smaller L quantum number is filled. In particular, no F^σ functions are required until the D^σ set is fully utilized. This is not possible for all classes of polyhedra, however. For example, for three-connected polyhedra and bipyramids, the F^σ shell is utilized before D^σ is filled (see the histograms in Figure 4.3). Deltahedra provide the most efficient way of covering a spherical surface, and their spherical harmonic expansions appear to reflect this characteristic. Other classes of deltahedra are less spherical;

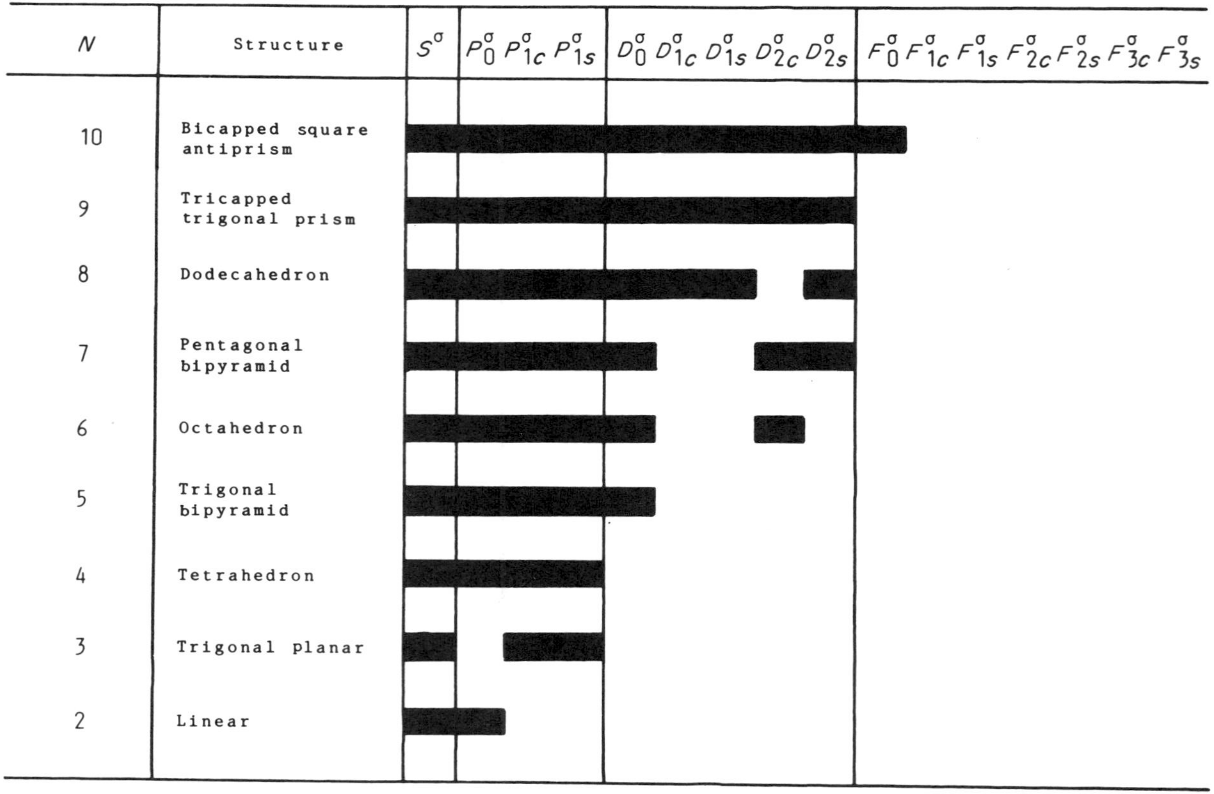

N	Structure	S^σ	$P_0^\sigma\ P_{1c}^\sigma\ P_{1s}^\sigma$	$D_0^\sigma\ D_{1c}^\sigma\ D_{1s}^\sigma\ D_{2c}^\sigma\ D_{2s}^\sigma$	$F_0^\sigma\ F_{1c}^\sigma\ F_{1s}^\sigma\ F_{2c}^\sigma\ F_{2s}^\sigma\ F_{3c}^\sigma\ F_{3s}^\sigma$
10	Bicapped square antiprism	■	■■■	■■■■■	■
9	Tricapped trigonal prism	■	■■■	■■■■■	
8	Dodecahedron	■	■■■	■■■■■	
7	Pentagonal bipyramid	■	■■■	■■■	
6	Octahedron	■	■■■	■■	
5	Trigonal bipyramid	■	■■■	■	
4	Tetrahedron	■	■■■		
3	Trigonal planar	■	■■		
2	Linear	■	■		

Figure 4.1 Summary of the appropriate σ cluster orbital basis sets for some deltahedral clusters.

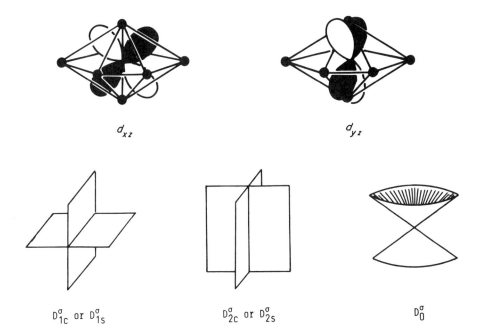

Figure 4.2 Nodal characteristics of D^σ atomic orbitals.

they may have atoms located on the nodal planes of some of the spherical harmonics which give rise to null functions.

(4) The energies of σ cluster orbitals depend primarily upon the L quantum number. Molecular orbitals with the fewest nodes also have the fewest antibonding nearest-neighbor interactions and consequently are the most bonding. It can be shown that the average energy for an L^σ manifold of orbitals, W_L^σ, in a spherical cluster is approximately equal to:[3]

$$W_L^\sigma \approx \alpha^\sigma + \frac{2e}{n} \beta^\sigma P_L(\cos \omega), \qquad (4.2)$$

where $P_L(x)$ is the Legendre polynomial of degree L, ω is the average angular separation between atoms, and e is the number of edges. α^σ and β^σ are the Coulomb and resonance integrals as normally defined in Hückel theory. The Legendre polynomial, $P_L(x)$, is the solution of Legendre's equation

$$(1 - x^2)\frac{d^2 P_L}{dx^2} - 2x\frac{dP_L}{dx} + L(L+1)P_L = 0. \qquad (4.3)$$

The quantity $-P_L(\cos \omega)$ is plotted for $L = 0$ to 5 in Figure 4.4 with ω ranging from 180° down to about 40°, which corresponds to a deltahedron with about 32 vertices. The S^σ cluster orbital with no angular nodes is always

N	Structure	S^σ	P_0^σ P_{1c}^σ P_{1s}^σ	D_0^σ D_{1c}^σ D_{1s}^σ D_{2c}^σ D_{2s}^σ	F_0^σ F_{1c}^σ F_{1s}^σ F_{2c}^σ F_{2s}^σ F_{3c}^σ F_{3s}^σ
9	Heptagonal Bipyramid	■	■	■ ■ ■	■
8	Hexagonal Bipyramid	■	■	■ ■ ■	■
7	Pentagonal Bipyramid	■	■	■ ■ ■	
6	Octahedron	■	■	■ ■	
5	Trigonal Bipyramid	■	■	■	

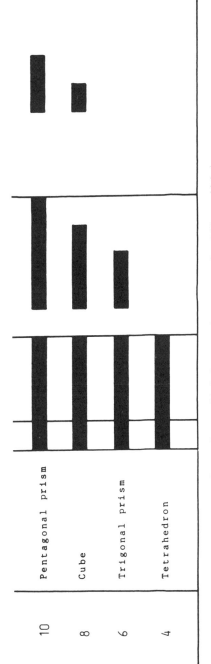

Figure 4.3 The L^σ cluster orbitals for some three-connected and bipyramidal clusters.

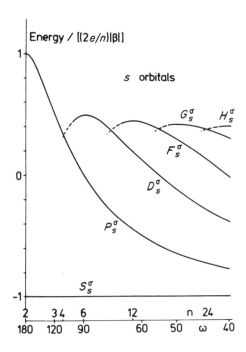

Figure 4.4 A plot of $-P_L(\cos\omega)$ against ω as a guide to L^σ cluster orbital energies. Reproduced with permission from Stone, A. J., *Inorg. Chem.*, **20**, 563 (1981).

the most stable. P^σ orbitals with a single nodal plane become bonding only when the number of vertices $n > 6$, and the D^σ orbitals become bonding only when $n > 16$. If the specific geometric details of the cluster are ignored, then the following important generalizations can be derived from the results shown in Figure 4.4. Filled electronic shells are associated with the following numbers of electrons:

2	S^σ filled ($n > 2$)
8	S^σ and P^σ shells filled ($n > 6$)
18	S^σ, P^σ, and D^σ shells filled ($n > 16$)

Clusters in which all the occupied subshells are precisely filled are expected to have spherical topologies. They should also be more stable than the adjacent members of the series, and will in general have higher ionization potentials. Clusters with intermediate numbers of electrons will not necessarily be spherical and may take up oblate or prolate spheroidal geometries. Prolate spheroidal clusters resemble a rugby ball (or American football) and oblate spheroids resemble a discus. The change in geometry will be particularly favorable if it results in the opening up of a significant HOMO-LUMO gap. This aspect is discussed in more detail in Section 4.3.

Although we have emphasized the spherical nature of the bonding problem, the tensor surface harmonic methodology is also useful for discussing specific structural aspects. For example, the factor of $2e/n$ in Equation (4.2) suggests that the average stabilization energy, W_L^σ, associated with a manifold of σ cluster orbitals is maximized when the polyhedron has as many edges connecting nearest neighbors as possible. For three-dimensional structures, deltahedra maximize the number of edges because they have exclusively triangular faces and we expect the maximum delocalization energies to be achieved for such geometries. While this is a reasonable generalization, it does not allow for the splittings of the L^σ shells in nonspherical environments. This point can be illustrated by reference to a specific example. In the previous chapter, it was noted that within the Hückel approximation the orbital energies of the radial molecular orbitals in an octahedral cluster relative to $\alpha^\sigma = 0$ are

$$S^\sigma(A_{1g}) \qquad 4\beta$$
$$P^\sigma(T_{1u}) \qquad 0$$
$$D^\sigma(E_g) \qquad -2\beta$$

The trigonal prism also has a pseudospherical geometry, and using the appropriate spherical harmonic expansion formulas (from Chapter 3), the radial σ orbitals are found to be:

$S^\sigma(A_1')$ $\qquad \sqrt{\frac{1}{6}}[\sigma(1) + \sigma(2) + \sigma(3) + \sigma(4) + \sigma(5) + \sigma(6)]$

$P_0^\sigma(A_2'')$ $\qquad \sqrt{\frac{1}{6}}[\sigma(1) + \sigma(2) + \sigma(3) - \sigma(4) - \sigma(5) - \sigma(6)]$

$P_{\pm 1}^\sigma(E')$ $\qquad \sqrt{\frac{1}{12}}[2\sigma(1) - \sigma(2) - \sigma(3) + 2\sigma(4) - \sigma(5) - \sigma(6)];$
$\qquad\qquad \frac{1}{2}[\sigma(2) - \sigma(3) + \sigma(5) - \sigma(6)]$

$D_{\pm 1}^\sigma(E'')$ $\qquad \sqrt{\frac{1}{12}}[2\sigma(1) - \sigma(2) - \sigma(3) - 2\sigma(4) + \sigma(5) + \sigma(6)];$
$\qquad\qquad \frac{1}{2}[\sigma(2) - \sigma(3) - \sigma(4) + \sigma(5)]$

These molecular orbitals are illustrated schematically in Figure 4.5 and compared with those of the octahedron. (Note that $P_{\pm 1}^\sigma \equiv P_{1c,1s}^\sigma$, etc.) The most stable molecular orbital, S^σ, always has energy $\alpha^\sigma + \lambda\beta^\sigma$ in the Hückel approximation for a cluster in which all the vertices are equivalent and have λ nearest neighbors. Consequently, for a cluster with two valence electrons, there is a clear preference for a deltahedral geometry, because this maximizes $\lambda = 2e/n$; i.e., the total stabilization energy for the octahedron would be $8\beta^\sigma$, whereas for the trigonal prism it would be $6\beta^\sigma$. For clusters with four or more electrons, the geometric preferences are not so clear-cut. The splitting of the P^σ set in the trigonal prism results in $P_z(A_2'')$ being stabilized by β and compensates for the smaller stabilization energy associated with $S^\sigma(A_1')$.

A significant and general feature of the molecular orbital splitting diagram in Figure 4.5 is the occurrence of a "barycenter" (or center-of-gravity) rule for all the molecular orbital energies relative to $\alpha^\sigma = 0$ when taken as

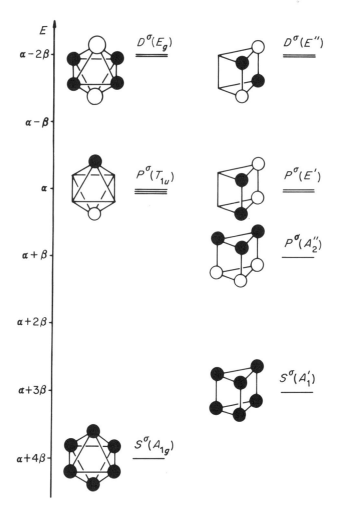

Figure 4.5 A comparison of the L^σ cluster orbitals of a trigonal prism and an octahedron.

a whole; e.g., for the octahedron, $4\beta^\sigma + 2 \times (-2\beta^\sigma) = 0$, and for the trigonal prism, $3\beta^\sigma + \beta + 2 \times (-2\beta^\sigma) = 0$. Hence, in simple cases, knowledge of the degeneracies of the irreducible representations for the L^σ functions and the number of edges of the polyhedron can lead immediately to the Hückel cluster orbital energies. For example, for the tetrahedron, the orbital energies are $S^\sigma(A_1) = \alpha^\sigma + 3\beta^\sigma$ (from the three-connected nature of the polyhedron) and $P^\sigma(T_2) = \alpha^\sigma - \beta^\sigma$ (from the center-of-gravity rule). By taking in-phase and out-of-phase combinations of these orbitals, the following cluster orbital energies for the cube may be derived; this is possible because the cube may be

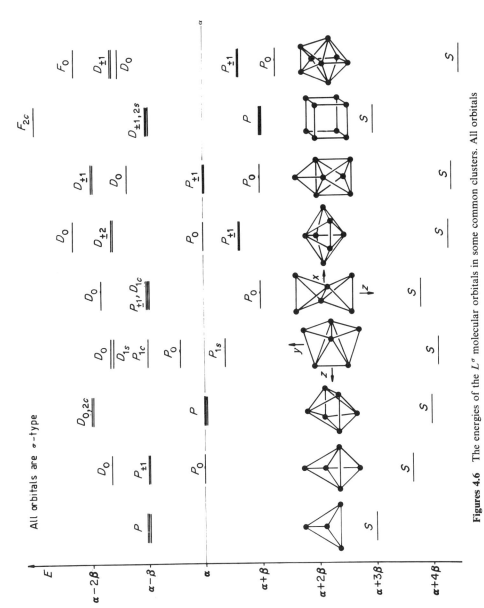

Figures 4.6 The energies of the L^σ molecular orbitals in some common clusters. All orbitals are σ-type.

regarded as a composite of two tetrahedra with all the nearest-neighbor interactions between atoms from different sets:

$$S_0^\sigma(A_{1g}) \qquad \alpha^\sigma + 3\beta^\sigma$$
$$P_{0,1c,1s}^\sigma(T_{1u}) \qquad \alpha^\sigma + \beta^\sigma$$
$$D_{1c,1s,2s}^\sigma(T_{2g}) \qquad \alpha^\sigma - \beta^\sigma$$
$$F_{2c}^\sigma(A_{2u}) \qquad \alpha^\sigma - 3\beta^\sigma$$

For the cube, the molecular orbital energies not only conform to the center-of-gravity rule, but also have a mirror plane relative to $\alpha^\sigma = 0$. This occurs because the vertices of the cube define an alternant graph such that the atoms can be partitioned into two groups, starred and unstarred, in such a way that none of the starred or unstarred positions are adjacent. The L^σ molecular orbital splitting diagrams for some common symmetrical clusters are shown in Figure 4.6.[4]

4.3 OBLATE AND PROLATE CLUSTERS

The TSH methodology is most clearly appropriate for polyhedral molecules which belong to high-symmetry point groups. It is actually just as useful for many molecules of lower symmetry, where the degeneracies of the L^σ (and L^π) orbitals are further removed. The group theoretical aspects of the problem can be tackled using descent-in-symmetry tables such as Table 4.1. Besides belonging to a lower symmetry point group, these cluster molecules are also less spherical in appearance. In particular, the distribution of vertex atoms may be better described as either prolate (like a rugby ball or American football) or oblate (like a discus). This distinction can be appreciated if a trigonal bipyramid ($B_5H_5^{2-}$) or a tricapped trigonal prism ($B_9H_9^{2-}$) is constructed from equilateral triangles. Although both molecules belong to the D_{3h} point group, they are geometrically quite different. In the former case the distance of the axial atoms from the centroid of the cluster is longer than that of the equatorial atoms, and the molecule is a prolate symmetric top. For the tricapped trigonal prism the capping atoms on the equator are farthest from the centroid, and this species is described as an oblate symmetric top. These differences in geometry have important consequences for the splittings of the cluster orbitals in these molecules. For clusters with 4 to 8 atoms, the P^σ set is bonding while the D^σ set is antibonding and therefore splittings within the P^σ manifold (see Figure 4.6) are particularly important for defining ground-state properties.

For a prolate cluster, the P_0^σ orbital is stabilized relative to the $P_{\pm 1}^\sigma$ pair while the opposite is true for an oblate cluster. The splittings, which are illustrated in Figure 4.7, follow another approximate center-of-gravity rule such that the stabilization or destabilization of P_0^σ is roughly twice that of the

TABLE 4.1 SYMMETRY DESIGNATIONS FOR THE LIGAND LINEAR COMBINATIONS AND CENTRAL ATOM ORBITALS IN DELTAHEDRAL COORDINATION COMPOUNDS

N		2	3	4	5	6	7	8	9	10
Point group	M	$D_{\infty h}$	D_{3h}	T_d	D_{3h}	O_h	D_{5h}	D_{2d}	D_{3h}	D_{4d}
L_N										
S^σ	s	a_{1g}	a'_1	a_1	a'_1	a_{1g}	a'_1	a_1	a'_1	a_1
P^σ_0	p_z	a_{1u}	a''_2		a''_2		a''_2	b_2	a''_2	b_2
P^σ_{1c}	p_x	e_{1u}	e'	t_2	e'	t_{1u}	e'_1	e	e'	e_1
P^σ_{1s}	p_y									
D^σ_{1c}	d_{xz}	e_{1g}	e''	t_2	e''	t_{2g}	e''_1	e	e''	e_3
D^σ_{1s}	d_{yz}									
D^σ_{2c}	d_{xy}	e_{2g}	e'	e	e'		e'_2	b_2	e'	e_2
D^σ_{2s}	$d_{x^2-y^2}$							b_1		
D^σ_0	d_{z^2}	a_{1g}	a'_1		a'_1	e_g	a'_1	a_1	a'_1	a_1

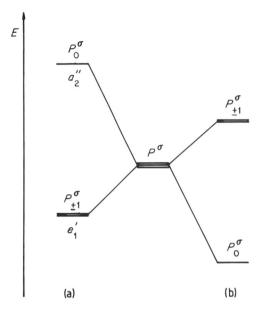

Figure 4.7 The splittings of the P^σ set in (a) oblate and (b) prolate geometries.

$P^\sigma_{\pm 1}$ orbitals. The origins of the relative orderings of P^σ_0 and $P^\sigma_{\pm 1}$ can be appreciated by considering how the cross-equator interactions change when the geometry is squashed along the z (principal) axis. The cross-equator antibonding interaction for P^σ_0 is increased for this oblate distortion, and therefore this orbital is destabilized relative to $P^\sigma_{\pm 1}$.[5]

The relevance of these arguments to real chemical situations can be appreciated by reference to some gold and lithium clusters. The lithium clusters, Li_n, have been observed in molecular beam experiments and are characterized primarily by mass-spectrometric measurements. In addition, their geometries have been calculated using accurate *ab initio* molecular orbital calculations, which can be considered to be reliable. The computed geometries for Li_8, Li_6, and Li_4 are illustrated in Figure 4.8.[6] The Li_8 cluster has a geometry based on a tetracapped tetrahedron and is closely related to the cube. The $S^\sigma(A_1)$ and $P^\sigma_{x,y,z}(T_2)$ molecular orbitals of the tetrahedron, which have been discussed in some detail in Section 4.2, can accommodate a total of eight valence electrons to give a particularly favorable closed-shell electronic configuration. These cluster orbitals are stabilized by mixing with the a_1 and t_2 orbitals of the capping atoms in much the same way as described above for the cube.

In contrast, for the Li_6 cluster, with a total of six valence electrons, an oblate geometry is preferred. This leads to the occupation of S^σ and two $P^\sigma_{\pm 1}$ orbitals of the P^σ set. This geometry maximizes the stabilization energy and leads to a large HOMO-LUMO gap by destabilization of the P^σ_0 component. The retention of a more spherical geometry would lead to an increase in the

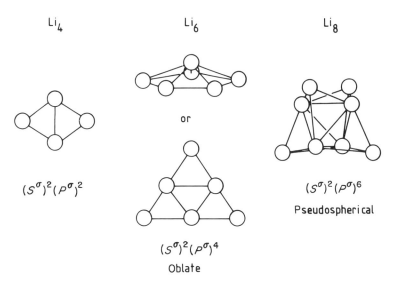

Figure 4.8 The *ab initio* calculated goemetries of Li$_8$, Li$_6$, and Li$_4$.

electronic energy for this molecule. *Ab initio* calculations indicate that two oblate geometries, a planar raft with D_{3h} symmetry and a nearly planar pentagonal pyramid with C_{5v} symmetry, have very similar energies.

For Li$_4$, with a total of four valence electrons, a prolate geometry is ideally required, because this results in the stabilization of P_0^σ at the expense of the $P_{\pm 1}^\sigma$ pair. In the observed geometry, the atoms are arranged in a diamond for which all three components of the P^σ set are split apart so that one member is stabilized at the expense of the other two. (One of the P^σ functions is actually null.)

In the gold clusters $[Au_m(PPh_3)_m]^{x+}$, the d orbitals of the gold atoms are effectively corelike, and bonding occurs primarily through the in-pointing s/p_z hybrids.[7] The out-pointing s/p_z hybrids are used to form the Au—PPh$_3$ bonds. The cluster $[Au_6(PPh_3)_6]^{2+}$ has a geometry based on two tetrahedra sharing a common edge.[8] This prolate geometry, illustrated in Figure 4.9, leads to a stabilization of P_0^σ at the expense of the other components of the P^σ manifold, and a stable closed-shell electronic configuration is achieved when four electrons occupy S^σ and P_0^σ. The related cluster $[Au_7(PPh_3)_7]^+$ has a very distinctive oblate geometry based upon a pentagonal bipyramid with the two axial gold atoms separated by a very short gold-gold bond (2.58 Å) (see Figure 4.9).[9] In this geometry the P^σ orbitals split to give $P_{\pm 1}^\sigma$ below P_0^σ, and a stable closed-shell electronic configuration is achieved when these orbitals are fully occupied by six valence electrons (see also Figure 4.6). Clearly, the alternative symmetrical geometry based on a capped octahedron for $[Au_7(PPh_3)_7]^+$ will not be favored for this electron count, since it has a

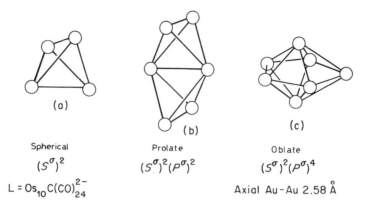

Figure 4.9 The gold skeletons of (a) $[Au_4(PPh_3)_3L]^{2+}$, (b) $[Au_6(PPh_3)_6]^{2+}$, and (c) $[Au_7(PPh_3)_7]^+$.

prolate geometry. Indeed, a simple consequence of this analysis is that the following electrochemical process should be accompanied by a change in cluster geometry:

$$Au_7(PPh_3)_7^+ \xrightarrow{-2e} Au_7(PPh_3)_7^{3+} \quad (4.4)$$

Oblate pentagonal bipyramid $\longrightarrow$ prolate capped octahedron.

A pseudospherical gold cluster with a closed-shell electronic configuration $(S^\sigma)^2(P^\sigma)^6$ has not yet been isolated, but recently a tetrahedral $[Au_4(PPh_3)_3^{2+}][L^{2-}]$ cluster, where L^{2-} is the unusual two-electron donor ligand $Os_{10}C(CO)_{24}^{2-}$, has been characterized with an $(S^\sigma)^2$ closed shell.[10]

4.4 CLUSTERS WITH INTERSTITIAL ATOMS

As the cluster increases in size, so does the size of the cavity in the center, and it becomes increasingly possible to incorporate atoms there. For 12-atom clusters with cuboctahedral and twinned-cuboctahedral geometries, close-packing considerations show that the incorporation of an interstitial atom with an identical radius to the peripheral atoms is possible. These, after all, are the central structural units of cubic close-packing and hexagonal close-packing. For transition metal atoms, the cavity at the center of an octahedron has the correct dimensions to incorporate first-row main group atoms such as B, C, and N. The presence of a central atom can contribute significantly to the stability of a cluster because its orbitals overlap strongly with the radial molecular orbitals of the peripheral atoms.

The number of bonding orbitals in such clusters can be readily understood as a development of the arguments developed above. The *s* and *p* valence orbitals of a main group central atom have the same transformation properties as the S^σ and P^σ linear combinations associated with the peripheral atoms, and there are generally strong overlaps between the two sets. The resultant strong bonding interactions reinforce the bonding pattern established for the parent cluster (Figure 4.10). If the latter has a spherical geometry, then a closed-shell stable electronic configuration is achieved when a total of eight valence electrons are available. Furthermore, the higher electronegativity of the central atom will ensure that all the P^σ components are occupied and a pseudospherical geometry will be preferred.

If the interstitial atom is a transition metal with nd, $(n+1)s$, and $(n+1)p$ valence orbitals, then the interaction diagram shown in Figure 4.11 is appropriate. The S^σ and P^σ linear combinations, which are already skeletal

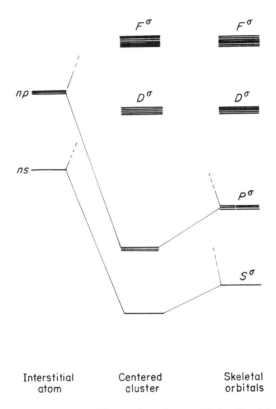

Figure 4.10 The interaction of the *s* and *p* valence orbitals of an interstitial main group atom with the S^σ and P^σ cluster orbitals of the outer cluster skeleton.

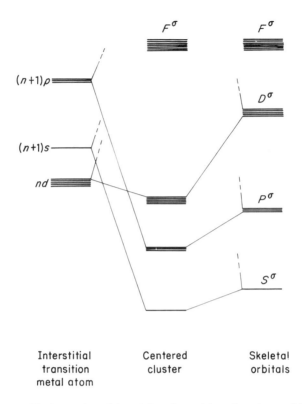

Figure 4.11 The interaction of the nd, $(n + 1)s$, and $(n + 1)p$ valence orbitals of an interstitial transition metal atom with the S^σ, P^σ, and D^σ cluster orbitals.

bonding, are stabilized by overlap with the higher-lying $(n + 1)s$ and $(n + 1)p$ valence orbitals and generate molecular orbitals which are bonding both within the spherical shell and between this shell and the interstitial atom. The D^σ linear combinations are doubly noded and for a 12-atom cage are weakly antibonding, but can overlap in a constructive fashion with the nd orbitals of the central atom. From the interaction diagram in Figure 4.11 it is apparent that the most stable electronic configuration for the centered cluster is achieved when 18 valence electrons are available to occupy the most strongly bonding molecular orbitals. With this number of valence electrons we would expect a spherical geometry to be retained, but if there are only 16 valence electrons then an oblate cluster geometry is anticipated.

Characterization of Sodium Clusters using esr

The great majority of cluster compounds described in this book are diamagnetic and therefore not amenable to electron spin resonance spectroscopy. However, this technique has been used to characterize sodium cluster ions which have been "trapped" in the cavities of zeolite structures. Initially these species were obtained by high-temperature and high-vacuum techniques, e.g., vapor phase deposition of sodium metal or pyrolysis of impregnated sodium azide. More recently they have been synthesized by adding solutions of $LiBu^n$ to NaX zeolite under inert atmosphere conditions. The resultant purple solid shows the well-resolved esr spectrum shown below. $\langle g \rangle = 2.0021$, and the 13-line hyperfine splitting pattern is attributed to Na_4^{3+} (^{23}Na, $I = 3/2$; 100% abundance).

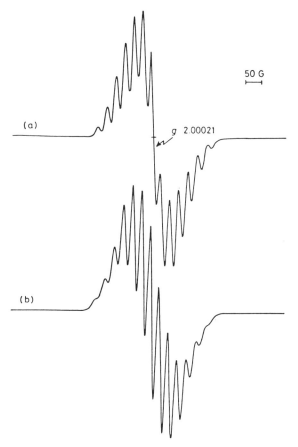

(a) ESR spectrum of Na_4^{3+} from the reduction of NaX zeolite by butyl-lithium in hexane at room temperature. (b) Simulated spectrum using Lorentzian rather than Gaussian linewidths.

The bonding in the cluster Na_6Mg, with a total of eight valence electrons, can also be understood by reference to Figure 4.10. It has, as anticipated, a spherical geometry based on an octahedron of sodium atoms with the more electronegative magnesium atom in the center (Figure 4.12).[11] In the octahedral point group the S^σ and P^σ cluster orbitals transform as A_{1g} and T_{1u}. A pseudospherical geometry is observed for the related ion Na_7Mg^+, which also has a total of eight valence electrons. In this instance the geometry

---

Mössbauer Data

In the area of cluster chemistry the nuclei most commonly used for Mössbauer studies are ^{57}Fe, ^{118}Sn, and ^{197}Au. The Mössbauer lines are sometimes not sufficiently well resolved to show all the chemically different environments in the cluster, particularly since the values for isomer shifts and quadrupole splittings are not large compared to common linewidths. Curve fitting methods are used to overcome the problem of overlapping peaks. A spectral analysis of this type for $[Au_{11}(CN)_3(PPh_3)_7]$ is illustrated in the figure.

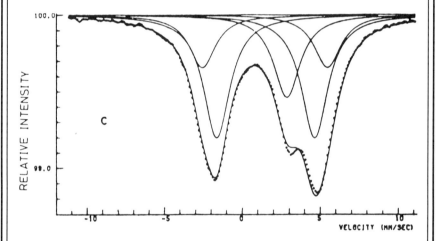

Mössbauer spectrum of $[Au_{11}(CN)_3(PPh_3)_7]$. The deconvolution analysis is based on three-site, five-line interpretation (two doublets and one singlet).

In spherical gold clusters with $12n_s + 18$ valence electrons, the central gold atom is observed as a singlet with a chemical shift of about 3 mm/s. The toroidal gold clusters with $12n_s + 16$ valence electrons have the central gold atom in a non-spherical environment and therefore a quadrupole doublet is anticipated; however, it is frequently lost in the quadrupole pairs associated with the peripheral gold (Au—L) fragments. The different chemical environments associated with the peripheral gold atoms are distinguishable, however. Some typical data are summarized below.

MÖSSBAUER DATA FOR GOLD PHOSPHINE CLUSTER COMPOUNDS						
	IS, mm/s			Qs. mm/s		
	Central Au	Au—L	Au—X	Central Au	Au—L	Au X
$[Au_{11}(CN)_3(PPh_3)_7]$	2.9	1.5	1.4	0	6.5	8.0
$[Au_{11}(SCN)_3(PPh_3)_7]$	2.7	1.4	0.6	0	6.7	4.5
$[Au_{11}(dppp)_5](SCN)_3$	2.8	1.7	—	0	6.6	—
$[Au_9(PPh_3)_8](PF_6)_3$	2.8	2.1	—	0	6.6	—
$[Au_9[P(p-C_6H_4Me)_3]_8]$ $(PF_6)_3$	2.8	2.3	—	0	6.8	—
$[Au_8(PPh_3)_8](PF_6)_2$	2.4	1.9	—	0	6.7	—
$[Au_8(PPh_3)_7](NO_3)_2$	2.4	2.42	—	0	6.88	—
$[Au_4(\mu\text{-}I)_2(PPh_3)_4]$		1.4			6.5	

is based on a pentagonal bipyramid of sodium atoms with an interstitial magnesium atom and the apical sodium atoms are well separated.

If the cluster has a total of six valence electrons, then an oblate geometry is preferred. Na_4Mg and Na_5Mg^+ illustrate this point, since the *ab initio* calculated geometries are based on a square of sodium atoms and a planar pentagon of sodium atoms, respectively. In both examples the magnesium atom resides at the center of the polygon. Similarly, $MgNa_2$ has a linear geometry which permits the stabilization of only one component of the P^σ set and therefore leads to the maximum stabilization energy for four valence electrons. It represents an extreme form of prolate geometry.

In contrast to the sodium and magnesium clusters described above, the cluster compounds of gold can be isolated as air-stable crystalline solids, and their structures can be studied in great detail using single-crystal X-ray

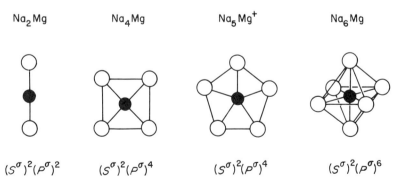

Figure 4.12 The *ab initio* calculated geometries of Na_xMg and Na_xMg^+ clusters.

diffraction techniques.[12] Stereochemical rearrangements can also be studied in solution. The structures of these molecular cluster compounds, illustrated in Figure 4.13, can be classified into two major groups. Those clusters with pseudospherical geometries are associated with a total of $12n_s + 18$ valence electrons, where n_s is the number of surface gold atoms. The spectrum of molecular orbitals in these species and their electronic requirements can be analyzed using Figure 4.11. Each surface gold atom has a filled $5d$ shell and a bonding orbital associated with the Au—PPh$_3$ bond, thereby accounting for the $12n_s$ component in the above formula. The skeletal molecular orbitals are those shown in Figure 4.11; i.e., S^σ, P^σ, and the $5d$ shell of the interstitial gold atom.

Structurally, the spherical clusters illustrated in Figure 4.13 can be derived from a centered chair of seven gold atoms with the additional atoms located on or about the principal axis, either singly or in triangular groups of three. This structural relationship is illustrated in Figure 4.14. The highest-nuclearity example is the icosahedron, which has triangles of gold atoms on the threefold axis above and below the centered chair. It was molecular orbital arguments of the type developed above which led to the prediction that gold clusters with the stoichiometry $[Au_{13}(PR_3)_{12}]^{5+}$ would have icosahedral geometries. This prediction was realized some years later when $[Au_{13}Cl_2(PMe_2Ph)_{10}]^{3+}$ was made and structurally characterized.[1]

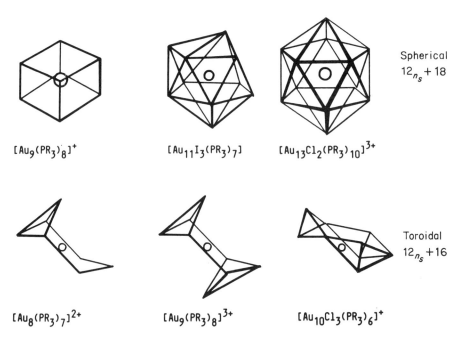

Figure 4.13 Geometries of pseudospherical and toroidal gold cluster compounds of the general form $[Au(AuPPh_3)_n]^{x+}$.

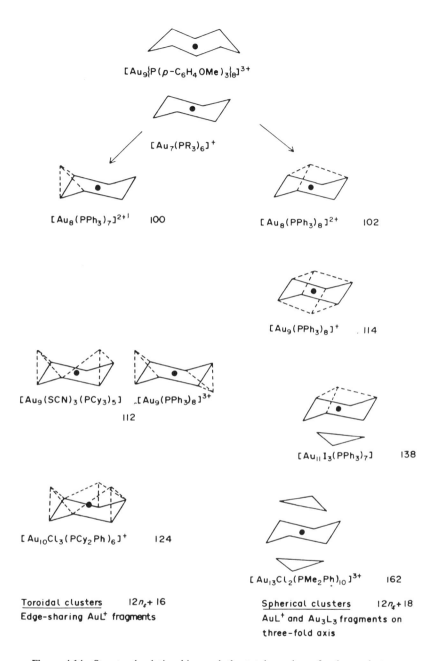

Figure 4.14 Structural relationships and the total number of valence electrons (pec) for pseudospherical and toroidal gold clusters of the form $[Au(AuPPh_3)_n]^{x+}$.

Mulliken overlap population analyses, which give an indication of the relative strengths of bonds within a molecule, have indicated that the radial gold-gold bonds are stronger than the tangential gold-gold bonds, emphasizing the important role of the interstitial atoms in stabilizing the structure. This is confirmed by the observation that the radial gold-gold bonds are consistently 0.2 to 0.3 Å shorter than the tangential bonds. A further consequence of the strong radial interactions is the ease with which the cluster skeletons can be deformed. Thus calculations on $[Au_{13}L_{12}]^{5+}$ indicate that although the icosahedron is the most stable geometry, because it maximizes the number of nearest neighbors on the spherical surface, the cuboctahedron and the anticuboctahedron are only about 0.8 eV less stable. Therefore, as long as the basic spherical topology is maintained, the skeleton can be distorted without a large energy loss being incurred. This lack of structural discrimination contributes to the stereochemical nonrigidity of gold cluster molecules in solution—an aspect which is discussed in more detail in Chapter 6.

The second class of gold cluster has been described as toroidal because the peripheral gold atoms define a ring or torus (see Figure 4.13 for some examples). These cluster compounds are characterized by $12n_s + 16$ valence electrons. They are clearly related to the oblate spheroids discussed in previous sections, and the electronic factor responsible for the change in electron count has an identical basis. In toroidal or oblate clusters the degeneracy of the P^σ cluster orbitals is split with $P^\sigma_{\pm 1}$ being stabilized relative to P^σ_0. The electronic origins of this splitting are illustrated in Figure 4.15. As a spherical Au_8 square antiprism is distorted into a toroidal crown, the bonding interactions associated with P^σ_0 around the square faces are lost, whereas the antibonding interactions between adjacent gold atoms are retained. Consequently, the crown geometry retains only two $P^\sigma_{\pm 1}$ bonding skeletal molecular orbitals.

These toroidal gold clusters also have easily deformable skeletons, and this is reflected not only in their stereochemical nonrigidity, but also in the fact that chemically very similar structures crystallize with different skeletal geometries; for example, the skeletal isomers of $[Au_9(P(C_6H_4OMe)_3)_8](NO_3)_3$ in Figure 4.16, where one isomer has a D_{4d} crown structure and the other a D_{2h} structure.[13] Except for $[Au_9(P(C_6H_4OMe)_3)_8]^{3+}$, which has a centered

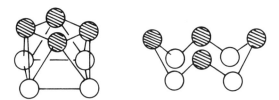

Figure 4.15 Illustration of P^σ_0 for a pseudospherical (square-antiprismatic) and a toroidal (crown) gold cluster. This orbital is antibonding and unavailable for bonding in the latter geometry.

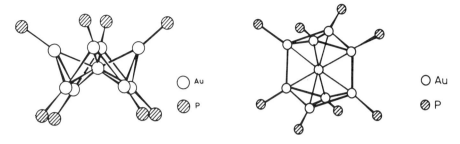

Figure 4.16 An example of skeletal isomerism in gold cluster chemistry.

crown structure, the remaining toroidal gold clusters have geometries based on a centered Au_7 chair with the additional gold atoms bonded to edges of the chair as in Figure 4.14. The stronger radial bonding relative to the tangential bonding is reflected in the observed bond lengths.

There are also a few examples of gold clusters with hemispherical topologies. In such compounds the P_0^σ cluster orbital is destabilized, but not sufficiently to become inaccessible. Consequently, such clusters are also characterized by $12n_s + 18$ valence electrons. Recently, examples of gold clusters with interstitial platinum and rhodium atoms have been isolated; their structures are illustrated in Figure 4.17.[14–15] In addition, Teo has shown that icosahedral silver-gold fragments can be linked to generate the high-nuclearity clusters illustrated in Figure 4.18.[16]

The occurrence of two major classes of gold cluster compounds with $12n_s + 16$ and $12n_s + 18$ valence electrons has some interesting chemical consequences, since the addition of an electron pair, either electrochemically or by ligand addition, should lead to a change in topology from toroidal to spheroidal. Two reactions illustrating this dramatic structural rearrangement are shown in Figure 4.19.[17]

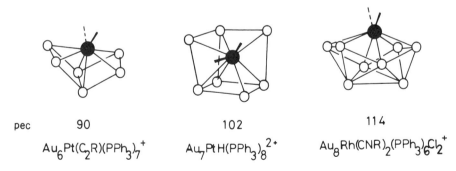

pec   90   102   114

$Au_6Pt(C_2R)(PPh_3)_7^+$ $Au_7PtH(PPh_3)_8^{2+}$ $Au_8Rh(CNR)_2(PPh_3)_6Cl_2^+$

Figure 4.17 Some hemispherical gold-platinum and gold-rhodium heteronuclear clusters.

Sec. 4.4 Clusters With Interstitial Atoms

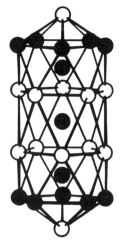

Figure 4.18 A gold-silver cluster based upon linked icosahedra; the gold atoms are shaded.

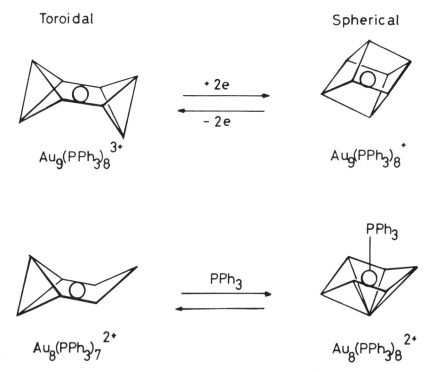

Figure 4.19 Examples of the pseudospherical to toroidal skeletal rearrangement observed for gold clusters.

4.5 MULTISPHERICAL CLUSTERS

The discussion above has concentrated on small clusters of the alkali and coinage metals, but mass spectroscopic studies of alkali metals in molecular beams show that clusters with hundreds of atoms are also formed (see Chapter 1). Obviously, the atoms in these large clusters do not lie on a single spherical surface. Instead the structures may be described as multispherical. The geometries adopted may be based upon close-packed arrangements or on shells with icosahedral point group symmetry. The molecular beam experiments reveal that certain cluster nuclearities are formed in greater abundance—these nuclearities are often called "magic numbers" and are taken to indicate the high relative stability of these fragments. For example, the following magic numbers are observed in a molecular beam experiment for sodium:[18] 8, 20, 34, 40, 58, and 92. The stability of these particular clusters appears to be explicable in terms of the completion of closed electronic subshells in a similar manner to that described above for the small clusters.[18]

In this case we cannot expect the particle on-a-sphere eigenfunctions to be appropriate, because there is generally more than one shell of atoms. Instead, the eigenfunctions for a free electron in a sphere may be used in the same way as the spherical harmonics are in TSH theory to give a "tensor solid harmonic" theory.[19] The solutions of the free-electron problem may be written

$$\psi = j_L(kr) Y_{L,M}, \qquad (4.5)$$

where $j_L(kr)$ is a spherical Bessel function and k determines the number of radial nodes. If we assume that only s orbitals are important, then we need only consider forming σ-type cluster orbitals from these expansion functions. The energy levels for the free-electron problem are

$$1s < 2p < 3d < 2s < 4f < 3p < 5g < 4d < 6h < 3s \cdots \qquad (4.6)$$

where $4d$, for example, has two angular nodes and one radial node in a notation analogous to that for atomic orbitals. The same ordering is expected from the tensor solid harmonic LCAO cluster orbitals, and we find that all the observed magic numbers do indeed occur for filled electronic shells. A practically identical pattern emerges from solving the one-electron problem using a spherically symmetric potential (a "jellium" potential).[18] There is also evidence that the larger alkali metal clusters adopt prolate and oblate geometries when they have a partially filled electronic subshell; Clemenger has provided a theoretical analysis of the problem within the jellium framework.[20] We might also expect such considerations to apply to colloidal metal particles of gold and silver where the atoms have a single s electron outside a filled d shell.[21]

EXERCISES

4.1 Derive the L^σ cluster orbitals for a cube and a square antiprism using the TSH methodology. (You may check that your linear combinations transform in the right way using projection operators.) Compare the relative stabilities of the two geometries for 2, 4, 6, and 8 skeletal valence electrons.

4.2 The following minimum-energy structures have been calculated for Li_n^+ clusters using *ab initio* techniques: Li_9^+ (centered square antiprism), Li_7^+ (pentagonal bipyramid), and Li_5^+ (trigonal bipyramid). Are these structures consistent with the ideas developed in this chapter?

4.3 Using TSH theory methodology, derive the L^σ linear combinations of inwardly directed radial hybrids for a pentagonal bipyramid and a cube (for the two layers of atoms in the latter structure, $\theta = \pi/2 \pm \frac{1}{2}\cos^{-1}\sqrt{\frac{1}{3}}$). Assuming that these are the only orbitals which need to be considered, construct qualitative energy-level diagrams for the two clusters. Would you expect these structures to share the same number of skeletal valence electrons?

4.4 Derive the L^σ linear combinations for a cluster in which two tetrahedra share an edge, treating all six vertex atoms at if they lie on the same spherical surface. By constructing a qualitative energy-level diagram for this system, predict when such a geometry will be favored over an octahedron.

4.5 Schmidbaur, H. et al. (*Angew. Chem. Int. Ed.*, **27**, 1544 (1988); **28**, 463 (1989)) have demonstrated that the diamagnetic octahedral cluster $[Au_6(Ptol_3)_3]^{2+}$ should be reformulated as $[Au_6C(Ptol_3)_6]^{2+}$ with an interstitial carbido-ligand. Use the molecular orbital arguments developed in this chapter to derive qualitative molecular orbital diagrams for $[Au_6L_6]^{2+}$ and $[Au_6CL_6]^{2+}$ based on idealized octahedral geometries and confirm that the carbido-formulation provides a better account of the closed shell requirements for the cluster (see also Mingos, D. M. P., *J.C.S. Dalton*, 1163 (1976)).

4.6 Predict the charges ($x+$) for the following series of gold compounds with interstitial main group atoms: $[Au_5C(PPh_3)_5]^{x+}$, $[Au_4N(PPh_3)_4]^{x+}$, $[Au_3S(PPh_3)_3]^{x+}$ and $[Au_2Cl(PPh_3)_2]^{x+}$ using the molecular orbital arguments developed in this chapter.

REFERENCES

1. Briant, C. E., Theobald, B. R. C., White, J. W., Bell, L. K., and Mingos, D. M. P., *J. Chem. Soc., Chem. Comm.*, 201 (1981).
2. Mingos, D. M. P., and Hawes, J. C., *Struct. Bond.*, **63**, 1 (1985).
3. Stone, A. J., *Molec. Phys.*, **41**, 1339 (1980).
4. Evans, D. G., and Mingos, D. M. P., *J. Organometallic Chem.*, **295**, 389 (1985).
5. Wales, D. J., and Mingos, D. M. P., *Inorg. Chem*, **28**, 2748 (1989).
6. Koutecký, J., and Fantucci, P., *Z. Physik D*, **3**, 147 (1986).
7. Mingos, D. M. P., *J. Chem. Soc., Dalton Trans.*, 1163 (1976).
8. Evans, D. G., and Mingos, D. M. P., *J. Organometallic Chem.*, **232**, 171 (1982). Briant, C. E., Hall, K. P., and Mingos, D. M. P., *J. Organometallic Chem.*, **254**, C18 (1983).

9. van der Velden, J. W. A., Beurskens, P. T., Bour, J. J., Bosman, W. P., Noordik, J. H., Kolenbrander, M., and Buskes, J. A. K. M., *Inorg. Chem.*, **23**, 146 (1984).
10. Dearing, A., Drake, S. R., Johnson, B. F. G., Lewis, J., McPartlin, M., and Powell, H. R., *J. Chem. Soc., Chem. Comm.*, 1331 (1988).
11. Koutecký, J., and Fantucci, P., *Chem. Rev.*, **86**, 539 (1986).
12. Hall, K. P., and Mingos, D. M. P., *Prog. Inorg. Chem.*, **32**, 237 (1984).
13. Briant, C. E., Hall, K. P., and Mingos, D. M. P., *J. Chem. Soc., Chem. Comm.*, 290 (1984).
14. Bour, J. J., Kanters, R. P. F., Schlebos, P. P. J., and Steggarda, J. J., *Recl. Trav. Chim. Pays-Bas*, **107**, 211 (1988).
15. Bott, S. J., Mingos, D. M. P., and Watson, M. J., *J. Chem. Soc., Chem. Comm.*, 1192 (1989).
16. Teo, B. K., and Keating, K., *J. Amer. Chem. Soc.*, **106**, 2224 (1984).
17. van der Linden, J. G. M., Paulissen, M. L. H., and Schmitz, J. E. J., *J. Amer. Chem. Soc.*, **105**, 1903 (1983).
18. Knight, W. D., Clemenger, K., de Heer, W. A., Saunders, W. A., Chou, M. Y., and Cohen, M. L., *Phys. Rev. Lett.*, **52**, 2141 (1984).
19. Wales, D. J., *Some Theoretical Aspects of Cluster Chemistry*, Ph.D. Thesis, University of Cambridge, 1988; Oxford Academic Publishers, Oxford, 1989.
20. Clemenger, K., *Phys. Rev. B*, **32**, 1359 (1985).
21. Wales, D. J., Kirkland, A. I., and Jefferson, D. A., *J. Chem. Phys.*, **91**, 603 (1989).

5

Clusters Where σ and π Interactions Are Important

5.1 INTRODUCTION

In the previous chapter we examined some clusters where only the radial bonding needed to be considered. Such species are exemplified by gold and alkali metal clusters, where the valence shell *p* orbitals are often too high-lying in energy to contribute significantly to the bonding. For these molecules the bonding patterns are particularly simple, but most clusters cannot be described in this way. In this chapter we will show how TSH theory may be used to treat clusters where the valence shell *p* orbitals must also be considered, and we will pay particular attention to the boranes and carboranes, which provide the most familiar examples of such species. We will also develop the TSH theory pairing principle in more detail, and show how this leads to predictions of deviations from the usual electron counting rules, and to descriptions of *nido* and *arachno* clusters. The relationship of the frontier molecular orbital energy levels in the *closo* boranes to their shape will also be discussed, as this is important for considerations of kinetic and thermodynamic stability. Our discussion will also extend to three-connected and four-connected clusters and to simple transition metal cluster carbonyls.

5.2 BONDING IN THE BORON HYDRIDES

For a long time the structures of even the simplest boranes were unknown, and it was not until the 1940s that the geometry of diborane (B_2H_6) was elucidated experimentally. Structure determination is today a fairly routine procedure involving X-ray and neutron diffraction and spectroscopic methods, such as ^{11}B nmr, for less stable species. Some of the first theoretical work in this area was performed by Longuet-Higgins and Roberts[1] in 1954 with some calculations on metallic borides of the general formula MB_6. In the following year the same authors predicted the existence of icosahedral $B_{12}H_{12}^{2-}$ (using a parameterized LCAO-MO approach[2]) before it was synthesized.

The next steps were taken in the early 1970s, when Wade,[3] Mingos,[4] Rudolph,[5] and Williams[6-7] noticed the structural patterns which are now known as the *debor* and capping principles. Lipscomb had previously conjectured that all the boranes' structures were icosahedral fragments, but the pattern recognized in the *closo*, *nido*, *arachno* nomenclature is now generally preferred. In this scheme the geometries of boranes and carboranes are considered to be based upon deltahedra (with all faces triangular). Species with a boron (or carbon) atom at each vertex of the deltahedron are *closo* compounds and are typified by the anions $B_nH_n^{2-}$ and the carboranes $(CH)_\alpha(BH)_\beta$ in which all the hydrogen atoms are terminally bound. A *nido* cluster is based upon a deltahedron with one missing vertex, and *arachno* clusters have two missing vertices. In general the (car)borane $(CH)_\alpha(BH)_\beta H_\gamma^{\delta-}$ adopts *closo*, *nido*, and *arachno* geometries when $\alpha + \gamma + \delta = 2$, 4, or 6, respectively. Examples were given in the opening chapters.

To derive general electron counting rules for boranes, carboranes, etc., we must consider the effect of the valence shell p orbitals in detail. As usual, we choose local axes at each cluster vertex such that the z axis points radially outward from the center of the cluster, the y axis points in the direction of θ increasing, and the x axis points in the direction of ϕ decreasing. The p_z orbitals are therefore classified as σ (no nodal plane containing the radius vector), and the p_x and p_y orbitals are π-type (one nodal plane containing the radius vector). Hence the p_z (or p^σ) orbitals span the same representation as the set of s orbitals, and the same linear combinations can be used with $\sigma(r)$ representing a p_z orbital rather than an s orbital. Of course, we could equally well consider any two independent linear combinations of the corresponding σ cluster orbitals composed of s and p_z functions. For example, we could form one in-pointing and one out-pointing hybrid at each vertex and construct one set of σ orbitals from the in-pointing orbitals and another from the out-pointing. The out-pointing hybrids will then form two-center two-electron σ bonds with the terminal hydrogen atoms, and need not be considered further. The in-pointing σ cluster orbitals, however, will clearly give rise to an energy-level pattern just like those discussed in Chapter 4.

From the $2n$ tangential p_x and p_y (p^π) orbitals, we can generally form n even and n odd π cluster orbitals, as described in Chapter 3. These are denoted L_μ^π (or $\psi_{L\mu}^\pi$) and $\bar{L}_\mu^\pi$ (or $\bar{\psi}_{L\mu}^\pi$), where $\mu = 0, 1c, 1s, 2c, \ldots$

We also noted in Chapter 3 that $\bar{V}_{L,\mu}^\theta = -V_{L,\mu}^\phi$ and $\bar{V}_{L,\mu}^\phi = V_{L,\mu}^\theta$ (and the same is true for any two orthogonal tangential directions at any point). Hence an odd π cluster orbital may be obtained from its even partner (except perhaps for a trivial change of phase) by rotating the tangential p orbitals at each vertex through 90°, all in the same sense. The converse is also true. Furthermore, this rotation interchanges bonding and antibonding characteristics, so that the odd π cluster orbitals are antibonding if their even partners are bonding, and vice versa. In deltahedral clusters we find that the even π orbitals are generally all bonding in character, which means that their odd partners must be antibonding. This can be understood qualitatively in pictorial form from the behavior of the even and odd orbitals in relation to the nodal surfaces and the maxima and minima of the parent spherical harmonic from which the expansion coefficients are obtained (Figure 5.1). Remember that the gradient vector, $\nabla Y_{L,\mu}$, points in the direction of the greatest rate of increase of $Y_{L,\mu}(\theta, \phi)$.

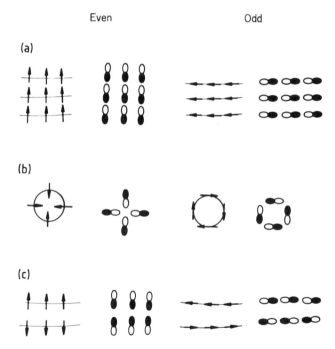

Figure 5.1 The behavior of the even (*left*) and odd (*right*) vector harmonics in regions of space (a) where $Y_{L,\mu}$ varies slowly, (b) near a maximum point, and (c) near a line where $Y_{L,\mu}$ has a minimum.[8] Notice that around the extrema the even functions are bonding and the odd functions antibonding, while in case (a) both are fairly nonbonding.

We therefore expect to find $n + 1$ strongly bonding orbitals consisting of S^σ and the even π set. Although some of the other in-pointing σ cluster orbitals may have bonding characteristics, they generally mix with bonding π orbitals which transform according to the same irreducible representation. This leaves the predicted skeletal electron count unchanged at $2n + 2$. Note that there are $2n - 1$ unoccupied skeletal orbitals which may be described as "inaccessible" because they are formed from in-pointing orbitals and are high in energy. For a cluster with $12 > n > 6$, the S^σ and P^σ cluster orbitals and the $n L^\pi$ cluster orbitals are all bonding in character. Even in deltahedra, the σ/π separation is only approximate and mixing occurs between orbitals of the same symmetry belonging to the two sets. For example, corresponding components of L^σ and L^π transform in the same way, and consequently their mixing is symmetry-allowed. The following descent-in-symmetry correlations arise for the σ and π cluster orbitals of the octahedron:

L^σ	O_h	L^π	O_h
S^σ	A_{1g}		
P^σ	T_{1u}	P^π	T_{1u}
D^σ	E_g	D^π	T_{2g}

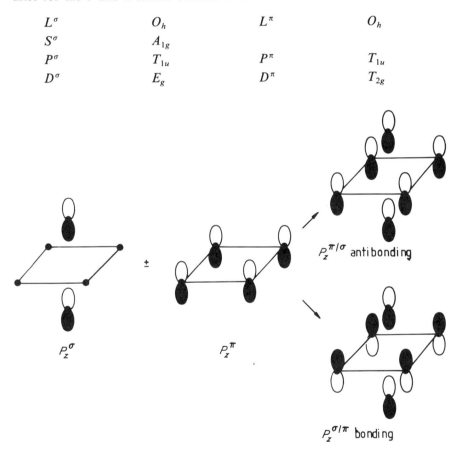

Figure 5.2 The mixing of P_z^σ and P_z^π orbitals in an octahedron gives strongly bonding and strongly antibonding combinations.

Hence P^σ/P^π mixing will be significant, but not D^σ/D^π, because different components of the latter sets are present. The P^σ/P^π mixing is actually quite strong because of the derivative nature of the relationship between these sets of orbitals; P_z^σ is concentrated on the z axis, while P_z^π is concentrated on the atoms in the xy plane. The interaction of these two functions is illustrated schematically in Figure 5.2. The in-phase P^σ/P^π combination is strongly bonding, whereas the out-of-phase combination is strongly antibonding and inaccessible. The occupied π cluster orbitals for a variety of clusters were summarized in Chapter 3 where a specific example ($B_6H_6^{2-}$) was given.

5.3 SHAPES AND PATTERNS IN BORANES AND CARBORANES

5.3.1 HOMO-LUMO Gaps in the $B_nH_n^{2-}$ Series

First we shall consider the trends which appear for the variation of the HOMO-LUMO gap in the *closo* boranes $B_nH_n^{2-}$. A plot of the HOMO and LUMO energies (calculated using the Fenske-Hall method[9]) for those clusters with $n+1$ skeletal electron pairs is shown in Figure 5.3. The structures on which the calculations for $n \geq 13$ are based are hypothetical.[10] Although the HOMO-LUMO gap clearly decreases with increasing nuclearity, this change is not smooth, and it often happens that relatively large gaps occur for clusters of even nuclearity compared to their neighbors with 1 more or 1 less vertex. The detailed variation is important because the HOMO-LUMO gap gives us a feel for how stable a cluster will be. More precisely, if the HOMO-LUMO gap is small, it is more likely that the cluster will be kinetically or thermodynamically unstable. Thermodynamic instability arises because the structure is not a true minimum of total energy or is subject to Jahn-Teller distortions.[11] Kinetic instability with respect to chemical reaction arises because the relatively high-lying HOMO and low-lying LUMO may interact favorably with electrophiles or nucleophiles, respectively. A summary of the HOMO-LUMO gaps observed is given in Table 5.1.

The bonding π cluster orbitals of $B_nH_n^{2-}$ are represented as bars in Figure 5.4, where the length of the bar indicates the spread of the L^π molecular orbitals. For the high-symmetry octahedral and icosahedral clusters, much of the $(2L+1)$-fold degeneracy is retained, but the splitting of the L^π manifolds becomes more significant in the less symmetrical point groups. However, the P^π, D^π, and F^π groups of orbitals are still clearly discernible. The filling of the D^π and F^π orbitals in deltahedra was discussed in Chapter 3.

Three factors are primarily responsible for the trends observed in Table 5.1, namely, the point group, nuclearity, and shape of the cluster. The effect of the order of the point group is clearly discernible in Figure 5.3. In low

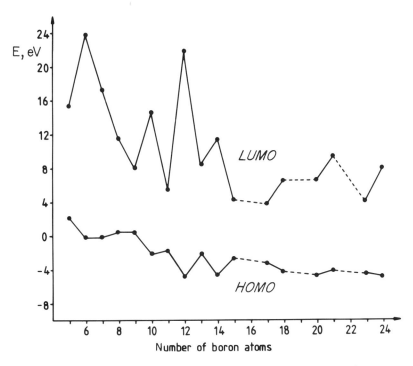

Figure 5.3 HOMO and LUMO energies (in eV) for *closo* boranes $B_nH_n^{2-}$ with n ranging between 5 and 24. Clusters with systematic deviations from $n+1$ skeletal electron pairs are not included.

symmetry, the mixing of frontier orbitals with other orbitals of the same symmetry species is more likely to raise the HOMO in energy and lower the LUMO and thereby reduce the HOMO-LUMO gap. The effect of cluster size on reducing the gap will be obvious to anyone who is familiar with the development of band structure in infinite solids. As the basis set increases in size, the average separation between the energy levels decreases.

What about the shape of the cluster? In Chapter 4 we showed how the prolate or oblate nature of a radially bonded cluster can have profound effects upon its electronic structure. For the clusters considered in this chapter, where σ and π interactions are both important, it appears that similar effects are present.[12] First, we observe that the large HOMO-LUMO gaps observed for "spherical" clusters of high symmetry, such as $B_6H_6^{2-}$ and $B_{12}H_{12}^{2-}$, can probably be ascribed largely to the point group and nuclearity. However, large gaps are also observed when only the $L_{\pm 1}^{\pi}$ members of the outer L^{π} manifold (the one with the largest L) are occupied; for example, in $B_5H_5^{2-}$ and $B_{10}H_{10}^{2-}$. Clusters with only a single member of the outer L^{π} set unoccupied also have relatively large gaps, e.g., $B_7H_7^{2-}$ ($D_{\pm 1}^{\pi}$ and $D_{\pm 2}^{\pi}$ occupied) and $B_{16}H_{16}^{2-}$ ($F_{\pm 1}^{\pi}$, $F_{\pm 2}^{\pi}$, and $F_{\pm 3}^{\pi}$ occupied). These observations are all suggestive

TABLE 5.1 THE CORRELATION BETWEEN POINT GROUP AND HOMO-LUMO GAP IN THE $B_nH_n^{2-}$ SERIES

n	5	6	7	8	9	10	11	12
Point group	D_{3h}	O_h	D_{5h}	D_{2d}	D_{3h}	D_{4d}	C_{2v}	I_h
HOMO-LUMO gap, eV	14	26	18.7	12	8	18	8	28.7

n	13	14	15	17	18	20	21	23	24
Point group	C_{2v}	D_{6d}	D_{3h}	C_{2v}	D_3	D_{6h}	D_3	D_3	T
HOMO-LUMO gap, eV	11.5	17.3	7.7	7.5	11.7	12	14.5	9.3	14

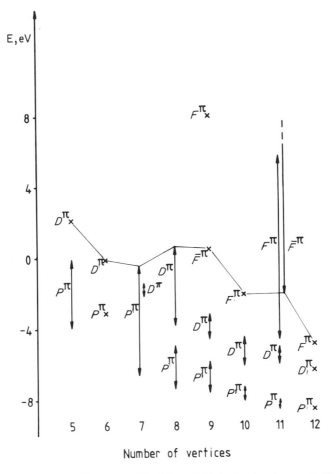

Figure 5.4 The spread in energy of the L^π manifolds in the *closo* boranes $B_nH_n{}^{2-}$. Note that the classifications are not always unambiguous, because of mixing. The HOMOs are linked by the thin line.

of a splitting pattern with a high-lying L_0^π orbital, as shown in Figure 5.5. Furthermore, the overlap of F_0^π and $\bar{F}_0^\pi$ for the tricapped trigonal prism ($n = 9$) also fits into this pattern, and the HOMO of $B_{11}H_{11}{}^{2-}$ is also $\bar{F}_0^\pi$. The occupation of the D^π and F^π orbitals from $B_5H_5{}^{2-}$ to $B_{15}H_{15}{}^{2-}$ was summarized in Chapter 3.

It is apparent that the TSH cluster orbitals provide a useful first approximation to the molecular orbitals of deltahedral boranes where the P^π, D^π, etc., orbitals are discernible in actual calculations and the odd and even π orbitals are well separated in energy. This occurs because the coverage of a spherical surface by a deltahedral array of atoms provides a reasonable

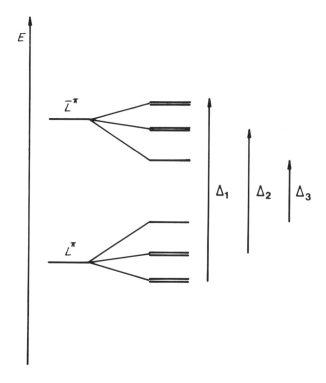

Figure 5.5 Splitting pattern of the outer L^π set suggested by the relatively large HOMO-LUMO gaps observed for $B_5H_5^{2-}$, $B_7H_7^{2-}$, $B_{10}H_{10}^{2-}$, and $B_{14}H_{14}^{2-}$.

representation of the surface itself. In $B_9H_9^{2-}$, the approximation begins to break down and the F_0^π and $\bar{F}_0^\pi$ orbitals overlap. In subsequent sections we will discuss some more severe examples where the usual TSH pattern of the occupied and unoccupied orbitals breaks down. One immediate consequence of these effects is that the *closo* boranes with a high-lying HOMO (which usually results where the L^π and $\bar{L}^\pi$ sets overlap) may show deviations from the simple $4n + 2$ electron counting rule. For example, B_8Cl_8 and B_9Cl_9 have *closo* deltahedral skeletons like $B_8H_8^{2-}$ and $B_9H_9^{2-}$, but have 2 fewer electrons. In these cases the D_0^π and $\bar{F}_0^\pi$ orbitals are depopulated, respectively. Similarly, the tricapped trigonal prismatic molecule Bi_9^{5+} has 2 electrons more than the usual *closo* count.

Experimentally, the clusters for which large HOMO-LUMO gaps are calculated tend to be the most stable. For example, $B_{10}H_{10}^{2-}$ and $B_{12}H_{12}^{2-}$ are unusually resistant to oxidation and are also the most resistant to acid hydrolysis.[13]

5.3.2 Charge Distributions

The charge distribution on the vertex atoms in a cluster depends upon the vertex connectivities and the cluster orbital occupations. In $B_8H_8^{2-}$, for example, the π cluster orbitals consist of the fully occupied P^π and D^π sets, and hence the observed charge distribution must be largely due to differences in connectivity. It is noteworthy that the low-connectivity vertices bear more negative charge. Assuming that the filled shells lead to a fairly uniform charge distribution, then partly occupied L^π shells, or "holes" in filled shells, must be of greatest importance. Actually this is simply a restatement of the problem, since the π cluster orbital set is actually defined by the representation spanned by the p^π orbitals. However, it can provide a useful picture; for example, in $B_7H_7^{2-}$, where only D_0^π of the D^π set is not present, there is a buildup of electron density in the equatorial plane relative to the apices (where the "missing" function has greatest amplitude). This result is typical of the buildup of charge at vertices of lower connectivity, and can be used to predict the most stable substitutional isomers for carboranes, for example.[14] The Mulliken atomic charges calculated by the Fenske-Hall method[9] for $B_7H_7^{2-}$, $B_8H_8^{2-}$, $B_9H_9^{2-}$, and $B_{10}H_{10}^{2-}$ are shown in Figure 5.6.

Intuitively we might expect the substitution of more electronegative atoms to be most energetically favorable at vertices where there is a charge buildup in the homonuclear cluster skeleton, and this is indeed the case. The reverse also holds for substitution by more electropositive atoms. These results

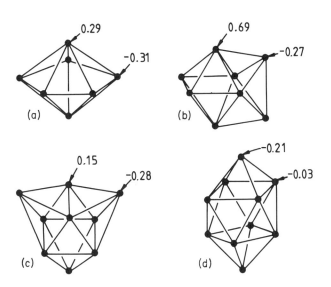

Figure 5.6 Mulliken atomic charges for the boron atoms in (a) $B_7H_7^{2-}$, (b) $B_8H_8^{2-}$, (c) $B_9H_9^{2-}$, and (d) $B_{10}H_{10}^{2-}$.

can be described more formally using first-order perturbation theory,[15] where the perturbation is the change in atomic number of a vertex atom, ΔZ_α:

$$\hat{\mathscr{H}}' = -\sum_{\alpha,i} \frac{\Delta Z_\alpha e^2}{4\pi\varepsilon_0 r_{i\alpha}}. \tag{5.1}$$

The subscripts α and i refer to the nuclei and electrons, respectively, and $r_{i\alpha}$ is the distance of electron i from nucleus α. The first-order change in the energy, E_0^1, of the ground-state wavefunction, ψ_0, is therefore

$$E_0^1 = \langle \psi_0 | \hat{\mathscr{H}}' | \psi_0 \rangle. \tag{5.2}$$

Since the electron density $\rho = \psi_0^* \psi_0$ and $\hat{\mathscr{H}}'$ is a simple multiplicative operator, the energy will clearly be lowered most when increases in Z_α match positions of greatest electron density. Hence electronegative substituents (such as the carbon atoms in carboranes) should be most stable at positions of negative charge (and less electronegative substituents at positions of positive charge) in the homonuclear cluster skeletons. In agreement with this principle, the most stable carborane isomers of $B_7H_7^{2-}$ have both carbon atoms in equatorial positions (see the description above of the charge distribution in $B_7H_7^{2-}$). Williams[6] has made extensive use of patterns such as this (i.e., the predilection of carbon atoms and bridging hydrogens for sites of low connectivity) in his empirical "coordination number pattern recognition" approach to the structures of boranes and carboranes.

5.4 THE PAIRING PRINCIPLE

The relationship between the even and odd π cluster orbitals is defined by the TSH theory pairing principle,[16-17] as described in Chapter 3. In this section we will discuss the pairing principle in more detail, and show how it leads to systematic deviations from the $2n + 2$ skeletal electron count derived above. The pairing principle is also of paramount importance for our considerations of cluster skeletal rearrangements, which follow in Chapter 6.

First we must deduce the transformation properties of the TSH theory parity operator in any given point group. Since the parity operation for a π orbital corresponds to a 90° rotation about the radius vector, it is clear that its character is simply 1 for any proper rotation. Furthermore, since a reflection changes the sense of the rotation, the character under any reflection or other improper operation (such as the inversion) must be -1. This particular irreducible representation is called the "pseudoscalar representation" and is denoted by Γ_u^o. Hence $\Gamma_{\hat{\mathscr{P}}} \equiv \Gamma_u^o$, where $\Gamma_{\hat{\mathscr{P}}}$ is the representation spanned by the parity operator, $\hat{\mathscr{P}}$. In fact it can also be shown that $\Gamma_u^o = \Gamma_z \otimes \Gamma_{\hat{R}_z}$, that is, the pseudoscalar representation is equivalent to the direct product of the representations spanned by z and $\hat{R}_z$ (a rotation about the global z axis).[18] Hence it is a simple matter to find the irreducible representation corresponding to $\mathscr{P}$ for any point group.

It is of some importance in what follows to know the relationship between the representations spanned by even cluster orbitals, Γ_e, and their odd partners, Γ_o. Clearly, for any particular pair of corresponding even and odd π orbitals, $\Gamma_o = \Gamma_e \otimes \Gamma_{\mathscr{P}}$. Γ_o is not always distinct from Γ_e; for example, in a point group with no improper operations, the character of $\Gamma_{\mathscr{P}}$ is 1 for every operation and $\Gamma_o = \Gamma_e$. There are actually three classes of point group which must be considered:[17]

(1) Γ_o and Γ_e are always distinct. Groups of this type contain a reflection, an improper rotation, or the inversion operation and include C_i, C_s, C_{2v}, $D_{(2n+1)d}$, C_{nh}, S_{4n-2}, D_{nh}, T_h, O_h, and I_h.
(2) Γ_o and Γ_e are always equivalent, because $\Gamma_{\mathscr{P}}$ is symmetric under every operation of the point group. Examples are C_n, D_n, T, O, and I, and molecules belonging to these groups are optically active.
(3) Some Γ_o and Γ_e are distinct but others are not. Such groups must contain an improper operation, as in class 1 above, but there are some irreducible representations with zero character under all the improper operations for which $\Gamma_o = \Gamma_e \otimes \Gamma_{\mathscr{P}} = \Gamma_e$. The E representations of C_{nv} ($n \geq 3$), S_{4n}, D_{2nd}, and T_d are of this type.

When a cluster belongs to point groups of type 2 or 3 above which are not Abelian (commutative) and cannot be written as the direct product of two simpler groups, the even and odd sets of cluster orbitals are not necessarily split into two distinct sets. (The proof of this assertion is rather complicated.[18]) Examples of such groups are T, T_d, and C_{nv} with $n \geq 3$. If the representation spanned by the p^π orbitals of a cluster belonging to such a point group has an odd number of E components, then the cluster is forced to deviate from the usual $n + 1$ skeletal electron pairs. This arises because one of the E components must be self-conjugate, and is converted into itself (except perhaps for a change of phase) by the parity operator. Because the parity operator also interconverts bonding and antibonding characteristics, it follows that this pair of orbitals must be nonbonding, with energy α^p. Hence there will be n strongly bonding orbitals, consisting of S^σ and $n - 1$ even π orbitals (with varying admixtures of L^σ orbital character), plus the nonbonding e pair. We therefore expect to find n or $n + 2$ skeletal electron pairs in such clusters, corresponding to the nonbonding pair being empty or fully occupied. Intermediate occupations of the nonbonding pair, such as the usual $n + 1$ electron pairs, would lead to Jahn-Teller distortions.

An example is easily found in the tetrahedron. Here the p^π orbitals span $E \oplus T_1 \oplus T_2$ and $\Gamma_{\mathscr{P}} = A_2$. Hence we assign T_2 as $P^\pi_{x,y,z}$ and T_1 as $\bar{P}^\pi_{x,y,z}$, leaving the E pair. If one component of this pair is even, then its partner must be odd in character, and since the two orbitals are degenerate they must be nonbonding. In B_4Cl_4 the nonbonding e pair is unoccupied and there are four (n) skeletal electron pairs. Some other examples will be discussed later in this chapter; a systematic study has been performed by Johnston and Mingos.[19] In fact, an odd number of E representations can arise only for systems with a single atom on a principal rotation axis of order 3 or more (for a cluster with a single shell of atoms), as demonstrated in the next chapter, where this issue arises again. Hence, for deltahedral clusters, deviations arise only in T, T_d, and C_{3v} symmetry. It is necessary to go to the hypothetical T_d-symmetry $B_{16}H_{16}$ molecule[17] to find an example of such a cluster in the deltahedral boranes (other than the tetrahedron itself).

The photoelectron spectrum of P_4 in the gas phase (Figure 5.7) confirms the relative ordering of the energy levels discussed above, i.e.,

$$\begin{array}{ccc} e & > \quad t_2 \quad > & a_1 \\ D^\pi/\bar{D}^\pi & P^\pi & S^\sigma \end{array}$$

The bands in the spectrum around 10 eV have mainly p orbital character, and the broad band between 15 and 18 eV (which shows evidence of Jahn-Teller splitting into three components) is largely derived from the $3s$ orbitals. The e ($D^\pi/\bar{D}^\pi$) and t_2 (P^π) bands also show Jahn-Teller splitting, which arises

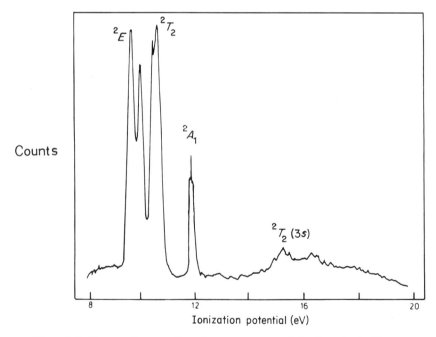

Figure 5.7 The gas-phase uv photoelectron spectrum of the P_4 molecule. Reproduced with permission from Brundle, C. R., et al., *Inorg Chem.*, **11**, 2011 (1972).

because the tetrahedral P_4^+ ion formed by the ionization of electrons from these orbitals would have a degenerate electronic ground state.

This above analysis clearly demonstrates how the application of quantum mechanics and group theory enriches the simple rules discussed in Chapter 2. The tetrahedron and other polar deltahedra with $3p + 1$ vertices and a single atom on a threefold rotation axis can achieve an electronic closed shell with $2n$ or $2n + 4$ skeletal electrons, but not $2n + 2$. B_4Cl_4 and P_4 provide examples of polar deltahedra with $p = 1$; for $p = 2$ and 3, there are no examples with boron atoms at all the vertices, but there do exist some transition metal clusters of this form which are discussed in Section 5.10.

The deviations from $n + 1$ skeletal electron pairs described above are systematic because they are forced by group theory alone. However, there are other, signficantly rarer deviations which cannot be explained in this way.[19] These are also mentioned in Section 5.10.

5.5 *nido* AND *arachno* CLUSTERS

The TSH method also provides a useful description of *nido* and *arachno* clusters.[20] Consider a *closo* deltahedron with $n + 1$ skeletal electron pairs and a corresponding *nido* structure with axial symmetry. The p^π orbitals of the "missing" vertex (which would lie on the principal rotation axis) span an E-type representation if the principal axis is of order 3 or more, and we know from Section 5.4 that there must be an even number of such representations in the p^π set of the *closo* cluster or it would exhibit a deviation from the $n + 1$ rule. Since one of these e pairs of orbitals has been lost, it follows that an odd number remain, and this entails the presence of a nonbonding e pair by the pairing principle. Furthermore, these nonbonding orbitals generally have significant amplitude around the open face because they would interact strongly with the e pair of the missing vertex to give n bonding L^π orbitals as usual. Usually this pair is stabilized by protonation of the open face, and $n + 2$ skeletal electron pairs result. However, there is scope for deviation from this rule if any of the assumptions made above do not hold. This is most likely if the *nido* cluster does not have a threefold principal axis or higher, in which case the nonbonding e pair will be split. If the splitting is small, then both components may still be occupied to give $n + 2$ skeletal electron pairs, but this need not always be true. The loss of one in-pointing σ orbital associated with the "missing" vertex does not affect the number of skeletal bonding orbitals, because the S^σ orbital is still strongly bonding and the other L^σ cluster orbitals generally mix with L^π functions and do not correlate with occupied orbitals.

This analysis also explains why a vertex of highest connectivity is usually "lost" to give *nido* boranes and carboranes. The nonbonding e pair must have a single nodal plane passing through the open face so as to match the p^π

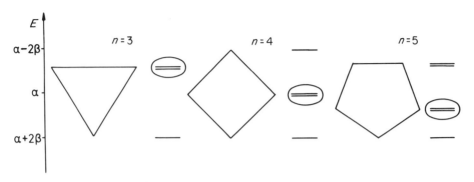

Figure 5.8 The Hückel π energy levels of C_nH_n for $n = 3$, 4 and 5. Notice that the singly noded e pair of orbitals (encircled) becomes steadily lower in energy. These energy-level diagrams may be constructed very simply by taking a point at $\alpha + 2\beta$ and rotating around $\alpha = 0$ in steps of $2\pi/n$, as indicated.

orbitals of the missing vertex. These orbitals are therefore stabilized when the open face is as large as possible,[20] as the antibonding interactions across the node are then balanced by the maximum number of bonding interactions for edges which the node does not cross. This effect can be observed in the Hückel π energy levels of cyclic polyenes. The energy of the singly noded e pair decreases as we go from three- to five-membered C_nH_n rings (Figure 5.8).

For *arachno* clusters, a similar argument is used, except that in this case a dimeric unit is "removed." This fragment has C_{2v} symmetry, while the *arachno* cluster will generally have fairly low symmetry. Hence, it is necessary in this case to use an idealized symmetry analysis, and to classify the relevant orbitals in $C_{\infty v}$. Again we assume that the hypothetical *closo* cluster has $n + 1$ skeletal electron pairs. The two in-pointing σ orbitals will not affect the electron count, for the same reason as above. The four p^π orbitals which are lost span the representations[20] shown in Table 5.2 (the approximate $C_{\infty v}$ classifications are obtained by inspection of the $C_{\infty v}$ character table). These combinations are illustrated in Figure 5.9; a range of related *closo*, *nido*, and *arachno* geometries was illustrated in Chapter 2.

Hence, in $C_{\infty v}$, an e_1 and an e_2 pair of orbitals are lost, and an odd number of e_1-type and e_2-type pairs are left. We therefore expect to find two

TABLE 5.2 CLASSIFICATION OF THE p^π ORBITALS OF A DIMERIC FRAGMENT IN C_{2v} AND IDEALIZED $C_{\infty v}$ SYMMETRY

Combination	C_{2v}	$C_{\infty v}$	Cartesian form
$p^x(1) + p^x(2)$	B_1	E_1	x
$p^y(1) + p^y(2)$	B_2	E_1	y
$p^x(1) - p^x(2)$	A_1	E_2	$x^2 - y^2$
$p^y(1) - p^y(2)$	A_2	E_2	xy

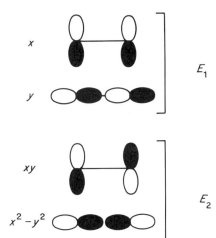

Figure 5.9 Linear combinations of the p^π orbitals of a "missing" dimeric fragment in an *arachno* cluster and their idealized $C_{\infty v}$ classifications.

nonbonding pairs of orbitals with large amplitudes in the open face, one e_1 and one e_2. If all these orbitals are occupied, then $n + 3$ skeletal electron pairs result. Of course, the idealized e orbitals will be split in lower symmetry, but such splittings appear to be generally small. The fact that two adjacent vertices are usually lost (at least for main group clusters) is explained by the greater stabilization of the nonbonding e orbitals in a larger open face, as above. The *arachno* electron counts of clusters with two nonadjacent vertices missing may be rationalized using an idealized symmetry analysis for each open face along the lines of the argument in the foregoing paragraphs. However, deviations from this rule must be anticipated because of the approximations made. An example is B_8H_{12}, which assumes an *arachno*-type structure although it has a *nido* electron count.[6]

The pattern which emerges is basically that when one or more vertices are "removed," the π cluster orbitals of the open structure which would interact strongly with the fragment π orbitals are roughly nonbonding. The analysis is rigorous for some high-symmetry cases with a threefold principal rotation axis or higher, but otherwise it is not. Hence we should be wary of trying to extend this analysis to *hypho* or even more open clusters where exceptions are likely to arise. In such cases the protonation patterns around the open faces are likely to be crucial. Usually protonation occurs around the open face(s), where the roughly nonbonding skeletal orbitals have greatest amplitude. This produces optimum stabilization of these orbitals, and *nido-* or *arachno*-type electron counts are especially likely if the additional protons span the same irreducible representations as the "missing" p^π orbitals.

5.6 THREE-CONNECTED CLUSTERS

The bonding in three-connected clusters such as prismane, C_6H_6, and cubane, C_8H_8, is normally described in terms of two-center two-electron bonds with $3n/2$ skeletal electron pairs. Why, then, do these clusters have a different electron count from the deltahedral clusters discussed above? The answer is basically that one of the principal assumptions which we have implicitly used does not hold for these structures: the L^π and $\bar{L}^\pi$ sets cannot be identified with bonding and antibonding functions anymore. Instead there is a region of significant overlap between these two sets which results in a higher electron count. This can be related to the low coordination number in these molecules,[21] or to our intuitive feeling that such structures do not provide such a good representation of the spherical surface.

An "equivalent orbital" method can be used to express the cluster orbitals in terms of linear combinations of edge-localized bonds. All the linear combinations of localized σ bonds correspond to occupied, bonding molecular orbitals. Linear combinations of these $3n/2$ σ bonds can be taken to give $n/2 + 2$ face-bonding combinations (with all the edges in phase around a face) and $n - 2$ face-antibonding (or nonbonding) combinations in which there are one or more changes of phase around each face. (*Note*: A three-connected cluster has $3n/2$ edges, $n/2 + 2$ faces, and n vertices.) In a TSH analysis we find that the $n/2 + 2$ strongly bonding orbitals consist of S^σ and $n/2 + 1$ orbitals of the L^π set. The remaining $n/2 - 1$ L^π orbitals and their parity-related $\bar{L}^\pi$ partners correspond to the other $n - 2$ members of the bonding set.

TABLE 5.3 THE IRREDUCIBLE REPRESENTATIONS FOR STRONGLY BONDING (L^π), NONBONDING, AND STRONGLY ANTIBONDING ($\bar{L}^\pi$) SKELETAL MOLECULAR ORBITALS IN SOME THREE-CONNECTED CLUSTERS

Clusters	Strongly bonding $(n/2+1)L^\pi + S^\sigma$	Nonbonding $(n-2)$	Strongly antibonding $(n/2+1)\bar{L}^\pi$
Tetrahedron $C_4H_4(T_d)$ $\Gamma^\sigma_u = A_2$	S^σ, $P^\pi(T_2)$	$D^\pi/\bar{D}^\pi(E)$	$\bar{P}^\pi(T_1)$
Trigonal prism C_6H_6 (D_{3h}) $\Gamma^\sigma_u = A''_1$	S^σ, $P^\pi(A''_2 + E')$ $D^\pi_0(A'_1)$	$D^\pi_{\pm 1}(E'')$ $\bar{D}^\pi_{\pm 1}(E')$	$\bar{P}^\pi_2(A'_2 + E'')$ $\bar{D}^\pi_0(A''_1)$
Cube $C_8H_8(O_h)$ $\Gamma^\sigma_u = A_{1u}$	S^σ, $P^\pi(T_{1u})$ $D^\pi_{0,2s}(E_g)$	$D^\pi_{\pm 1,2c}(T_{2g})$ $\bar{D}^\pi_{\pm 1,2c}(T_{2u})$	$\bar{P}^\pi(T_{1g})$ $\bar{D}^\pi_{0,2s}(E_u)$
Dodecahedron $C_{20}H_{20}(I_h)$ $\Gamma^\sigma_u = A_u$	S^σ, $P^\pi(T_{1u})$, $D^\pi(H_g)$ $F^\pi(T_{2u})$	$F^\pi(G_u)/\bar{F}^\pi(G_g)$ $G^\pi(H_g)/\bar{G}^\pi(H_u)$	$\bar{P}^\pi(T_{1g}), \bar{D}^\pi(H_u)$ $F^\pi(T_{2g})$

The representations spanned by the various sets of orbitals, faces, and edges can be summarized as follows:

$$\Gamma_{\text{strongly bonding}} = \Gamma_{\text{face}},$$
$$\Gamma_{\text{strongly antibonding}} = \Gamma_{\text{face}} \otimes \Gamma_{\mathscr{P}}, \qquad (5.3)$$
$$\Gamma_{\text{weakly bonding}} = \Gamma_{\text{edge}} - \Gamma_{\text{face}}.$$

The symmetries and TSH classification of the molecular orbitals in some common three-connected clusters are given in Table 5.3. Characteristic energy-level diagrams for three-connected and deltahedral clusters are compared in Figure. 5.10.

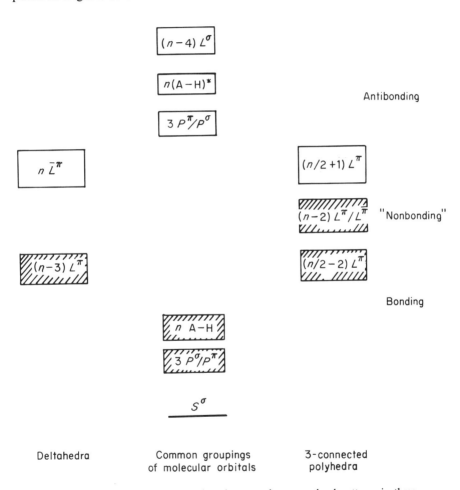

Figure 5.10 A schematic comparison between the energy-level patterns in three-connected clusters and deltahedra. The hatched boxes represent filled molecular orbitals.

5.7 FOUR-CONNECTED CLUSTERS

Delocalization is inevitable for four-connected main group clusters with terminal ligands, because there are only three valence hybrids per vertex available to participate in skeletal bonding. Furthermore, the connectivity is generally high enough for the L^π and $\bar{L}^\pi$ to be separated into distinct bonding and antibonding sets to give $n + 1$ skeletal electron pairs, as for deltahedral clusters. Hence, despite some open faces, such clusters may be regarded as *closo*.[22] The same is generally true when a cluster has vertices with a range of connectivities. Some examples of four-connected cluster skeletons are illustrated in Figure 5.11, the TSH classification of their bonding and nonbonding skeletal orbitals is given in Table 5.4.

The square antiprism (Figure 5.11b) is particularly interesting because it might also be described as an *arachno* bicapped square antiprism. In this structure there is no single atom on the principal rotation axis, and therefore the TSH pairing principle does not require there to be any strictly nonbonding

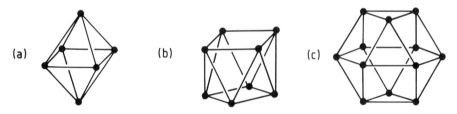

Figure 5.11 Examples of four-connected cluster skeletons: (a) octahedron, (b) square antiprism, (c) cuboctahedron.

TABLE 5.4 SUMMARY OF BONDING AND NONBONDING MOLECULAR ORBITALS FOR THE FOUR-CONNECTED POLYHEDRAL $B_nH_n{}^{2-}$ MOLECULES

		Radial and tangential skeletal molecular orbitals	
n	Geometry	Bonding	Nonbonding
6	Octahedral	$S^\sigma(A_{1g})P^\sigma(T_{1u})D^\pi_{1s,1c,2s}(T_{2g})$	
8	Square antiprism	$S^\sigma(A_1)P^\sigma(A_2E_1)D^\pi_0(A_1)D^\pi_{1s,1c}(E_2)$	$D^\pi_{2s,2c}(E_3)$ $\bar{D}^\pi_{2s,2c}(E_1)$
10	Bicapped cube	$S^\sigma(A_{1g})P^\sigma(A_{2u}E_u)D^\pi_0(A_{1g})D^\pi_{1s,1c}(E_g)$ $D^\pi_{2s}(B_{2g})F^\pi_{1s,1c}(E_u)$	$D^\pi_{2c}(B_{1g})\bar{D}^\pi_{2c}(B_{1u})$
12	Cuboctahedron	$S^\sigma(A_{1g})P^\sigma(T_{1u})D^\pi(E_gT_{2g})$ $F^\pi_{0,1s,1c}(T_{1u})F^\pi_{2c}(A_{2u})$	
12	Twinned cuboctahedron	$S^\sigma(A_1')P^\sigma(A_2''E')D^\pi(A_1'E'E'')$ $F^\pi_{0,1s,1c}(A_2''E')F^\pi_{3c}(A_1')$	

orbitals. Instead the HOMO and LUMO are parity-related e_3 ($D_{\pm 1}^{\pi}$) and e_1 ($\bar{D}_{\pm 1}^{\pi}$) pairs of orbitals. The electron count observed depends upon whether both pairs of orbitals or only the e_3 pair are occupied, and this in turn depends upon the HOMO-LUMO gap. In fact, transition metal square antiprisms provide examples of both possible electron counts.[23]

5.8 TRANSITION METAL CLUSTERS

The bonding in transition metal clusters is complicated by the presence of the valence-shell d orbitals, and we find that there is a much wider range of possibilities than for main group clusters. The following analysis applies to single-shell clusters in which the valence ns, np, and $(n + 1)d$ orbitals are all sufficiently close in energy to participate in bonding. In clusters with π-acceptor ligands, typified by carbon monoxide, there may be substantial stabilization of metal orbitals by the ligands, and such species usually fit in with the analysis of this section.

The d orbitals provide an additional basis function of σ symmetry, two of π symmetry, and two of δ symmetry. Hence there may be substantial mixing of these functions with the s and p orbitals, just as p_z and s orbitals mix in main group clusters. Because of the larger number of orbitals, it is convenient to count the number of inaccessible orbitals rather than those which are expected to be occupied. When we do this, a fairly simple division emerges between three in-pointing hybrids (one σ and two π), three out-pointing hybrids (again one σ and two π), and the three remaining orbitals (one σ and two δ). The latter three orbitals correspond to the t_{2g} set, consisting of d_{z^2}, d_{xy}, and $d_{x^2-y^2}$, which are important in transition metal complexes of octahedral symmetry. These three orbitals and the three out-pointing orbitals generally correlate formally with accessible, occupied orbitals with varying amounts of ligand character, so long as there are sufficient π-acceptor-type ligands. For example, clusters composed of $M(CO)_3$ "conical" fragments generally conform to this pattern, but those with a lower ratio of carbonyls to metal atoms may not. This implies that the precise arrangement of the ligands is not particularly important for electron-counting purposes, so long as there are enough of them. This observation fits in well with the known fluxionality of the carbonyl ligands in many transition metal clusters. These accessible orbitals include all the metal-ligand σ bonds and the t_{2g} set, which is usually stabilized by overlap with, for example, carbonyl π^* orbitals.

The three in-pointing hybrids per vertex interact to give $n + 1$ strongly bonding orbitals and $2n - 1$ strongly antibonding orbitals, just as for a main group cluster. The latter $2n - 1$ orbitals usually account for the only inaccessible orbitals correlating formally with metal valence hybrids. They are too high in energy to be occupied by metal electrons, and are too isolated in space (because they are in-pointing) to mix with, and be stabilized by, the ligand

Clusters and Color

The great majority of cluster compounds are highly colored as a result of strong absorption bands in the visible region. If related compounds in a column of the transition series of the periodic table are compared, they become less highly colored, presumably because the HOMO-LUMO gap increases as a result of stronger metal-metal bonding.

Examples

$Fe_3(CO)_{12}$	Dark-green	$Co_4(CO)_{12}$	Black	$Fe_5C(CO)_{15}$	Black
$Ru_3(CO)_{12}$	Orange	$Rh_4(CO)_{12}$	Red	$Ru_5C(CO)_{15}$	Red
$Os_3(CO)_{12}$	Pale-yellow	$Ir_4(CO)_{12}$	Yellow	$Os_5C(CO)_{15}$	Orange

The color intensifies also as the nuclearity increases, because the number of bonding molecular orbitals and antibonding molecular orbitals increases and the HOMO-LUMO gap diminishes as the "band" structures develop.

Examples

$Os_5C(CO)_{15}$	Orange	$Rh_6C(CO)_{15}^{2-}$	Yellow
$Os_7C(CO)_{19}H_2$	Brown	$Rh_8C(CO)_{19}$	Black
$Os_8C(CO)_{21}$	Purple		

The color changes are less dramatic in the following series of columnar platinum clusters:

$[Pt_6(CO)_{12}]^{2-}$	Orange-red	$[Pt_{15}(CO)_{30}]^{2-}$	Yellow-green
$[Pt_9(CO)_{11}]^{2-}$	Violet-red	$[Pt_{18}(CO)_{36}]^{2-}$	Olive-green
$[Pt_{12}(CO)_{14}]^{2-}$	Blue-green		

The color can also change with interstitial atom: $[Co_6C(CO)_{15}]^{2-}$ (black) and $[Co_6N(CO)_{15}]^-$ (orange-red).

---

orbitals. Note that the δ orbitals always correlate with accessible, occupied levels for these species, and need not be considered further. The skeletal electron count for such clusters (including all the metal-ligand bonds) is therefore $12n + 2(n + 1) = 14n + 2$, in agreement with the pattern noted in Chapter 2. This is a highly significant result, since it means that it is not necessary to calculate the full spectrum of molecular orbitals for such clusters. The identification of the inaccessible orbitals, which can be achieved from a calculation on an isostructural main group cluster, suffices to define these orbitals for a particular geometry.

In Figure 5.12 the results of a simple molecular orbital calculation for $[Ru_6(CO)_{18}]^{2-}$ are illustrated; they confirm the important features of the bonding model developed above. It should be noted that the skeletal molecular orbitals do not follow the same ordering as that described previously for

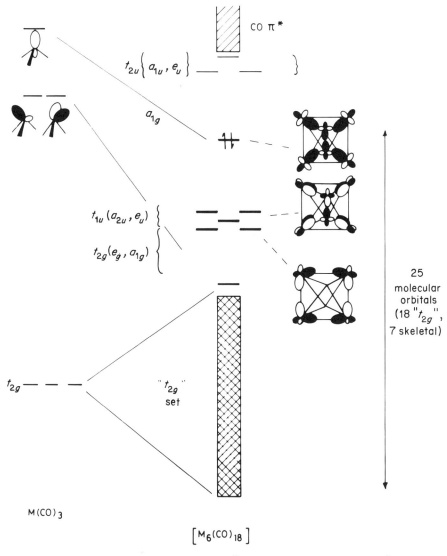

Figure 5.12 The results of an extended Hückel calculation for $[Ru_6(CO)_{18}]^{2-}$.

borane clusters. This results from the different relative orderings of the metal valence orbitals $(n + 1)d < ns \ll np$ and the mixings which occur within the skeletal orbitals.

The photoelectron spectrum of the tetrahedral cluster $Mo_4S_4(\eta\text{-}C_5H_5)_4$ (which is isostructural to P_4 and can also be described using six edge bonds) is illustrated in Figure 5.13. The relative ordering of the highest occupied

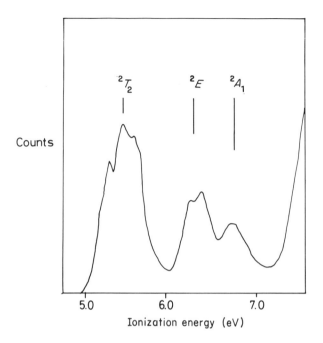

Figure 5.13 The gas-phase uv photoelectron spectrum of the tetrahedral cluster $Mo_4S_4(\eta\text{-}C_5H_4Pr^i)_4$. Reproduced with permission from Green, J. C., University of Oxford.

skeletal orbitals is $t_2 > e > a_1$; compare P_4, where $e > t_2 > a_1$ as shown in Figure 5.7.

Lauher, and Ciani and Sironi, have reported calculations on a wide range of bare metal clusters which support the above generalizations.[24] Consequently, the electronic closed-shell requirements deduced for *closo*, *nido*, and *arachno* deltahedra and three-connected and four-connected clusters can be carried over to the isostructural transition metal cluster carbonyls so long as there are no additional inaccessible orbitals (see Table 5.5). $Fe_5C(CO)_{15}$, for example, has a square-based pyramid of iron atoms and a *nido* electron count. Many other examples of *closo*, *nido*, and *arachno* metal clusters which conform to this pattern were given in Chapter 2. There are also a number of transition metal clusters with atoms on a threefold rotation axis which have $7n$ or $7n + 2$ valence electron pairs rather than $7n + 1$. For example, the tetrahedra $Re_4H_4(CO)_{12}$ and $Ru_4H_4(CO)_{12}$ have $7n$ and $7n + 2$ valence electron pairs, respectively, while $Os_7(CO)_{21}$ (capped octahedron) has $7n$ pairs.[19]

The above analysis emphasizes the similarity between these simple transition metal clusters and boranes and carboranes. For example, both the

Infrared Data for Carbonyl, Carbido, and Hydrido Clusters

Carbonyls

Terminal, edge-bridging, and face-bridging carbonyls can generally be distinguished from one another on the basis of their infrared spectral data. The following frequency ranges are helpful for neutral carbonyl clusters:

	$v(CO)$, cm^{-1}
Terminal CO	2150–1900
μ^2-CO	1850–1750
μ^3-CO	1800–1600

In cluster anions these frequencies move down, and bands 210 to 150 cm^{-1} below the terminal $v(CO)$ bands are indicative of edge-bridging and 210 to 250 cm^{-1} below indicative of face-bridging carbonyls.

π-bonded CO's, whereby the terminal CO on one metal forms a π bond to an adjacent metal atom (see below), have $v(CO)$ at 1650 to 1330 cm^{-1}.

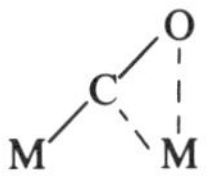

Hydrides

The infrared frequencies of metal-hydrogen stretching modes are very sensitive not only to the mode of coordination but also to the angle subtended at hydrogen by the metal atoms. Deuterium substitution results in a shift by a factor of about $1/\sqrt{2}$, which is useful for confirming the assignment.

	$v(M-H)$, cm^{-1}	$v(M-D)$, cm^{-1}
Terminal M—H	1950–1750	1325–1270

Bridging M—H—M

Compound	$v_{asym}(M_2H)$, cm^{-1}	$v_{sym}(M_2H)$, cm^{-1}	$\theta(M-H-M)$, deg.
$H_2Os_3(CO)_{10}$	1228	1177	92.6
$H_2Os_3(CO)_9(C=CH_2)$	1392	1286	101.5
$HW_2(CO)_{10}^-$	1520	1022	123.3
$H_3Mn_3(CO)_{12}$	1660 av.	875 av.	131 av.
$HCr_2(CO)_{10}^-$	1720	818	158.9

The broad range of frequencies observed for edge-bridging hydrides occurs because the v_{sym} and v_{asym} are very sensitive to the angle subtended at the hydrogen.

For interstitial hydrogens in octahedral cavities, the following frequencies have been observed:

$[Ru_6H(CO)_{18}]^-$ 825 cm^{-1}; $[Ru_6D(CO)_{18}]^-$ 600 cm^{-1}

Carbido clusters

For interstitial carbido clusters, bands assigned to $v(M-C)$ in the range 650 to 800 cm^{-1} are observed. The following represent typical data:

Compound	$v(M-C)$, cm^{-1}
$[Fe_5C(CO)_{15}]$	770, 790
$[Co_6C(CO)_{15}]^{2-}$	719, 772
$[Rh_6C(CO)_{15}]^{2-}$	653, 689
$[Os_{10}C(CO)_{24}]^{2-}$	753

For the trigonal prismatic clusters, the two bands have been assigned to A_2'' and E' modes. Could you confirm this assignment using the D_{3h} point group character table?

octahedral clusters $B_6H_6^{2-}$ and $Co_6(CO)_{14}^{4-}$ have $2n - 1 = 11$ inaccessible orbitals (Figure 5.14), and the electron counts are

$B_6H_6^{2-}$ $2 \times [(4 \times 6) - 11] = 26$
$Co_6(CO)_{14}^{4-}$ $2 \times [(9 \times 6) - 11] = 86$.

Other examples of octahedral clusters with $14n + 2 = 86$ electrons are $Co_6(CO)_{16}$ and $Ru_6C(CO)_{17}$.

TABLE 5.5 SUMMARY OF ELECTRONIC CHARACTERISTICS OF TRANSITION METAL CLUSTERS WHERE RADIAL AND TANGENTIAL BONDING EFFECTS ARE STRUCTURALLY IMPORTANT

Polyhedral description	Characteristic electron count	Example
closo deltahedron	$14n + 2$	$Os_5(CO)_{15}^{2-}$ (trigonal bipyramid), $Rh_6(CO)_{16}$ (octahedron), $[Rh_{12}Sb(CO)_{27}]^{3-}$ (icosahedron)
nido deltahedron	$14n + 4$	$Ru_5C(CO)_{15}$ (square pyramid)
arachno deltahedron	$14n + 6$	$Fe_4H(CO)_{13}^-$ (butterfly)
Four-connected polyhedron	$14n + 2$	$[Co_8C(CO)_{18}]^{2-}$ (square antiprism)
Three-connected polyhedron	$15n$	$[Rh_6C(CO)_{15}]^{2-}$ (trigonal prism)
Ring compounds	$16n$	$Os_3(CO)_{12}$ (triangle)

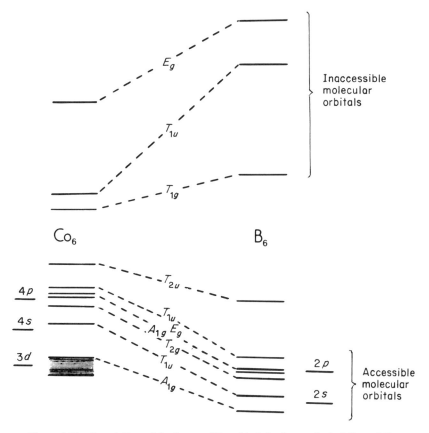

Figure 5.14 Correlation of the inaccessible orbitals in the octahedral B_6 and Co_6 clusters.

The isolobal principle (see Chapter 2) also finds application in this area. For example, if a metal fragment isolobal to a BH vertex (such as $Fe(CO)_3$) is substituted in a borane or carborane, then we expect the resulting cluster to be isostructural to the parent main group cluster, but with 10 more valence electrons. These 10 extra valence electrons represent the difference between the six accessible orbitals, which correlate with the three out-pointing and the three "t_{2g}" orbitals for the transition metal cluster, and the BH σ bond.

5.9 SYMMETRY-FORCED DEVIATIONS IN ELECTRON COUNT

The distinction between accessible and inaccessible orbitals for main group clusters of boron and carbon is generally a good one, and few deviations from the rules described in Chapter 2 are observed. For transition metal carbonyls

Location of Hydrido Ligands

The direct location of hydrido ligands from Fourier electron density maps derived from X-ray data is problematical because of the presence of heavy atoms in the structure. Ideally, neutron diffraction data should be used to resolve this structural problem, but large single crystals are not always available and the procedure is expensive! The locations of edge-bridging, face-bridging, and interstitial hydrides in clusters have been found with great accuracy using neutron data and, for example, have shown that the hydrido ligands in $[Ni_{12}(CO)_{21}H_2]^{2-}$ are not in symmetrical octahedral interstitial sites, but lie much closer to the central hexagonal layer of nickel atoms (0.73 Å) than the outer triangular plane (1.69 Å). In contrast, in $[Co_6H(CO)_{15}]^-$ the hydride is located symmetrically at the center of the octahedron.

Sometimes X-ray and neutron diffraction data are combined so that information is obtained about the light atoms without having to have a complete set of neutron data, e.g., if the crystal does not diffract well in the neutron beam.

If crystals which are suitable for neutron data cannot be obtained, then circumstantial evidence from the X-ray data has to be used to propose the location of the hydrogen ligands. These depend on the stereochemical positions of the ligands about the metal, expansion of metal-metal bond lengths, and inspection of the van der Waals packing arrangement in the molecule. Potential energy calculations have also been used to confirm the positions of the hydrides.

Very beautiful illustrations of the molecular packing diagrams for metal clusters are to be found in McPartlin, M., *Angew. Chem., Int. Ed.*, **25**, 853 (1986).

---

there are also surprisingly few deviations. However, we have already alluded on several occasions to the fact that there are systematic symmetry-forced deviations from the usual number of skeletal electron pairs, i.e., $n + 1$. Using the pairing principle and group theory, we can analyze the following three classes of deltahedra.[19]

Polar deltahedra have a single atom on the principal threefold rotation axis and have $3p + 1$ vertices. As discussed above, such species have a nonbonding pair of e-type orbitals and either n or $n + 2$ occupied skeletal orbitals. A range of examples of polar deltahedra with increasing values of p is illustrated in Figure 5.15. In polar deltahedra with only a C_2 principal rotation axis, there are no degenerate irreducible representations, and hence the idealized nonbonding e pair of orbitals must be split. The degree of splitting will determine the preferred electron count of the cluster; e.g., $B_{11}H_{11}^{2-}$ has the usual $n + 1 = 12$ skeletal electron pairs.

Bipolar deltahedra,[19] on the other hand, are structures with two atoms on the principal axis and include bipyramids such as $B_5H_5^{2-}$. Some other examples are given in Table 5.6. A few bipolar transition metal deltahedra are not particularly spherical and do not follow the $n + 1$ rule because there is only a small energy difference between two sets of parity-related e-type

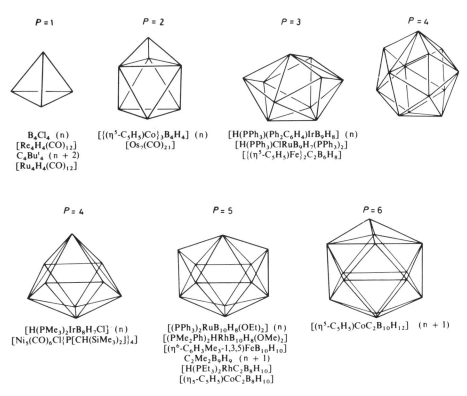

Figure 5.15 A range of polar clusters of increasing nuclearity: (a) C_{3v} polar clusters, and (b) C_{2v} polar clusters at the bottom.

frontier orbitals. In such cases either $n-1$, $n+1$, or $n+3$ skeletal electron pairs may be accommodated, corresponding to successive filling of these two pairs. $n+3$ pairs are found for the trigonal bipyramids $Ni_5(CO)_{12}^{2-}$ and $Rh_5(CO)_{14}^{2-}$ and the pentagonal bipyramid $(\eta\text{-}C_5H_5)Ni(\eta\text{-}C_5H_5)Ni(\eta\text{-}C_5H_5)$ (references 25, 26, and 27, respectively).

Similar deviations can also occur for *nonpolar* deltahedra,[19] which have no atoms on the principal rotation axis (Tables 5.7 and 5.8), and we have already mentioned the tricapped trigonal prism (Figure 5.6). This structure is unusual because the HOMO is actually odd in character ($\bar{F}_0^\pi$) while the LUMO is its even partner. This results in a fairly small HOMO-LUMO gap, a relatively high-lying HOMO and a relatively-low-lying LUMO (F_0^π). Hence the number of skeletal electron pairs is usually $n+1=10$ ($B_9H_9^{2-}$), but can be $n=9$ (B_9Cl_9) or $n+2=11$ (Bi_9^{5+}). See also Exercise 5.3.

TABLE 5.6 BIPYRAMIDAL CLUSTERS

Cluster	Skeletal electron pairs	Reference
Trigonal bipyramidal		
$C_2B_3H_5$	6 ($n+1$)	a
$[Os_5(CO)_{16}]$	6	b
$[Sn_5]^{2-}$	6	c
$[Ni_5(CO)_{12}]^{2-}$	8 ($n+3$)	d
$[Rh_5(CO)_{14}I]^{2-}$	8	e
Pentagonal bipyramidal		
$[B_7H_7]^{2-}$	8 ($n+1$)	f
$[(CO)_3FeC_2B_4H_6]$	8	g
$[(\eta^5\text{-}C_5H_5)Co(MeC_2B_3H_4)Co(\eta^5\text{-}C_5H_5)]$	8	h
$[(\eta^5\text{-}C_5H_5)Ni(\mu\text{-}C_5H_5)Ni(\eta^5\text{-}C_5H_5)]$	10 ($n+3$)	i

[a] Williams, R. E., *Adv. Inorg. Chem. Radiochem.*, **18**, 67 (1976).
[b] Eady, C. R., Johnson, B. F. G., Lewis, J., Reichert, B. E., and Sheldrick, G. M., *J. Chem. Soc., Chem. Comm.*, 271 (1976).
[c] Edwards, P. A., and Corbett, J. D., *Inorg. Chem.*, **16**, 903 (1977).
[d] Longoni, G., Chini, P., Lower, L. D., and Dahl, L. F., *J. Amer. Chem. Soc.*, **97**, 5034 (1975).
[e] Martinengo, S., Ciani, G., and Sironi, A., *J. Chem. Soc., Chem. Comm.*, 1059 (1979).
[f] Lipscomb, W. N., *Boron Hydrides*, Benjamin, New York, 1963.
[g] Grimes, R. M., *J. Amer. Chem. Soc.*, **93**, 261 (1971).
[h] Beer, D. C., Miller, V. R., Sneddon, L. G., Grimes, R. N., Mathew, M., and Palenik, G. J., *J. Amer. Chem. Soc.*, **95**, 3046 (1973).
[i] Salzer, A., and Werner, H., *Angew. Chem., Int. Ed. Engl.*, **17**, 869 (1978).

TABLE 5.7 TRICAPPED TRIGONAL PRISMATIC CLUSTERS

Cluster	Skeletal electron pairs	Reference
B_9Cl_9	9 (n)	a
$[B_9H_9]^{2-}$	10 ($n+1$)	b
$[Ge_9]^{2-}$	10	c
$[TlSn_8]^{3-}$	10	d
$[(\eta^5\text{-}C_5H_5)CoC_2B_6H_8]$	10	e
$[(PMe_3)_2PtC_2B_6H_8]$	10	f
$[Bi_9]^{5+}$	11 ($n+2$)	g

[a] Hursthouse, M. B., Kane, J., and Massey, A. G., *Nature (London)*, **228**, 659 (1970).
[b] Guggenberger, L. J., *Inorg. Chem.*, **7**, 2260 (1968).
[c] Belin, C. H. E., Corbett, J. D., and Cisar, A., *J. Amer. Chem. Soc.*, **99**, 7163 (1977).
[d] Burns, R. C., and Corbett, J. D., *J. Amer. Chem. Soc.*, **104**, 2804 (1982).
[e] Dustin, D. F., Evans, W. J., Jones, C. J., Wiersma, R. J., Gong, H., Chan, S., and Hawthorne, M. F., *J. Amer. Chem. Soc.*, **95**, 4565 (1973).
[f] Welch, A. J., *J. Chem. Soc., Dalton Trans.*, **225** (1976).
[g] Friedman, R. M., and Corbett, J. D., *Inorg. Chem.*, **12**, 1134 (1973).

TABLE 5.8 (TRIANGULATED) DODECAHEDRAL CLUSTERS

Cluster	Skeletal electron pairs	Reference
B_8Cl_8	8 (n)	a
$[\{(\eta^5\text{-}C_5H_5)Co\}_4B_4H_4]$	8	b
$[B_8H_8]^{2-}$	9 ($n+1$)	c
$[(\eta^5\text{-}C_5H_5)CoSnC_2B_4Me_2H_4]$	9	d
$[\{(\eta^5\text{-}C_5H_5)Ni\}_4B_4H_4]$	10 ($n+2$)	e

[a] Jacobson, R. A., and Lipscomb, W. N., *J. Chem. Phys.*, **31**, 605 (1959).
[b] Pipal, J. R., and Grimes, R. N., *Inorg. Chem.*, **18**, 257 (1979).
[c] Guggenberger, L. J., and Muetterties, E. L., *Inorg. Chem.*, **8**, 2771 (1969); Klanberg, F., Eaton, D. R., Guggenberger, L. J. and Muetterties, E. L., *Inorg. Chem.*, **6**, 1271 (1967).
[d] Wong, K. S., and Grimes, R. N., *Inorg. Chem.*, **16**, 2053 (1977).
[e] Bowser, J. R., Bonny, A., Pipal, J. R., and Grimes, R. N., *J. Amer. Chem. Soc.*, **101**, 6229 (1979).

5.10 PARTIAL INVOLVEMENT OF THE TANGENTIAL ORBITALS

The analysis presented above concentrates on situations where the vertex atoms have high local symmetry and both p_x and p_y contribute to the skeletal π cluster orbitals. In this section we consider the consequences of replacing these symmetric fragments with angular fragments which have only C_{2v} local symmetry. In such clusters the tangential x and y directions are necessarily different.

If CH vertices in a cluster are replaced by CH_2 units, then only one component of the p^π set at each vertex is available for skeletal bonding. The group theoretical aspects of this problem can be highlighted by comparing the triangular $C_3H_3^+$ cyclopropyl cation with cyclopropane, $[CH_2]_3$, which has the perpendicular ($\perp$) conformation shown in Figure 5.16 but could also exist in a hypothetical parallel ($\parallel$) conformation. In Chapter 3 it was shown that

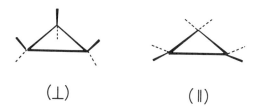

Figure 5.16 The two possible conformations of cyclopropane, $(CH_2)_3$: perpendicular ($\perp$) and parallel ($\parallel$).

Mass Spectrometry for Characterization of Molecular Clusters

Neutral carbonyl and cyclopentadienyl cluster compounds are frequently sufficiently volatile for conventional mass spectrometry data to be available. Information can be obtained not only from the parent ion but also from the fragmentation pattern, which enables the carbonyls to be counted off. This technique is useful also for determining the number of hydrido ligands associated with the cluster.

Cationic and anionic clusters are generally insufficiently volatile for these measurements, but are amenable to analysis using FABMS (fast atom bombardment mass spectrometry) techniques. Well-resolved spectra have been obtained for clusters with molecular weights up to 4000. The positive ion and cluster fragments are observed with well-resolved fine structures. An analysis of the isotopic ion distributions of the peaks clearly establishes the molecular formulas of the clusters. The cluster samples are generally dissolved in *meta*-nitrobenzyl alcohol to form a matrix, and ions are generated from impact on the matrix of a xenon atom beam derived from an Xe^+ ion beam accelerated to 8 kV.

An FABMS for $[Au_6Pt(PPh_3)_7](BPh_4)_2$ is illustrated in Figure (A). Peak A in the above spectrum has been expanded and is shown in Figure (B). The observed (solid line) and simulated (dashed line) spectra are in excellent agreement for $[Au_6PtP_7BC_{150}H_{125}]^+$.

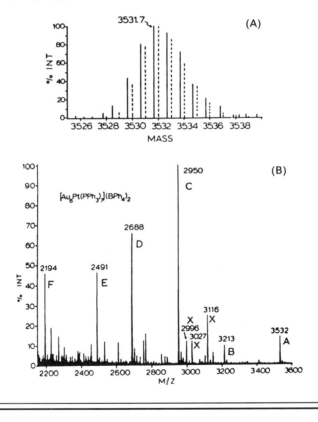

the p^θ and p^ϕ orbitals of a triangular cluster span different irreducible representations and do not mix at all. Therefore the molecular orbitals for $C_3H_3^+$ separate into distinct sets consisting of either p^θ or p^π linear combinations. In addition to the S^σ orbital, there are skeletal bonding π orbitals of symmetries A_2'' (P_0^π) and E' ($P_{\pm 1}^\pi$). The TSH theory pairing principle defines the relationship between the even and odd π orbitals as follows:

$$a_2'' = P_0^\pi(\theta) \xrightarrow{\hat{\mathscr{P}}} a_2' = \bar{P}_0^\pi(\phi),$$

$$e' = P_{\pm 1}^\pi(\phi) \xrightarrow{\hat{\mathscr{P}}} e'' = \bar{P}_{\pm 1}^\pi(\theta).$$

In the planar ($\parallel$) form of cyclopropane, only the p^θ orbitals are available for skeletal bonding and only the bonding a_2'' (P_0^π) and antibonding e'' ($\bar{P}_{\pm 1}^\pi$) orbitals are generated. In the perpendicular form only the p^ϕ are available and lead to bonding e' ($P_{\pm 1}^\pi$) and antibonding a_2' ($\bar{P}_0^\pi$) orbitals. Therefore the original $L^\pi/\bar{L}^\pi$ sets in $C_3H_3^+$ are partitioned in different ways in the two conformers. The perpendicular conformer has the most skeletal bonding orbitals and is clearly the favored geometry for $(CH_2)_3$ with the S^σ (A_1') and $P_{\pm 1}^\pi$ (E') orbitals occupied.

These observations have direct analogues in transition metal chemistry.[28-29] The $Os(CO)_4$ fragment is isolobal with CH_2 (see Figure 5.17), and the $M_3(CO)_{12}$ clusters (M = Os, Ru) are characterized by a total of 48 valence electrons. The angular PtL_2 fragment has an inaccessible p orbital perpendicular to the plane and is therefore also isolobal to CH_2 (Figure 5.17). Since this orbital is high in energy, only the in-plane d_{xz}/p_x hybrids participate in skeletal bonding in triangular platinum clusters such as $Pt_3(\mu_2\text{-H})_3(PPr_2^iPh)_3H_3$, which has a planar geometry. The presence of only 42 valence electrons in this cluster is explained by the inaccessibility of the p_y orbitals and the fact that the in-plane π-type valence orbitals, which will also be used for skeletal bonding, are stabilized by metal-ligand bonding in this cluster. Strong π-acceptor ligands can stabilize the in-phase a_2'' combination of the p_y orbitals to give a

Isolobal fragments

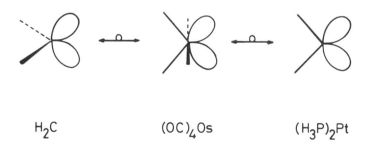

Figure 5.17 The $Os(CO)_4$ and PtL_2 fragments are isolobal to CH_2.

44-electron cluster such as $[Pt_3(CO)_6]^{2-}$. This molecule provides a transition metal analogue of the hypothetical cyclopropane geometry in which all the hydrogen atoms lie in the plane of the carbon triangle. The lowest-lying orbital generated from the three out-of-plane tangential hybrids is P_0^π (A_2'') in both cases.

In contrast, if the p^θ (or d_{yz}/p_y) orbitals are required for ligand bonding, then only the in-plane π orbitals can be considered for skeletal bonding in $Os_3(CO)_{12}$ and the perpendicular conformation of cyclopropane. Since the TSH pairing operation interconverts the in-plane and out-of-plane tangential hybrids, it follows that the bonding characteristics of the two parity-related orbitals in different sets are inverted. Hence the doubly degenerate e-type orbital is bonding for the in-plane hybrids but antibonding for the out-of-plane hybrids. Similarly, the singly degenerate a_2-type orbital is bonding for the out-of-plane hybrids but is antibonding for the in-plane hybrids. There are therefore two accessible skeletal orbitals formed from tangential hybrids which lie in the plane of the triangle in $Os_3(CO)_{12}$, as well as S^σ.

Deviations from the usual electron counting rules occur only if one of the tangential hybrids at each vertex that normally contribute to skeletal bonding is inaccessible for some reason. Such cases may arise when there are extra ligands (as for $Os(CO)_4$ vertices) or when one of the sets of tangential π orbitals in a transition metal cluster is inaccessible by virtue of a large $nd - (n+1)p$ energy separation. The following generalizations hold for clusters of the above type based upon ML_2 fragments:

(1) The bonding, nonbonding, and antibonding π cluster orbitals for the two alternative conformations of $[ML_2]_n$, which are related by 90° rotations of the ML_2 fragments, correlate with the complete sets of L^π and $\bar{L}^\pi$ orbitals before metal-ligand bonding is considered.

(2) The skeletal molecular orbitals of one conformer are one subset of the $L^\pi/\bar{L}^\pi$ set and are not related to one another by the THS theory parity operation. The total number of tangential orbitals in each conformer equals the number of vertices.

(3) The parity operation relates the tangential orbitals of one conformer, A, to those of the other, B, as follows:

$$\Gamma_\pi(A) \text{ (bonding)} \otimes \Gamma_{\mathscr{P}} \to \Gamma_{\bar\pi}(B) \text{ (antibonding)}$$

and $$\Gamma_{\bar\pi}(A) \text{ (antibonding)} \otimes \Gamma_{\mathscr{P}} \to \Gamma_\pi(B) \text{ (bonding)},$$

where $\Gamma_{\mathscr{P}}$ is the irreducible representation spanned by the TSH parity operator, $\mathscr{P}$. Nonbonding orbitals of A are converted into nonbonding orbitals of B by this process.

(4) Since the tangential skeletal orbitals of one conformer are not related to one another by the parity operation, there is no mirror relation for the energies of these orbitals. The most stable conformer is the one in which all the bonding orbitals are precisely filled.

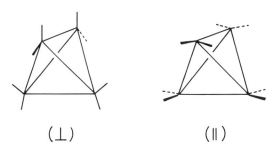

Figure 5.18 The two D_{2d} symmetry conformers of $[ML_2]_4$.

In a tetrahedral $[ML_2]_4$ cluster, the two D_{2d} conformations illustrated in Figure 5.18 are possible. If only the π orbitals parallel to the plane of the ML_2 fragment are available for skeletal bonding at each vertex then the skeletal π sets correlate as follows:

	T_d	D_{2d} ‖	D_{2d} ⊥
$\bar{P}^\pi$ (T_1)		A_2	E
$D^\pi/\bar{D}^\pi$ (E)		B_1	A_1
P^π (T_2)		E	B_2

Of the five occupied π cluster orbitals originally associated with the tetrahedron (t_2 and e), three are available in the parallel conformer and two in the perpendicular conformer. Consequently, the two conformers are associated with 54 and 56 valence electrons instead of the 60 usually found for a tetrahedron. $Pt_4(\mu_3\text{-}H)_4(PPr_2^iPh)_4$, which has 56 valence electrons, provides an example of such a compound.

TABLE 5.9 EXAMPLES OF PLATINUM CLUSTER COMPOUNDS

Skeletal geometry	Example	Electron count	
Triangle	$[Pt_3(CO)_3(PR_3)_3]$	42	
	$[Pt_3(CO)_3(PR_3)_4]$	44	(48)
Tetrahedron	$[Pt_4H_8(PR_3)_4]$	56	(60)
Butterfly	$[Pt_4(CO)_5(PR_3)_4]$	58	(62)
Trigonal bipyramid	$[Pt_5H_8(PR_3)_5]$	68	(72)
Edge-bridged tetrahedron	$[Pt_5(CO)_6(PR_3)_4]$	70	(74)
Trigonal prism	$[Pt_6(CO)_{12}]^{2-}$	86	(90)

Numbers in parentheses are valence electron counts for clusters of the manganese, iron, and cobalt groups where both p_x and p_y participate in bonding.
Source: Mingos, D. M. P., and Wardle, R. W. M., *Transition Met. Chem.*, **10**, 441 (1985).

Similar effects are observed in other platinum clusters, and in general the total valence electron count is reduced by 4 compared to that predicted from the usual PSEPT rules (Table 5.9). Similar considerations are required for the δ orbitals in clusters with π-donor ligands (Chapter 8).

EXERCISES

5.1 Using Euler's rule (faces + vertices = edges + 2) and simple geometric arguments, deduce the number of faces, vertices, and edges in a general n-vertex (a) deltahedron, (b) three-connected structure, (c) four-connected structure. What are the relationships between these three classes?

5.2 Which of the bicapped square-antiprismatic carboranes $C_2B_8H_{10}$ would you expect to be the most stable?

5.3 Sketch the $\bar{F}_0^\pi$ and F_0^π orbitals for a tricapped trigonal prism ($Y_{3,0} \propto z(5z^2 - 3r^2)$). Given that neither orbital is occupied in B_9Cl_9, $\bar{F}_0^\pi$ is occupied in $B_9H_9^{2-}$, and both are occupied in Bi_9^{5+}, how would you expect the bond lengths of the central trigonal prism to vary?

5.4 Rationalize the electron counts of the following clusters: (a) cubane, C_8H_8; (b) octahedral $Rh_6(CO)_{16}$; (c) Bi_5^{3+}; (d) $Rh_{12}Sb(CO)_{27}^{3-}$ (icosahedron of rhodium atoms with an interstitial antimony atom[30]); (e) $Ni_9C(CO)_{17}^{2-}$ (capped square antiprism[31]); (f) $B_{11}H_{11}S$ (icosahedron); and (g) $Fe(CO)_3B_4H_8$ (square pyramid with the iron atom at the apex).

5.5 Consider the bicapped square antiprism $B_{10}H_{10}^{2-}$. Identify the L^σ, L^π, and $\bar{L}^\pi$ basis set required for the skeletal valence orbitals in this molecule. Draw qualitative energy-level diagrams for these three sets of orbitals and identify the accessible and inaccessible orbitals after mixing has occurred between them.

5.6 Consider the photoelectron spectrum of $Mo_4S_4(\eta\text{-}C_5H_5)_2$ (Figure 5.13). What are the ground-state electronic configurations of the tetrahedral ion generated by ionization of an electron from (a) the t_2 HOMO, (b) the neighboring e orbital, and (c) the a_1 orbital? By considering the symmetries of vibrations of the metal tetrahedron, sketch the appropriate Jahn-Teller distortions for cases (a) and (b) and deduce the resulting point group symmetry of the metal atom cage in each case.

5.7 In Section 5.10 we discussed some clusters where each vertex does not contribute the usual one σ and two π orbitals to skeletal bonding. Explain why this is the case for (a) PtL_2 and (b) $Os(CO)_4$ fragments. In (b) you may find it helpful to form and reduce the representation spanned in C_{2v} by the four CO σ donor orbitals and compare this result with the symmetries of the valence-shell orbitals of the metal atom. Note that we assume, as in Figure 5.17, that the metal δ orbitals are essentially filled in d^8 osmium and hence are not available for metal-ligand bonding.

REFERENCES

1. Longuet-Higgins, H. C., and Roberts, M. de V., *Proc. Roy. Soc. A*, **224**, 336 (1954).
2. Longuet-Higgins, H. C., and Roberts, M. de V., *Proc. Roy. Soc. A*, **230**, 110 (1955).
3. Wade, K., *J. Chem. Soc., Chem. Comm.*, 792 (1971).
   Wade, K., *Adv. Inorg. Chem. Radiochem.*, **18**, 67 (1976).
4. Mingos, D. M. P., *Nature Phys. Sci.*, **236**, 99 (1972).
5. Rudolph, R. W., and Pretzer, W. R., *Inorg. Chem.*, **11**, 1974 (1972).
   Rudolph, R. W., *Accts. Chem. Res.*, **9**, 446 (1976).
6. Williams, R. E., *Adv. Inorg. Chem. Radiochem.*, **18**, 67 (1976).
7. Williams, R. E., *Inorg. Chem.*, **10**, 210 (1971).
8. Stone, A. J., *Polyhedron*, **3**, 1299 (1984).
9. Hall, M. B., and Fenske, R. F., *Inorg. Chem.*, **11**, 768 (1972).
10. Brown, L. D., and Lipscomb, W. N., *Inorg. Chem.*, **16**, 2989 (1977).
    Bicerano, J., Marynick, D. S., and Lipscomb, W. N., *Inorg. Chem.*, **17**, 2041 (1978).
11. Burdett, J. K., *Molecular Shapes*, Wiley, New York, 1980.
12. Wales, D. J., and Mingos, D. M. P., *Inorg. Chem.*, **28**, 2748 (1989).
13. Todd, L. J., *Prog. Boron Chem.*, **2**, 10 (1970).
14. Gimarc, B. M., and Ott, J. J., *J. Amer. Chem. Soc.*, **108**, 4298 (1986).
15. See, for example, Atkins, P. W., *Molecular Quantum Mechanics*, Oxford University Press, London, 1983.
16. Stone, A. J., *Molec. Phys.*, **41**, 1339 (1980).
17. Fowler, P. W., *Polyhedron*, **4**, 2051 (1985).
18. Ceulemans, A., *Molec. Phys.*, **54**, 161 (1985).
19. Johnston, R. L., and Mingos, D. M. P., *J. Chem. Soc., Dalton Trans.*, 647 (1987).
20. Stone, A. J., and Alderton, M. J., *Inorg. Chem.*, **21**, 2298 (1982).
21. Stone, A. J., and Wales, D. J., *Molec. Phys.*, **61**, 747 (1987).
22. Wales, D. J., *Molec. Phys.*, **67**, 303 (1989).
23. Mingos, D. M. P., and Johnston, R. L., *Struct. Bonding*, **68**, 29 (1987).
24. Lauher, J. W., *J. Amer. Chem. Soc.*, **100**, 5305 (1978).
    Ciani, G., and Sironi, A., *J. Organometallic Chem.*, **197**, 233 (1980).
25. Longoni, G., Chini, P., Lower, L. D., and Dahl, L. F., *J. Amer. Chem. Soc.*, **97**, 5034 (1975).
26. Martinengo, S., Ciani, G., and Sironi, A., *J. Chem. Soc., Chem. Comm.*, 1059 (1079).
27. Salzer, A., and Werner, H., *Angew. Chem., Int. Ed.*, **17**, 689 (1978).
28. Evans, D. G., and Mingos, D. M. P., *J. Organometallic Chem.*, **240**, 321 (1982).
29. Evans, D. G., and Mingos, D. M. P., *Organometallics*, **2**, 435 (1983).
30. Vidal, J. L., and Troup, J. M., *J. Organometallic Chem.*, **213**, 351 (1981).
31. Longoni, G., Ceriotti, A., Della Pergola, R., Manassero, M., Perego, M., Piro, G., and Sansoni, M., *Phil. Trans. Roy. Soc. Lond.*, **A308**, 47 (1982).

6

Skeletal Rearrangements in Clusters

6.1 INTRODUCTION

In this chapter we will review the theoretical study of cluster vertex rearrangements, and demonstrate the usefulness of TSH theory in this area.[1-2] TSH theory is found to provide some very powerful selection rules for distinguishing orbital symmetry–allowed from orbital symmetry–forbidden processes.

The discussion falls conveniently into two halves. The first deals with rearrangements of main group clusters, such as boranes and carboranes, while the second deals with transition metal clusters. There are rather fewer data available for the rearrangement rates of transition metal cluster skeletons when compared to borane and carborane clusters, so that much of the analysis presented for the former molecules has yet to be tested experimentally. This probably reflects, at least in part, the greater advancement of ^{11}B nmr relative to nmr spectroscopy for transition metals. Nonetheless, some important conclusions may be reached concerning the interpretation of future experimental data for transition metal cluster rearrangements. For example, if different trends are observed for isostructural main group and transition metal clusters, then it follows that some of the alternative mechanisms must be operating.

6.2 THE DIAMOND-SQUARE-DIAMOND REARRANGEMENT

In the diamond-square-diamond (DSD) mechanism for the skeletal rearrangement of boranes, due to Lipscomb,[3] an edge common to two triangular faces of the cluster skeleton breaks and a new edge is formed perpendicular to it (Figure 6.1). DSD rearrangements, or combinations of several concerted DSD processes, have been proposed to rationalize fluxional processes and isomerizations of boranes, carboranes, and metalloboranes.[4-5] King[6] has used topological considerations to distinguish between inherently rigid clusters (which contain no *degenerate* edges) and those for which one or more DSD processes are *geometrically* possible. An edge is said to be degenerate if a DSD rearrangement in which it is broken leads to a product having the same cluster skeleton as the starting material. For example, the three equatorial edges of $B_5H_5^{2-}$ are degenerate and the remainder are not. Alternative DSD processes are distinguished by the connectivities of the four vertices around the diamond in question. Hence the DSD process for the trigonal bipyramid (shown in Figure 6.2) is described as 44(33) because an edge connecting two four-coordinate equatorial vertices is broken while an edge is made between the two three-coordinate apical vertices. A general DSD process is denoted by $\alpha\beta(\gamma\delta)$, although this does not always specify the process unambiguously. For a degenerate DSD rearrangement to be geometrically possible, the connectivities of the new cluster vertices must match those of the original skeleton and we must have $\alpha + \beta = \gamma + \delta + 2$. In the *closo*-borane series, only $B_5H_5^{2-}$, $B_8H_8^{2-}$, $B_9H_9^{2-}$, and $B_{11}H_{11}^{2-}$ satisfy these requirements.[7] The highest point group symmetry that the transition state for a DSD process in a deltahedron can have is C_{4v} if the cluster has $4p + 1$ vertices, as, for example, $B_5H_5^{2-}$ and $B_9H_9^{2-}$. For $B_8H_8^{2-}$ and $B_{11}H_{11}^{2-}$, the highest-symmetry geometry achievable during the single DSD process is C_{2v}.

King's approach was partially successful in that all the structures he predicted to be rigid are found experimentally to be nonfluxional. However, some of the above molecules of the $B_nH_n^{2-}$ series with topologically possible

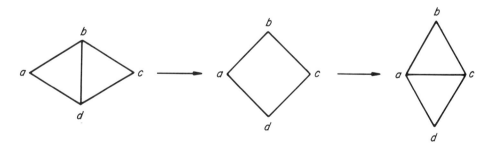

Figure 6.1 The diamond-square-diamond (DSD) process.

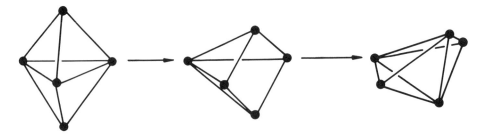

Figure 6.2 The single DSD process for $B_5H_5^{2-}$. Notice the square-based pyramidal transition state.

single DSD rearrangements are not fluxional on the nmr time scale. They are the ones which have $4p + 1$ vertex atoms.

6.3 THE CONSERVATION OF ORBITAL SYMMETRY

The most important principle used in this chapter is the "conservation of orbital symmetry" as pioneered by Woodward and Hoffmann.[8] We assume that the reader has some familiarity with this concept already, from the viewpoint of either organic or theoretical chemistry.

Consider a single-step reaction mechanism which converts reactants, R, into products, P. The commonly encountered reactions in textbooks are the Diels-Alder process and the cyclization of polyenes such as butadiene, but for clusters we are usually interested in single-molecule rearrangements which lead back to some permutation of the initial skeletal geometry. Using a molecular orbital treatment, one may classify the orbitals of the reactants and products according to the irreducible representations of the point group of symmetry elements which are retained throughout the process. A correlation diagram is then drawn between orbitals which are expected to evolve into one another during the reaction, with the proviso that orbitals having the same symmetry species cannot cross. Orbitals which are correlated must necessarily transform in the same manner in the point group of the symmetry operations which are retained along the reaction coordinate. If it happens that one of the occupied orbitals of the reactants correlates with an unoccupied orbital in the product, then the process is expected to be unfavorable if the reactants are in their ground electronic states. Such mechanisms are generally called "orbitally forbidden," although the large energy barriers do not rigorously mean that the process cannot occur. As an example, the correlation diagram for the disrotatory ring closure of butadiene is illustrated in Figure 6.3. The classifications are with respect to the symmetry plane, which is retained throughout the process. In constructing the correlation diagram it may happen that one

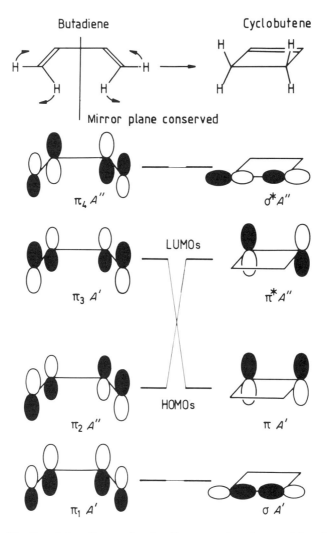

Figure 6.3 Correlation diagram for the disrotatory ring closure of butadiene, which is "forbidden" by orbital symmetry for this molecule in its electronic ground states. The orbitals are classified according to irreducible representations of the point group C_s because only a single mirror plane is retained throughout in the disrotatory process.

cannot connect two orbitals for which a correspondence is expected, because of the noncrossing rule; this is known as an "avoided crossing."

In fact, we should really consider the correlation diagram for the electronic states of the reactants and products, but this need not concern us here. The main apparent weakness of the above method is that low-symmetry processes may have no symmetry elements retained, in which case there can be no crossings. However, one may argue that the chemical groups which lower the molecular symmetry play no part in the process, and hence make use of the correlation diagram for reactants and products of idealized symmetry. This approach will be followed below, although the effects of low symmetry will also be discussed explicitly in some cases.

6.4 TSH ANALYSIS OF SKELETAL REARRANGEMENTS

The main use of TSH theory in considering skeletal rearrangements is concerned with the identification of symmetry-forbidden processes, which exhibit a crossing of occupied and virtual molecular orbitals. We know that the occupied cluster orbitals of deltahedral boranes basically consist of the even L^π set and the S^σ orbital. Hence, the identification of the symmetries of the occupied molecular orbitals is a straightforward matter for both starting material and product. Reduction in symmetry may then be performed to the point group containing the symmetry elements retained throughout the process. If the symmetries of the sets of bonding orbitals referred to this point group are not equivalent, then a crossing must occur and the rearrangement will be energetically unfavorable.

Of course, not all orbital symmetry–allowed rearrangements will be energetically favorable; clearly, if no symmetry elements are preserved throughout a given process, then there cannot be a crossover, but this is no guarantee that the energy barrier to the process will be small.

Detailed considerations allow us to deduce two theorems for identifying degenerate rearrangements which necessarily involve orbital crossings.

Theorem I

> A crossing is found to occur if the proposed transition state has a single atom lying on a principal rotation axis of order 3 or more.

For a single DSD process, this is the case if the degenerate edge lies opposite a single atom, as it does in $B_5H_5^{2-}$ and $B_9H_9^{2-}$. In fact, we noted above that the DSD process could only pass through a C_{4v} "transition state" for *closo* boranes, $B_nH_n^{2-}$, with $n = 4p + 1$ atoms, i.e., $B_5H_5^{2-}$ and $B_9H_9^{2-}$. These crossings were first noted by Gimarc and Ott.[9]

The proof of the above theorem is straightforward. The p^π orbitals of the unique atom lying on the principal axis will span an E irreducible representation if the molecular point group is axial and the rotation axis is of order 3 or more. It can easily be shown that the total number of E representations in the $L^\pi/\bar{L}^\pi$ set is then odd[10] and one of them must be nonbonding. We will illustrate this result for molecules belonging to C_{nv} point groups by considering the transformation properties of the p^θ and p^ϕ orbitals of each set of n equivalent atoms lying equidistant to the n-fold principal rotation axis. These orbitals each clearly span the regular representation of C_{nv} (where the character equals the order of the point group for the identity operation and zero otherwise), so that any E-type irreducible representation must occur an even number of times in the representation spanned by the nonpolar atoms. The TSH pairing principle then implies that one pair must be self-conjugate and nonbonding. Such structures therefore have a doubly degenerate nonbonding pair of e orbitals at the HOMO-LUMO level which corresponds to the crossover of two molecular orbitals at this geometry, as illustrated in Figure 6.4 for the single DSD processes in $B_5H_5^{2-}$ and $B_9H_9^{2-}$.

In the transition state for the single DSD process (Figure 6.1), the frontier orbitals of the open face (Figure 6.5) are self-conjugate; i.e., they are interchanged by the TSH parity operation. If the cluster has $4p + 1$ atoms ($p = 1, 2, \ldots$), these orbitals are degenerate, but otherwise they need not be. Theorem I enables us to identify mechanisms where there must be an orbital crossing due to a degeneracy at the HOMO-LUMO level.

A more detailed examination enables us to identify cases where there must be a crossing in lower point group symmetries. These are forced not so much by a degeneracy of the frontier orbitals in the prevailing point group, but rather by the pairing principle causing a change in the number of occupied orbitals with a given parity under reflection. We have seen that the TSH theory parity operation corresponds to the pseudoscalar irreducible representation of the point group, which is symmetric to all proper rotations and antisymmetric to improper rotations and reflections. Therefore L^π and $\bar{L}^\pi$ orbitals related by the parity operation have opposite parities with respect to reflection. The high-symmetry case (above) can also be discussed in terms of the parity-under-reflection argument, which proceeds as follows. The frontier orbitals in the transition state (Figure 6.5) have complementary characteristics with respect to reflection in the mirror planes which pass through opposite vertices of the open square face. If the open face is squeezed along one of these mirror planes, then the two parity-related orbitals are split apart. Squeezing across the other mirror plane, however, causes splittings of the two components in the opposite sense, and hence there is an orbital crossing in the correlation diagram (Figure 6.6) if a mirror plane is retained throughout a DSD process. If instead a C_2 axis is retained, then there is an avoided crossing, because the frontier orbitals are both antisymmetric under this operation. Note that these symmetry elements must be present for the whole cluster skeleton, not just the open face. Hence we arrive at our second theorem:

Theorem II

A degenerate DSD process in which a mirror plane is retained throughout involves an orbital crossing and is therefore orbitally "forbidden."

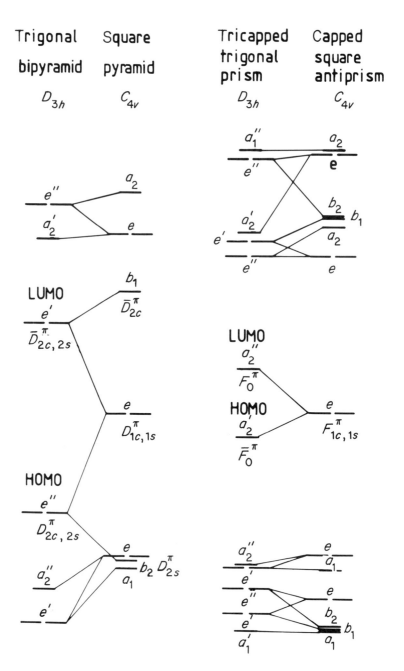

Figure 6.4 Correlation diagrams for the forbidden single DSD processes in $B_5H_5^{2-}$ (*left*) and $B_9H_9^{2-}$ (*right*). Only the L^π cluster orbitals are considered, and the units of energy are parameterized so that the energy scales and origins of the two diagrams are different.[11]

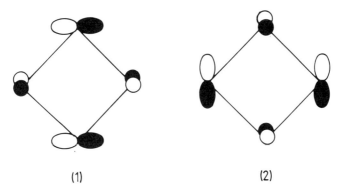

Figure 6.5 The frontier orbitals in the open face of the single DSD transition state. Both orbitals are antisymmetric with respect to a C_2 rotation about an axis through the center of the face. Under reflection in vertical and horizontal mirror planes, (1) is antisymmetric and symmetric, respectively, while (2) behaves in the opposite fashion.

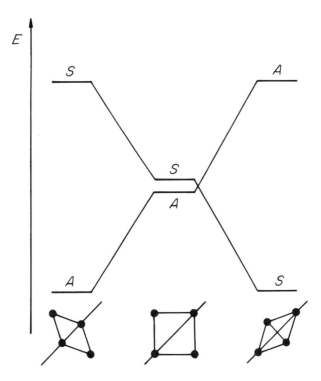

Figure 6.6 Orbital correlation diagram for a single DSD process in which a mirror plane through two of the critical atoms is retained throughout. An orbital crossing occurs due to the change in bonding character of the parity-related orbitals.

Sec. 6.4 TSH Analysis of Skeletal Rearrangements

In fact, we can generalize this result further to identify the conditions under which the orbitals belonging to the L^π frontier set can cross their $\bar{L}^\pi$ partners. The point groups which may be conserved throughout actually fall into three distinct classes[12] as mentioned in Chapter 5:

(1) L^π and $\bar{L}^\pi$ always transform according to different irreducible representations. Groups of this type include the inversion and/or a mirror plane, e.g., C_i, C_s, C_{2v}, O_h, etc.
(2) L^π and $\bar{L}^\pi$ always transform according to the same irreducible representations. This is the case for point groups such as C_n and D_n where the parity operator transforms as A_1.
(3) Some L^π and $\bar{L}^\pi$ transform in the same way while others do not. Groups of this type include C_{nv} ($n \geq 3$) and S_{4n}.

For a DSD process, mirror planes or a C_2 axis may be conserved; and in the point groups C_s and C_2, the parity operator transforms as A'' and A, respectively, Hence, if a mirror plane is conserved, parity-related L^π and $\bar{L}^\pi$ always span different IRs (and can cross), while if a C_2 operation only is conserved they must span the same IR (and cannot cross). We would also expect these results to extend to cases where the symmetry elements are only approximate, but in a less rigorous manner. Only a C_2 axis is retained in the single DSD processes for $B_8H_8^{2-}$ and $B_{11}H_{11}^{2-}$, and consequently these molecules are stereochemically nonrigid.

We may also apply these orbital symmetry selection rules to nondegenerate rearrangements in which the starting and finishing clusters are different, so long as they both have $n+1$ skeletal electron pairs. For example, in the single DSD process which interconverts 1,2-$C_2B_3H_5$ to 1,5-$C_2B_3H_5$, a mirror plane is retained throughout and there is a crossing.[9] The analysis may also be extended to multiple DSD processes, whether they be concerted or stepwise. For example, consider a degenerate, concerted double DSD process in which a mirror plane is preserved throughout and passes through both open faces in the transition state. If one edge is broken across the mirror plane and the other is broken simultaneously in the plane, then there is no orbital crossing. This follows because of the complementary nature of the frontier orbitals in the two open faces, which preserves the total number of occupied orbitals with S and A parity under the reflection by means of two avoided crossings. In contrast, if the two edges are made and broken parallel to one another (both in or both across the mirror plane), then the number of S or A parity orbitals in the occupied set changes by plus or minus 2.

We will refer to crossing of the above type as "TSH theory symmetry-forced." The identification of such processes is very important, as we demonstrate below, and the rules can also be applied to rearrangements of transition metal clusters in favorable cases.

The orbital symmetry rules are related to the number of skeletal bonding orbitals usually associated with a particular geometry. For example, we know from the previous chapters that a five-vertex square-based pyramidal cluster such as B_5H_9 should have $n + 2 = 7$ skeletal bonding orbitals. However, in the single DSD 44(33) "transition state," we encounter a $B_5H_5^{2-}$ square-based pyramid with only six skeletal electron pairs. In contrast, the single DSD process for $B_8H_8^{2-}$ probably leads to a distorted bicapped trigonal prismatic intermediate structure with C_{2v} symmetry. It can be shown from the "capping principle"[13] (see Chapters 2 and 7) that a bicapped trigonal prism can be associated with either 9 or 10 skeletal bonding molecular orbitals. Consequently, the starting and transition-state geometries have compatible closed-shell electronic requirements and are not anticipated to have very different energies. Furthermore, there is no threefold or higher principal rotation axis in the transition state, and no mirror plane is maintained throughout—hence there is no orbital crossing.

6.5 APPLICATION TO BORANES

As discussed above, the single DSD process for $B_5H_5^{2-}$ is "forbidden" by orbital symmetry. Some alternative mechanisms have been considered, but none are expected to be particularly favorable. For $B_6H_6^{2-}$ there is no topologically possible low-order DSD process, and this molecule is not fluxional on the nmr time scale. The two possible carborane isomers are both known,[14] which is also circumstantial evidence that there is a significant barrier to rearrangement. Otherwise we might expect only the most stable carborane to be isolatable.

In contrast, both $B_8H_8^{2-}$ and $B_{11}H_{11}^{2-}$ have orbital symmetry–allowed single DSD processes, and both exhibit fluxionality on the nmr time scale.[15–16] Only one carborane isomer has been prepared in each case,[17–18] although for $C_2B_9H_{11}$ this conclusion is based on nmr evidence alone. For $B_7H_7^{2-}$, $B_9H_9^{2-}$, and $B_{10}H_{10}^{2-}$ there is no evidence of fluxionality on the nmr time scale.[15] However, only one of the six possible carborane isomers $C_2B_7H_9$ has been isolated,[19] while three[20] of the $C_2B_8H_{10}$ and two[21] of the $C_2B_5H_7$ carborane isomers are known. For all three species there are orbital symmetry–allowed concerted double DSD processes, although this requires the two open faces to share an edge in the case of $B_{10}H_{10}^{2-}$ and $B_7H_7^{2-}$. Furthermore, in each case a mirror plane would be retained in the first step of a stepwise double DSD rearrangement. For the pentagonal bipyramid, for example, the 54(44) DSD process results in a capped octahedral C_{3v} geometry with a single atom on the principal axis, as Onak and his coworkers have noted.[22]

The case of $B_{12}H_{12}^{2-}$ and the related carboranes $C_2B_{10}H_{12}$ is also interesting. All three carboranes are known; the 1,2 isomer (carbon atoms

NMR and Infrared Data for Polyhedral Borane Anions ($B_nH_n^{2-}$)

The $B_nH_n^{2-}$ salts are colorless crystalline solids with substantial thermal and chemical stabilities; e.g., $[Et_4N]_2B_6H_6$ has m.p. > 300°, $[Et_3NH]_2B_{10}H_{10}$ has m.p. > 302°, and $[Et_3NH]_2B_{12}H_{12}$ has m.p. 298–300°. They are water-soluble and air-stable in the solid state. Their ^{11}B nmr and infrared characteristics are summarized below.

^{11}B nmr data and ν(B—H) from Infrared Data

Ion	^{11}B rel. to Et_2OBF_3, δ (ppm)	J(B—H), Hz	ν(B—H) cm^{-1}
$[B_6H_6]^{2-}$	13	122	2421
$[B_7H_7]^{2-}$	0.3 [5B]	120	2500
	22.7 [2B]	119	2450
$[B_8H_8]^{2-}$ *	6 [8B] (Room temp)	128	2480, 2450
Square antiprism, and bicapped-trigonal prism in solution suggested	-9.3 [2B] ⎫ 3.6 [4B] ⎬ < 4° 22.2 [2B] ⎭	128	
$[B_9H_9]^{2-}$	3.4 [3B]	133	2540, 2478
	21.5 [6B]	124	2445, 2415
$[B_{10}H_{10}]^{2-}$	0.7 [2B]	138	2532
	28.9 [8B]	125	2450, 2470
$[B_{11}H_{11}]^{2-}$ *	17.0 [11B]	130	
$[B_{12}H_{12}]^{2-}$	14.0 [12B]	115	2480
			Raman (2517, 2472)

*Stereochemically nonrigid.

adjacent) may be converted to the 1,7 by heating to 470°C, and the 1,7 isomer may be converted to the 1,12 (carbon atoms at opposite ends of a diameter) by heating to 630°C. The high temperatures are indicative of high activation energy barriers for these rearrangements. A sextuple DSD mechanism has been proposed to account for the first interconversion, and a relative rotation of layers, which is actually equivalent to a quintuple DSD, for the second.[4] Both processes involve several DSD rearrangements, and are therefore expected to be less favorable than the single DSD possible for $B_8H_8^{2-}$ and $B_{11}H_{11}^{2-}$, and the double DSD possible $B_9H_9^{2-}$. In fact, *ab initio* calculations indicate that neither geometry is a true transition state.[1]

The above theories may also be applied to the *nido* boranes and carboranes to predict the relative energy barriers to skeletal rearrangements in these species. If we assume that any rearrangement must leave the open face essentially unchanged, because of the bridging protons, then we can proceed as follows. The 7- and 8-vertex *nido* geometries have symmetry-allowed single

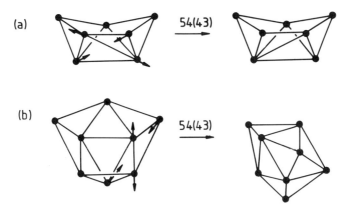

Figure 6.7 Single DSD processes for (a) seven- and (b) eight-vertex *nido* boranes.

DSD rearrangements, both described as 54(43) in King's notation[6] (Figure 6.7). The 9-vertex *nido* geometry has a symmetry-allowed double DSD rearrangement which is equivalent to concurrent 54(53) and 54(44) processes in which the two open faces share an edge (Figure 6.8). The 10- and 11-vertex *nido* geometries have no low-order topologically allowed DSD processes available, and are therefore expected to have larger barriers to rearrangement. However, the quintuple DSD which is topologically possible for the 11-vertex system does not have a TSH symmetry-forced crossing, because both starting material and transition state have a single atom on the principal axis. The prediction for the relative order of the energy barriers of these *nido* boranes and the corresponding carboranes to skeletal rearrangements is therefore:

7-, 8-vertex *nido* < 9-vertex *nido* ≪ 10-, 11-vertex *nido* geometries.

Figure 6.8 Concerted double DSD process for the nine-vertex *nido*-skeleton.

6.6 SINGLE EDGE-CLEAVAGE PROCESSES

Rodger and Johnson[23] have considered the geometric symmetry selection rules for the rearrangement of transition metal clusters with between 5 and 12 atoms. Their approach must be carefully distinguished from the method described in this chapter, as it is not concerned with the conservation of *orbital* symmetry, but with the fact that any reaction mechanism must follow a vibrational normal mode of the cluster at every point. The geometric selection rules which result are also more rigorous than the orbital symmetry rules, because the latter can only suggest when the energy barrier to a given process will be prohibitive. The above authors consider single edge-cleavage processes only, on the grounds that these are expected to be most favorable. Many of these mechanisms correspond to the diamond-square-diamond processes considered previously,[1] except that they are assumed to be stepwise if more than one edge must be broken.

Most interesting are the mechanisms which were not considered in the above sections because they involve tetrahedral caps in the intermediate structures. Such geometries were discounted for main group clusters because they are rarely observed in solid state structural analyses. For transition metal clusters, however, such species might play a more significant role. Incidentally, the fact that all the rearrangement rates of the boranes and carboranes could be rationalized in terms of DSD processes provides circumstantial evidence that the other structures are too high in energy to play any part for these species.

Since these other processes do not immediately lead back to the original geometry, we must be more careful in applying Theorems I and II. Some new criteria will also be developed in terms of the usual electron counting rules for *closo*, *nido*, and *arachno* clusters, plus the "capping"[24-26] and "condensation"[27] principles which were outlined in Chapter 2. The latter are used to determine whether the predicted electron count changes from the starting cluster to the suggested intermediate. Processes where the electron count changes are expected to be relatively unfavorable because there will be an unoccupied low-lying orbital or an occupied high-lying orbital. For example, if a DSD process converts a *closo* deltahedron into a capped deltahedron with $n - 1$ vertices and a principal rotation axis of order 3 or more passing through the cap, then the process will probably be unfavorable. This is because capped deltahedra of this type have a nonbonding $L^{\pi}/\bar{L}^{\pi}$ e pair and n skeletal bonding orbitals.[1] The electronic ground state of such a species would be a triplet and subject to a Jahn-Teller distortion.[28] In any case, the frontier orbitals are high-lying because they are nonbonding. For DSD processes which result in a capped deltahedron with lower symmetry, the "e" pair are split, but we still expect them to be approximately nonbonding and the HOMO in particular to be high-lying.[1] Pathways which involve such intermediates are therefore expected to entail appreciable energy barriers.

Note, however, that this empirical "skeletal electron count selection" (SECS) rule should not be placed upon the same footing as the orbital or geometric symmetry selection rules above. It would be quite wrong to describe mechanisms in which the predicted electron count changes as "forbidden," and we will not do so. Nonetheless we would still expect them to be relatively high activation energy processes, and predictions will be made on the basis of this assumption.

6.7 THE SDDS MECHANISM

The presence of a single square face in an otherwise deltahedral cluster opens up the possibility of a new low-energy rearrangement mechanism. Such a process may be written as a square → diamond, diamond → square, or SDDS, mechanism. This concerted process is illustrated for the square-based pyramid and the capped square antiprism in Figure 6.9a and b, respectively. If a mirror plane is retained throughout, then this mechanism involves two orbital crossings if one edge lies in the plane and the other lies across it. If the edges are parallel, as in Figure 6.9, then there are two avoided crossings and the process is orbital symmetry–allowed. There is no TSH symmetry-forced crossing in this case, since the starting and finishing molecules both have a single atom on the fourfold principal axis. Hence this is expected to be a favorable process for *nido* transition metal clusters. For main group clusters, however, the presence of bridging hydrogen atoms around the open face

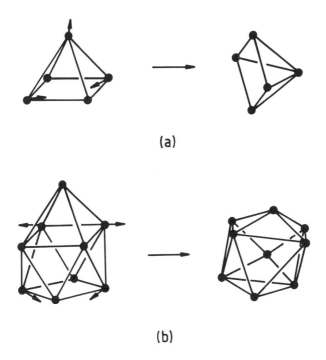

Figure 6.9 The square → diamond, diamond → square (SDDS) mechanism for a square-based pyramid (a), and a capped square antiprism (b).

means that this mechanism would involve a much greater perturbation to the structure of the cluster, and is therefore not likely to occur. It could, however, be significant in accounting for the behavior of $B_8H_8^{2-}$ in solution,[29] where a fluxional bicapped trigonal prismatic geometry may be present. The SDDS mechanism is allowed by orbital symmetry for the latter species, and may account for its nonrigidity. Furthermore,[30] for square-based pyramidal $C_5H_5^+$ there are no bridging protons around the square face, and hence the barrier to the symmetry-allowed SDDS mechanism should be lower than in, say, B_5H_9.

6.8 TRANSITION METAL CLUSTERS CONFORMING TO THE PSEPT

In transition metal clusters, the description of the frontier orbitals is more complicated than for main group clusters, and we usually find that an S^σ cluster orbital is the HOMO, with a set of even L^π cluster orbitals lying below it[31-32] (see, for example, Figure 5.12 in Chapter 5). The inwardly hybridized skeletal orbitals of π symmetry are d and p in character. Hence the predicted degeneracy for a system with a single atom on a principal axis of order 3 or more will not necessarily occur at the HOMO-LUMO level of the cluster. However, the crossover still represents a crossing between an occupied and virtual molecular orbital, and is therefore still expected to be an orbitally "forbidden" process.

Hence we reach the conclusion that the orbital symmetry selection rules developed for main group clusters are expected to be applicable to rearrangements of transition metal clusters, provided that:

(1) The ligands do not greatly reduce the energy barrier by lowering the symmetry.
(2) The metal fragments remain isolobal with BH throughout the rearrangement process.
(3) The metal cluster has a spectrum of skeletal molecular orbitals similar to those for the isostructural main group cluster. It should be noted, however, that since the splittings between L_{dp}^π and $\bar{L}_{dp}^\pi$ orbitals are much smaller for transition metal than main group clusters, the difference in activation energies between orbitally "allowed" and "forbidden" processes is correspondingly less.

The additional criterion that multiple DSD processes will generally be less favorable than a single DSD process is also expected to hold. In metal carbonyl clusters where there are bridging ligands which lower the symmetry of the cluster, the orbital symmetry selection rules are weakened and lower barriers for "forbidden" rearrangement processes are anticipated.

Low-nuclearity (4 to 8) metal atom clusters based on conical $M(CO)_3$ fragments, such as $[Os(CO)_3]_m$, are generally observed to be stereochemically rigid on the nmr time scale. Presumably in such clusters the spectra of skeletal molecular orbitals based on the in-pointing dp hybrids of the conical $M(CO)_3$ fragments resemble those of boranes sufficiently for the arguments developed above to be applicable. For example ^{13}C nmr evidence indicates that $Os_6(CO)_{18}$ (a bicapped tetrahedron), $Os_7(CO)_{21}$ (a capped octahedron), and $Os_8(CO)_{21}$ (a bicapped octahedron) are stereochemically rigid on the nmr time scale.[33] There is some evidence, however, that the barriers to rearrangement are lower than in deltahedral boranes which have only orbital symmetry–forbidden DSD processes available. For example, $Os_6(CO)_{16}(P(OMe)_3)_2$ can be isolated in the solid state in two isomeric forms based upon a bicapped tetrahedron with the phosphites occupying either one or both of the capping atoms. In solution the clusters exhibit nmr spectra consistent with these solid state data with no fluxionality observed. However, on standing, the more stable isomer is obtained.[33]

When asymmetry is introduced into the cluster by changing some of the metal atoms or replacing carbonyl ligands with, for example, hydrides, the orbital symmetry rules begin to lose their validity. For example, the clusters $FeOs_n Ru_{3-n}(\mu_2\text{-}H)_2(CO)_{13}$ ($n = 0, 1,$ or 2) have distorted tetrahedral geometries. Nmr data indicate that the isomers interconvert via a subtle "breathing" motion of the framework with concomitant movements of the carbonyls and hydrides.[34]

For higher-nuclearity carbonyl clusters with combinations of bridging and terminal carbonyls, the situation is again less clear-cut because the bridging carbonyls interact strongly with the L_{dp}^π skeletal molecular orbitals. The great majority of pseudospherical deltahedral rhodium clusters, such as $Rh_6(CO)_{16}$ (octahedral), $[Rh_7(CO)_{16}]^{3-}$ (capped octahedral), $[Rh_{13}H_{5-n}(CO)_{24}]^{n-}$ ($n = 1$ to 4, cuboctahedral), and $[Rh_{14}H_{4-n}(CO)_{24}]^{n-}$ (capped cuboctahedral), are stereochemically rigid on the nmr time scale.[35]

In the distorted icosahedral cluster $[Rh_{12}Sb(CO)_{27}]^{3-}$, which has an interstitial antimony atom, the ^{13}C nmr data suggest that all the rhodium atoms are equivalent at ambient temperatures. A cuboctahedral intermediate has been proposed to account for this observation, but the results described in this chapter suggest that a fluxional process involving the carbonyls alone is more likely to be responsible.[36] The bicapped square-antiprismatic clusters $[Rh_{10}E(CO)_{22}]^{n-}$ ($n = 3$, $E = P$ or As; $n = 2$, $E = S$), which also have interstitial atoms, provide the only well-established examples of metal cluster compounds which are isostructural to *closo*-borane skeletal geometries and are stereochemically nonrigid. The nmr data in the range $90°$ to $100°C$ indicate that a skeletal rearrangement is occurring in addition to the lower-energy carbonyl permutations. Since the corresponding $B_{10}H_{10}^{2-}$ molecule is nonfluxional on the nmr time scale in this temperature range it is probable that the double DSD process has a lower activation energy for the transition metal clusters.[36,37]

In summary, although low-symmetry environments created by ligands may lead to a decrease in the activation-energy barriers to rearrangements which are "forbidden" by orbital symmetry under high symmetry, the remaining barriers are probably still significant. Rearrangements are generally observed only at temperatures significantly above ambient.

The related *nido* clusters $[Rh_9E(CO)_{21}]^{2-}$ (E = P or As), which have capped square-antiprismatic geometries, become fluxional at room temperature,[38] indicating a somewhat smaller activation energy. The SDDS process, described above, may account for this behavior, and is illustrated in Figure 6.9b. There is no TSH symmetry-forced crossing.

6.9 RADIALLY BONDED METAL CLUSTERS

In the radially bonded clusters, discussed in Chapter 4, the inwardly hybridized tangential p^π orbitals are too high in energy to contribute any accessible skeletal orbitals.[39] In such clusters the TSH forced crossing rule is inapplicable, because it assumes that all the even π cluster orbitals are occupied. In these radially bonded clusters the number of skeletal bonding orbitals remains constant so long as all the skeletal atoms remain roughly disposed over the surface of a sphere.[40]

All pseudospherical gold clusters $[Au(AuPR_3)_n]^{x+}$ ($n = 7$ to 12) have a set of four radial skeletal bonding orbitals consisting of S^σ, P_x^σ, P_y^σ, and P_z^σ. Any rearrangement which proceeds along a pathway in which a roughly spherical topology is maintained is symmetry-allowed, because the molecular orbitals constitute a complete spherical set and therefore correlate smoothly. Furthermore, since the Au—Au radical bonding is stronger than the tangential bonding, the energy differences between alternative cluster geometries are small, and these compounds are expected to be stereochemically nonrigid. There is a significant quantity of ^{31}P nmr data in support of this conclusion.[40]

In toroidal gold clusters, there are three (S^σ, P_x^σ, and P_y^σ) skeletal bonding orbitals, and skeletal rearrangements involving alternative toroidal geometries with the same three bonding orbitals will not involve orbital crossings. Hence rearrangements in which the geometry remains oblate will be orbital symmetry–allowed. However, rearrangements which would interconvert the oblate starting geometry into a prolate spheroidal cluster—e.g., a pentagonal bipyramid such as $[Au_7(PPh_3)_7]^+$ into a capped octahedral cluster—are forbidden by orbital symmetry because the latter species would be characterized by occupied S^σ and P_z^σ orbitals. Since toroidal gold clusters are known to be stereochemically nonrigid we conclude that the rearrangements must involve transition states with oblate spheroidal geometries. This situation results from the way that the P^σ orbitals split in the two topologies, as discussed in Chapter 4. Circumstantial evidence for the soft nature of the potential energy surface in these gold clusters compounds is also provided by

High-Resolution Solid State nmr Using Magic Angle Sample Spinning

The majority of gold cluster compounds of the type $[Au_x(PR_3)_x]^{m+}$ and $[AuAu_x(PR_3)_x]^{m+}$ are stereochemically nonrigid in solution (even at low temperatures). As a consequence, their $^{31}P\{^1H\}$ spectra show only a single average signal and are structurally uninformative. In the solid state it has been possible to obtain reasonably high-resolution spectra (linewidths around 300 Hz) using magic angle sample spinning and cross polarization techniques. A typical spectrum is illustrated in the figure for $[Au_9\{P(C_6H_4OMe)_3\}_8](NO_3)_3$. The cluster geometry is stereochemically rigid in the solid state. It has effective D_{2h} symmetry and two chemically distinct ^{31}P environments which are discernible in the spectrum. The broad satellites are the result of spinning sidebands.

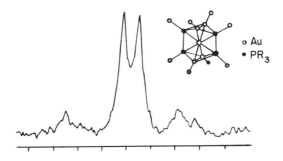

○ Au
● PR_3

The stereochemical nonrigidity of cluster compounds is not always frozen out in the solid state. There are examples of gold phosphine cluster compounds which remain fluxional in the solid state.

The ^{13}C spectra of $Fe_3(CO)_{12}$ have been measured using magic angle spinning techniques in the temperature range 31° to −121°. Below −95° the spectrum is consistent with the crystal structure, i.e., 10 terminal and 2 bridging CO's. At 24°C there are three pairs of resonances of equal intensity. These results have indicated that terminal-bridge carbonyl exchange is occurring in the solid state. It has been suggested that the carbonyls in $Fe_3(CO)_{12}$ occupy a close-packed icosahedral arrangement and the triangle of metal atoms may reorientate within this ligand polyhedron and thereby cause bridge-terminal exchange. It has been argued that in the solid state the intramolecular forces between the carbonyl ligands lead to a rigid carbonyl shell and the potential energy surface associated with reorganization of this shell is much steeper than that associated with the ring metal triangle tilting motions which result in the exchange process. B. F. G. Johnson et al.* have argued that a libration motion of the triangle by a few degrees about the twofold axis leads to a D_3 structure in which all carbonyls are terminally bound.

*J. Chem. Soc., Chem. Comm., 889 (1988).

the occurrence of skeletal isomers in the solid state.[40] We would expect the radially bonded clusters of the alkali metals to be stereochemically nonrigid for the same reasons.

6.10 CLUSTERS WITH CAPPING Ib METAL ATOMS

Recently a large number of mixed metal clusters containing $M(PPh_3)$ (M = Cu, Ag, Au) fragments have been characterized. In these compounds compact structures based upon face-sharing tetrahedra are generally formed with the $M(PPh_3)$ fragment capping the least hindered triangular face. When two or more of the $M(PPh_3)$ fragments are bonded to the cage, they are generally observed to undergo skeletal rearrangements which lead to their site exchange. The remaining part of the cluster skeleton remains rigid throughout this process. As we have previously pointed out,[41] this is a special case of skeletal stereochemical nonrigidity which results from the unusual bonding capabilities of the d^{10} $M(PPh_3)^+$ fragments. These fragments can form multicentered bonds through an out-pointing s/p_z hybrid with either two or three transition metal atoms. The fact that $M(PPh_3)^+$ fragments can function equally well as either face- or edge-bridging groups and in each case increment the total electron count by 12 provides a ready explanation for the stereochemical nonrigidity of such compounds.[42]

6.11 PLATINUM AND PALLADIUM CLUSTERS

Palladium and platinum clusters do not have electron counting rules analogous to those of the boranes and carboranes. The pattern of orbital energy levels characteristic of these species differs significantly from that observed for main group and transition metal carbonyl clusters which conform to the PSEPT (Chapter 2). As a consequence, some of the rearrangement processes which are unfavorable for $B_nH_n^{2-}$ and, for example, $[Os(CO)_3]_n$ become energetically feasible. This is illustrated by the observed fluxionality of $Pt_4(CO)_5(PPh_3)_4$ on the nmr time scale.[43] A tetrahedron-butterfly rearrangement has been proposed to account for this observation. For compounds which conform to PSEPT, such a process involves a change in the number of skeletal electron pairs from 6 to 7 when one of the metal-metal bonds is broken (Figure 6.10). However, for the platinum butterfly the increase in overlap between filled orbitals on the wingtip atoms as they are brought together is not sufficiently large to create an orbitally forbidden process.

There is also evidence that in the stacked platinum clusters $[Pt_3(CO)_3]_n^{2-}$ the platinum triangles rotate relative to one another by a low-energy trigonal twist mechanism. The process involves an intermediate trigonal prism for octahedra, and is not expected to be favorable for either main group clusters

Nmr Studies of Platinum Clusters: The usefulness of ^{195}Pt isotopomer effects

The ^{195}Pt nucleus ($I = \frac{1}{2}$) occurs with an abundance of 33.8% and has a relative sensitivity of 9.94×10^{-3} (compare the ^{1}H sensitivity of 1.0). Therefore ^{195}Pt spectroscopy can be used routinely to study platinum clusters. The presence of ^{195}Pt in less than 100% abundance results in complex spectra which represent the superposition of the spectra for clusters with all the possible permutations of the isotopomers. The ^{195}Pt nmr spectra, or the spectra of other nuclei associated with ligands coordinated to platinum (such as ^{31}P, ^{1}H, and ^{13}C), therefore yield direct information about the nuclearity and structure of the cluster. The ^{31}P($I = \frac{1}{2}$) nmr spectrum for a triangular platinum cluster (shown in the figure) illustrates the general principles underlying the spectral analysis.

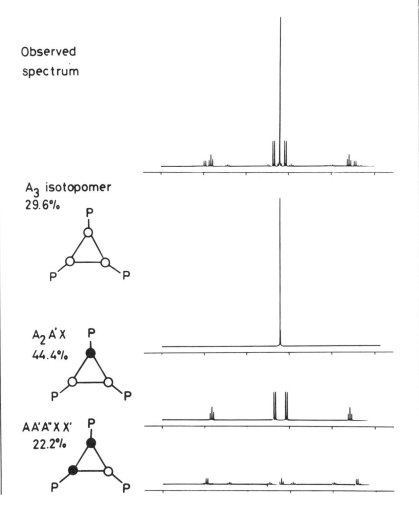

The A_3 isotopomer (with no ^{195}Pt nuclei) occurs with an abundance of 29.6% ($\approx \frac{2}{3} \times \frac{2}{3} \times \frac{2}{3}$) and gives rise to a single line resonance which defines the ^{31}P chemical shift. The most abundant isotopomer has a single ^{195}Pt nucleus and has an abundance of 44.4% ($\approx \frac{1}{3} \times \frac{2}{3} \times \frac{2}{3} \times 3$). The spin system in this case is written $A_2A'X$ (A, A' = P, X = ^{195}Pt). Although the phosphorus nuclei are chemically equivalent, they are no longer magnetically equivalent. The spectrum consists of a pair of outer triplets and a pair of inner doublets. The former result from direct coupling between ^{31}P and the adjacent ^{195}Pt [^1J(Pt—P) $\approx$ 4400 Hz; ^3J(Pt—P) $\approx$ 60 Hz], and the latter from coupling between ^{31}P and a nonadjacent ^{195}Pt [^2J(Pt—P) $\approx$ 426 Hz; ^3J(Pt—P) $\approx$ 60 Hz]. This subspectrum contains important information concerning the structure. The spectrum of a dimeric cluster containing the Pt_2P_2 fragment would contain a pair of outer and inner doublets. Similarly, a tetrahedral Pt_4P_4 compound would give an outer pair of quartets and an inner doublet. For additional examples see Mingos, D. M. P., and Wardle, R. W. M., *Transition Met. Chem.*, **10**, 441 (1985).

The isotopomer with two ^{195}Pt nuclei has an abundance of 22.2% ($\approx \frac{1}{3} \times \frac{1}{3} \times \frac{2}{3} \times 3$), and the associated spin system is written AA'A"XX'. The resultant second-order spectrum is complex, but the following features can be discerned. The inner triplet of triplets is associated with the phosphorus atoms not coordinated to ^{195}Pt nuclei. The outer lines result from second-order splittings associated with the ^{31}P nuclei directly bonded to ^{195}Pt atoms.

The remaining isotopomer with all ^{195}Pt nuclei occurs with only 3.7% abundance ($\approx \frac{1}{3} \times \frac{1}{3} \times \frac{1}{3}$) and has spin system AA'A"XX'X". The spectrum is very complex and requires computer simulations for its interpretation.

or transition metal clusters with the normal 86 valence electron count from the SECS rule. In the trigonal prismatic $[Pt_6(CO)_{12}]^{2-}$, however, the bonding is somewhat different. The valence-shell p orbitals are relatively high in energy for platinum, and this leads to rather weak interactions between the two platinum triangles and there is only one accessible orbital corresponding to bonding between them.[44] Six of the seven skeletal molecular orbitals are strongly bonding within the platinum triangles and only weakly bonding between them; their energies are not sensitive to relative rotation of the triangles. The other, a'_1, bonding orbital also shows little preference for either

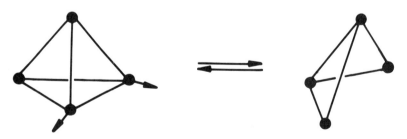

Figure 6.10 The tetrahedron-butterfly interconversion proposed for $Pt_4(CO)_5(PPh_3)_4$.

geometry. Hence, this cluster has only 86 valence electrons, the same as an octahedral cluster such as $[Os_6(CO)_{18}]^{2-}$, and rearrangements involving relative rotation of the $Pt_3(CO)_6$ triangles are therefore facile. The frontier orbitals for this cluster are given in Figure 6.11. This result also suggests that ligand disposition can be an important factor in transition metal rearrangements. Hence $[Pt_6(CO)_{12}]^{2-}$, where the carbonyls bridge only the triangular edges, is fluxional, but $[Os_6(CO)_{18}]^{2-}$ is not. In platinum clusters the metal-metal bonding across those edges which are not carbonyl-bridged is weak and consequently the potential energy surfaces for rearrangements involving these edges appear to be soft.

The related trigonal prismatic clusters $[PtRh_2(CO)_x]_2^-$ appear to undergo

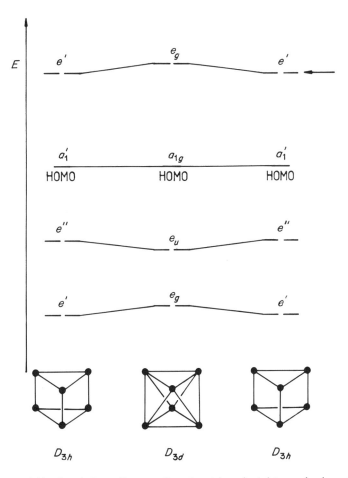

Figure 6.11 Correlation diagram for the trigonal twist mechanism for $[Pt_6(CO)_{12}]^{2-}$. The unoccupied e' orbitals indicated by the arrow correspond to an accessible pair in clusters which follow the PSEPT rules.

a similar fluxional process involving the rotation of PtRh$_2$ triangles. The mixed-metal cluster [PtRh$_8$(μ_3-CO)$_3$(μ_2-CO)$_9$(CO)$_7$]$^{2-}$ has a structure based upon two octahedra sharing a common face. Rotation of the two rhodium triangles about the pseudo-threefold axis has been proposed as a possible explanation of the observed ^{13}C nmr data.[45]

6.12 REARRANGEMENTS OF THE LIGAND SHELL[46-47]

In recent years ^{1}H and ^{13}C nmr studies have established that in many carbonyl and hydride clusters the ligands migrate over the surface of the cluster very readily. Indeed, there are very few carbonyl clusters which are stereochemically rigid above 60°C, Rh$_6$(CO)$_{16}$ being exceptional in this regard. At the other extreme the ligands in some clusters are still fluxional on the nmr time scale at -150°C. For intramolecular ligand migration to occur in a metal cluster, an intermediate bridging interaction is required. Therefore ligands such as H and CO, which retain the same donor characteristics independent of whether they occupy terminal or bridging sites, are more labile than ligands such as Cl and SR, whose donor characteristics are dependent upon coordination number. The theoretical analysis of cluster bonding developed in the previous chapters indicates that there are a large number of orbitals available for metal-metal and metal-ligand bonding. Specifically, for a deltahedron there are $7n + 1$ accessible orbitals with the appropriate radial and nodal characteristics. The number of ligands around the cluster is determined primarily by the d electron configuration of the metal, as the following examples for an octahedron illustrate:

Formal configuration	d^6	d^8	d^{10}
Accessible orbitals	43	43	43
Orbitals formally occupied by metal electrons	18	24	30
Orbitals available for M—L bonding	25	19	13

The symmetry of the ligand shell is not generally defined specifically by the symmetries of the accessible orbitals of the metal cage, because different arrangements of ligands can produce linear combinations of donor orbitals of matching symmetry. Consequently, the observed carbonyl arrangement in a crystal structure generally represents a shallow minimum on a very soft potential energy surface and there is no general theory for rationalizing the observed structures. The following factors are important:

(1) **Local Geometric Effects.** In the absence of bridging carbonyls, the metal-metal bonds and the metal-carbonyl bonds are usually found in

the orientation suggested by the inert gas rule for the metal vertices. For example, in $Os_3(CO)_{12}$ and $Ir_4(CO)_{12}$, the local geometries are octahedral.

(2) **Charge Effects.** The ratio of bridging to terminal carbonyls increases markedly as the negative charge on the cluster increases. Bridging carbonyls are more effective at accepting electron density through their π^* orbitals than terminal carbonyls, and consequently the cluster is able to achieve electroneutrality more effectively in this way. For example, $Co_6(CO)_{16}$ has 12 terminal and 4 face-bridging carbonyls, whereas $[Co_6(CO)_{14}]^{2-}$ has 6 terminal and 8 face-bridging carbonyls. The replacement of carbon monoxide by better σ-donor ligands can have a similar effect; hence, although $Ir_4(CO)_{12}$ has only terminal ligands, $Ir_4(CO)_9(PPh_3)_3$ has three edge-bridging carbonyls.

Even when there are bridging carbonyls, local geometric effects can still make a contribution, and in $Fe_3(CO)_{12}$ the geometries around the iron atoms are approximately octahedral if the bridged metal-metal bond is ignored. Similarly, the local geometry in platinum clusters is frequently based on fragments with a square-planar or trigonal-bipyramidal geometry.

(3) **Cluster Topology.** The number of edges and faces of the cluster places geometric and symmetry constraints on the locations of the carbonyls. For example, $Co_6(CO)_{16}$ has T_d point group symmetry with the bridging carbonyls occupying four of the eight faces of the metal octahedron in a symmetrical fashion, and $[Co_6(CO)_{14}]^{2-}$ has all eight faces of the octahedron occupied by bridging ligands. $[Co_6(CO)_{15}]^{2-}$, on the other hand, has a stoichiometry which cannot be accommodated within the T_d or O_h point groups, and a C_{3v} structure with three face-bridging carbonyls is adopted.

(4) **Steric Effects.** We noted in Chapter 1 that for a given cluster there is a geometric limit to the number of carbonyl ligands which can be accommodated. As this limit is approached there is a tendency for the ligands to adopt a pseudospherical deltahedral geometry with a higher proportion of bridges. These effects are particularly important for the lower-nuclearity clusters of the lighter transition metals. The approximately icosahedral arrangement of ligands in $Fe_3(CO)_{12}$, $Co_4(CO)_{12}$, and $Rh_4(CO)_{12}$ in preference to the cuboctahedral arrangements found in $Os_3(CO)_{12}$ and $Ir_4(CO)_{12}$ can be attributed to steric effects. However, as the nuclearity of the cluster increases, electronic rather than steric effects are likely to dominate.

Recently, Lauher[48] and Kepert[49] have attempted to put the stereochemical aspects of carbonyl arrangements in metal clusters on a more quantitative basis, using simple model potentials. The cluster metallic radii are used to

^{13}C nmr Data
For Metal Carbonyl Clusters

^{13}C data are obtained generally for enriched samples with 50 to 75% ^{13}CO. Terminal carbonyls generally have resonances at about 180 to 210 ppm, and bridging carbonyls, 220 to 260 ppm, to low field of TMS. If the carbonyls are bonded to ^{103}Rh, which has $I = \frac{1}{2}$, 100% abundance, then the nmr data at low temperatures can give precise information concerning the structure of the cluster in solution. Terminal carbonyls are observed as doublets with $J(Rh\text{---}C) \approx 70$ to 100 Hz, edge-bridging carbonyls as triplets with $J(Rh\text{---}C) \approx 35$ to 45 Hz, and face-bridging carbonyls as quartets with $J(Rh\text{---}C) \approx 20$ Hz. The latter values are, of course, valid only for symmetrically bridged compounds. As the temperature is raised, intramolecular carbonyl rearrangement processes can lead to an average signal and a binomial multiplet of n lines where n = number of rhodium atoms + 1 is observed. The table below gives some typical data. The coupling constant of the multiplet is the weighted mean of the coupling constants for the frozen structure if the rearrangement is intramolecular.

^{13}C NMR Studies on Metal Cluster Compounds

Compound	Temp. (°C)	Behavior	^{13}C chemical shift (ppm)
$Rh_4(CO)_{12}$	+63	Fluxional	189.5 quintet
$Rh_6(CO)_{16}$	+70	Not fluxional	231.5 quadruplet; 180.1 doublet
$Rh_6(CO)_{15}^{2-}$	−69	Fluxional	209.2 septet
$Rh_7(CO)_{16}I^{2-}$	−31	Fluxional	218.5 octet

define a nonspherical metal surface and the carbonyls are allowed to migrate over it. The equilibrium geometries are calculated on the basis of competing repulsive interactions between carbonyls and attractive interactions between the carbonyls and the metal skeleton.

(5) **Semibridging Effects.** The cluster topology and the steric effects noted above can lead to situations where the distinction between terminal and bridging carbonyls becomes blurred. There are many clusters which have such semibridging carbonyls with intermediate geometries. When there is a soft potential energy surface connecting alternative ligand arrangements, minor electronic and steric effects can "freeze out" intermediate and asymmetric structures. The potential energy surface is likely to be particularly soft when topological and symmetry factors do not exert a strong effect. For example, in $[Fe_4(CO)_{13}]^{2-}$ there are three semibridging carbonyls around the face of the tetrahedron which maintain C_3

symmetry. The unique Fe-CO distances are 1.8 and 2.2 Å, respectively, and the Fe—C—O bond angle is 155°. These semibridging carbonyls represent "snapshots" of the reaction coordinate for the interconversion of terminal and bridging ligands and underline the fact that low-energy transformations must involve concerted movements of the ligands.

A variable-temperature ^{13}C nmr spectrum of a metal carbonyl cluster can establish whether ligand rearrangement is intramolecular and also which nuclei are permuted by the fluxional process. It cannot, however, tell us directly what the mechanism for the rearrangements is. The latter must be established using additional circumstantial evidence based on chemical knowledge of the likely bond-making and bond-breaking processes and consideration of related studies on less symmetrical derivatives.

Nmr data can often discriminate between local and global fluxional processes. For example, the ^{13}C nmr spectrum of $Os_6(CO)_{18}$, which has a C_{2v} bicapped tetrahedral structure based upon $Os(CO)_3$ fragments, has three coalescence temperatures at $-100°$, $-10°$, and $40°C$. This has been attributed to the three local rotations of the distinct $Os(CO)_3$ units. Hence, if the local rotations of the $M(CO)_n$ units in a cluster have substantially higher activation energies than carbonyl terminal-bridge exchange, then global migrations are observed only at high temperatures. The activation energies for terminal-bridge exchange are reduced by those effects which promote the formation of carbonyl bridges as described above. These include an increased negative charge on the cluster; e.g., $[Rh_{13}(CO)_{24}H_2]^{3-}$ is nonrigid at $-80°C$ but the carbonyls in $[Rh_{13}(CO)_{24}H_3]^{2-}$ are fluxional only above $25°C$. The replacement of CO by better σ-donor ligands has the same effect; e.g., $Os_3(CO)_{11}(PEt_3)$ exhibits ligand fluxionality at $100°C$ below $Os_3(CO)_{12}$. Steric crowding also lowers the activation energies so long as the structure retains some "defects." The concept of "defects" stems from solid state chemistry, where ion migration is found to have a lower activation energy if there are vacancies in the structure to which the ions can migrate. If a cluster carbonyl has a vacant face or edge, then a concerted movement of the ligands can result in the occupation of this site and the concomitant creation of a new "defect" somewhere else. In $Rh_4(\eta\text{-}C_5H_5)_4(\mu_3\text{-CO})_2$, only two of the tetrahedral faces are bridged and the ligands are fluxional, whereas in $Fe_4(\eta\text{-}C_5H_5)_4(\mu_3\text{-CO})_4$, where each face is bridged, they are not. Similarly, $Rh_6(CO)_{16}$ is rigid even at $60°C$, but $[Rh_6(CO)_{15}]^{2-}$, with one less carbonyl, is fluxional at $-70°C$. Finally, $[Rh_6C(CO)_{15}]^{2-}$, which has carbonyls over all nine edges of the trigonal prism of rhodium atoms, is rigid at $25°C$ despite the larger negative charge.

If $M(CO)_n$ rotation and carbonyl terminal-bridge exchange processes both have low activation energies, then global ligand migration may occur. If the metal has a nuclear spin, this results in a simple multiplet with a ^{13}C—M coupling constant which is a weighted mean of the coupling constants found

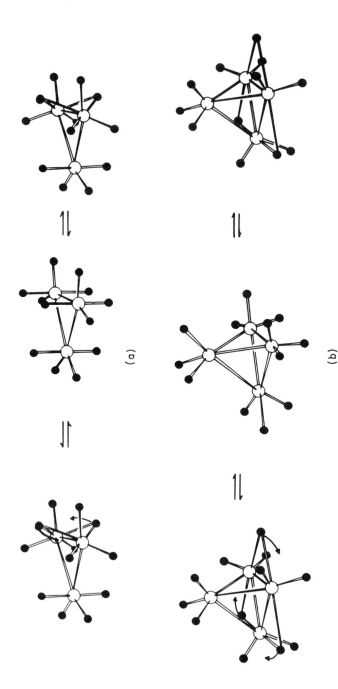

Figure 6.12 The mechanism proposed to explain the carbonyl fluxionality in (a) $Rh_4(CO)_{12}$ and (b) $Fe_3(CO)_{12}$. In each case the bridge-terminal exchange produces a high-symmetry intermediate which can reform the original cluster or rearrange it to a different permutational isomer.

in the "frozen" structure. The usual mechanisms proposed to explain the stereochemical nonrigidity of $Rh_4(CO)_{12}$ and $Fe_3(CO)_{12}$ are illustrated in Figure 6.12. The intermediate structure for $Rh_4(CO)_{12}$ has T_d point group symmetry, which results from the concerted opening up of the carbonyl bridges, and is identical to that observed for $Ir_4(CO)_{12}$ in the solid state. Another mechanism involving rotation of the metal tetrahedron within the icosahedral shell of ligands has been proposed by Johnson[47] to account for the observed nmr data.

6.13 SUMMARY

The orbital symmetry selection rules for cluster rearrangements developed using TSH theory are clearly very powerful, and the skeletal electron count selection (SECS) rule is also useful. The latter rule is not as rigorous as the geometric or orbital symmetry selection rule, as it is based upon the assumed conservation of the number of skeletal electron pairs. It is applied using the general principles of the PSEPT, such as the capping principle and the condensation principle, to predict the electron counts of suggested transition states or intermediates. An alternative justification of this approach is to reason that if several structures with the same skeletal electron count exist, then fluxionality is more likely to occur.

There are a number of possibilities for transition metal clusters which do not arise for main group clusters. For example, "radially" bonded clusters, exemplified by various gold and alkali metal clusters, are all expected to be readily fluxional. Palladium and platinum clusters may also exhibit fluxional behavior because deformations of nonbridged metal-metal bonds have very soft potential energy surfaces. *nido* transition metal clusters may also be able to rearrange by the square-diamond, diamond-square (SDDS) mechanism, where one face opens and another closes.

EXERCISES

6.1 Using the results of Chapter 4, confirm that the trigonal twist mechanism for the octahedron, which proceeds via the trigonal prism, is allowed by orbital symmetry for Na_6Mg, where the magnesium atom is interstitial.

6.2 With the aid of models, determine the polyhedra which are formed when a single DSD process occurs in (a) $B_6H_6^{2-}$, (b) $B_7H_7^{2-}$ (consider breaking an axial-equatorial edge), and (c) $B_{12}H_{12}^{2-}$. In each case find the symmetry elements which are retained throughout. Hence predict whether these cluster skeletons are fluxional on the nmr time scale at room temperature.

6.3 Using TSH theory methodology, construct qualitative energy-level diagrams for $B_5H_5^{2-}$ in trigonal bipyramidal and square-based pyramidal geometries. Hence show that the single DSD process for the trigonal bipyramid is "forbidden" by orbital symmetry.

6.4 Using the appropriate real spherical polar forms of the vector spherical harmonics based upon $Y_{2,\pm 1}$, sketch the form of the D_{xz}^π and D_{yz}^π orbitals in the square face of a square-based pyramid. Choose the z axis to coincide with the fourfold rotational axis. Show also that a suitable set of L^π cluster orbitals for this geometry is $P_{x,y,z}^\pi$, $\bar{P}_{x,y,z}^\pi$, D_{xz}^π, D_{yz}^π, $D_{x^2-y^2}^\pi$, and $\bar{D}_{x^2-y^2}^\pi$, where D_{yz}^π is equivalent to $\bar{D}_{xz}^\pi$.

6.5 By considering the changes in energy of the frontier orbitals in the two open faces (see Exercise 6.4), construct schematic correlation diagrams for the following double DSD processes: show that there are (a) two crossings if the edges which are made and broken are parallel with respect to a conserved mirror plane and (b) two avoided crossings if they are perpendicular.

6.6 Using the same approach as in Exercise 6.5 sketch orbital correlation diagrams for the following SDDS processes: show (a) that there are two crossings if the edges being made and broken lie perpendicular to one another with respect to a conserved mirror plane, and (b) that there are two avoided crossings if these critical edges are parallel.

6.7 For $B_5H_5^{2-}$ and $B_9H_9^{2-}$, use models to trace the structural effect of the degenerate single DSD process. The most symmetrical intermediate on each pathway has C_{4v} point group symmetry, and the starting and finishing geometries are related by a 90° "pseudorotation." What are the corresponding results for $B_8H_8^{2-}$ and $B_{11}H_{11}^{2-}$?

6.8 In Figure 6.6 the observed crossings involve $L_{1c,1s}^\pi$ in the intermediate structure. Can you explain why the L_0^π and $L_{2c,2s}^\pi$ orbitals do not give rise to such a crossing?

REFERENCES

1. Wales, D. J., and Stone, A. J., *Inorg. Chem.*, **26**, 3845 (1987).
2. Wales, D. J., Mingos, D. M. P., and Lin, Z., *Inorg. Chem.*, **28**, 2754 (1989). Wales, D. J., and Mingos, D. M. P., *Polyhedron*, **15**, 1933 (1989).
3. Lipscomb, W. N., *Science*, **153**, 373 (1966).
4. Wade, K., *Electron Deficient Compounds*, Nelson, London, 1971.
5. See, for example, Kennedy, J. D., in *Progress in Inorganic Chemistry*, Vol. 34, Wiley, New York, 1986.
6. King, R. B., *Inorg. Chim, Acta*, **49**, 237 (1981).
7. For this chapter in particular, and throughout the book, the reader will find a good set of models invaluable. We have found that the "Polydron" building blocks available from toy shops and Early Learning Centres in the United Kingdom are perfect for this purpose.
8. Woodward, R. B., and Hoffmann, R., *Angew. Chem., Int. Ed.*, **8**, 781 (1969). Woodward, R. B., and Hoffmann, R., *The Conservation of Orbital Symmetry*, Academic Press, London, New York, 1970.
9. Gimarc, B. M., and Ott, J. J., *Inorg. Chem.*, **25**, 83, 2708 (1986). Gimarc, B. M., and Ott, J. J., *J. Comp. Chem.*, **7**, 673 (1986).
10. Johnston, R. L., and Mingos, D. M. P., *J. Chem. Soc., Dalton Trans.*, 647 (1986).

11. Stone, A. J., and Wales, D. J., *Molec. Phys.*, **61**, 747 (1987).
12. Fowler, P. W., *Polyhedron*, **4**, 2051 (1985).
13. Forsyth, M. J., and Mingos, D. M. P., *J. Chem. Soc., Dalton Trans.*, 610 (1977).
14. McNeill, E. A., Gallaher, K. L., Scholer, F. R., and Bauer, S. H., *Inorg. Chem.*, **12**, 2108 (1973).
    Shapiro, I., Keilin, B., Williams, R. E., and Good, C. D., *J. Amer. Chem. Soc.*, **85**, 3167 (1963).
    Beaudet, R. A., and Poynter, R. L., *J. Chem Phys.*, **53**, 1899 (1970).
15. Muetterties, E. L., Hoel, E. L., Salentine, C. G., and Hawthorne, M. F., *Inorg. Chem.*, **14**, 950 (1975).
16. Muetterties, E. L., Wiersema, R. J., and Hawthorne, M. F., *J. Amer. Chem. Soc.*, **95**, 7520 (1973).
    Tolpin, E. I., and Lipscomb, W. N., *J. Amer. Chem. Soc.*, **95**, 2384 (1973).
17. Shapiro, I., Good, C. D., and Williams, R. E., *J. Amer. Chem. Soc.*, **84**, 3837 (1962).
18. Rogers, H. N., Lau, K., and Beaudet, R. A., *Inorg. Chem.*, **15**, 1775 (1976).
    Hart, H., and Lipscomb, W. N., *Inorg. Chem.*, **7**, 1070 (1968).
    Berry, T. E., Tebbe, F. N., and Hawthorne, M. F., *Tetrahedron Letters*, 715 (1965).
19. Koetzel, T. F., Scarborough, F. E., and Lipscomb, W. N., *Inorg. Chem.*, **7**, 1076 (1968).
20. Tebbe, F. N., Garrett, P. M., Youngs, D. C., and Hawthorne, M. F., *J. Amer. Chem. Soc.*, **88**, 609 (1966).
    Rietz, R. R., Schaeffer, R., and Walter, E., *J. Organometallic Chem.*, **63**, 1 (1973).
    Garrett, P. M., Smart, J. C., Ditta, G. S., and Hawthorne, M. F., *Inorg. Chem.*, **8**, 1907 (1969).
21. Rietz, R. R., and Schaeffer, R., *J. Amer. Chem. Soc.*, **95**, 6254 (1973).
    Grimes, R. N., *Carboranes*, Academic Press, New York, 1970.
    Onak, T. P., Gerhart, F. J., and Williams, R. E., *J. Amer. Chem. Soc.*, **85**, 3378 (1963).
22. Abdou, Z. J., Abdou, G., Onak, T., and Lee, S., *Inorg. Chem.*, **25**, 2678 (1986).
23. Rodger, A., and Johnson, B. F. G., *Polyhedron*, **7**, 1107 (1988).
24. Mingos, D. M. P., *Nat. Phys. Sci.*, **236**, 99 (1972).
25. Mason, R., Thomas, K. M., and Mingos, D. M. P., *J. Amer. Chem. Soc.*, **95**, 3802 (1973).
26. Mingos, D. M. P., *J. Chem. Soc., Dalton Trans.*, 610 (1977).
27. Mingos, D. M. P., *J. Chem. Soc., Chem. Comm.*, 706 (1983).
28. McIvor, J. W., and Stanton, R. E., *J. Amer. Chem. Soc.*, **94**, 8618 (1972).
    McIvor, J. W., and Stanton, R. E., *J. Amer. Chem. Soc.*, **97**, 3632 (1975).
29. Muetterties, E. L., Wiersema, R. J., and Hawthorne, M. F., *J. Amer. Chem. Soc.*, **95**, 7520 (1973).
30. Strohrer, W.-D., and Hoffmann, R., *J. Amer. Chem. Soc.*, **94**, 1661 (1972).
31. Evans, D. G., *Inorg. Chem.*, **25**, 4602 (1986).
32. Woolley, R. G., *Inorg. Chem.*, **27**, 430 (1987).

33. Lewis, J., Johnson, B. F. G., et al., *J. Chem. Soc., Dalton Trans.*, 149 (1988).
34. Gladfelter, W. L., and Geoffrey, G. L., *Inorg. Chem.*, **19**, 2579 (1980). Geoffrey, G. L., *Acc. Chem. Res.*, **13**, 469 (1980).
35. Salter, I. D., *Adv. Dynamic Stereochem.*, **2**, 57 (1988).
36. Heaton, B. T., Strona, L., Della Pergola, R., Vidal, J. L., and Schoening, R. C., *J. Chem Soc., Dalton Trans.*, 1941 (1983).
37. Garlaschelli, L., Fumagalli, A., Martinengo, S., Heaton, B. T., Smith, D. O., and Strona L., *J. Chem. Soc., Dalton Trans.*, 2265 (1982).
38. Vidal, J. L., Walker, W. E., Pruett, R. L., and Schoening, R. C., *Inorg. Chem.*, **18**, 129 (1979).
39. Mingos, D. M. P., and Johnston, R. L., *Structure and Bonding*, **68**, 30 (1987).
40. Hall, K. P., and Mingos, D. M. P., *Prog. Inorg. Chem*, **32**, 237 (1984).
41. Mingos, D. M. P., *Polyhedron*, **3**, 1289 (1984).
42. Salter, I. D., *Adv. Organometallic Chem.*, **29**, 249 (1989).
43. Moor, A., Presgasin, P. S., Venanzi, L. M., and Welch, A. J., *Inorg. Chim. Acta*, **85**, 103 (1984).
44. Underwood, D. J., Hoffmann, R., Tatsumi, K., Nakamura, A., and Yamamoto, Y., *J. Amer. Chem. Soc.*, **107**, 5968 (1985).
45. Fumagalli, A., Martinengo, S., Chini, P., Albinanti, A., Bruckner, S., and Heaton, B. T., *J. Chem Soc., Chem. Comm.*, 195 (1978).
46. Band, E., and Muetterties, E. L., *Chem. Rev.*, **78**, 639 (1978).
47. Johnson, B. F. G., and Benfield, R. E., in Johnson, B. F. G. (ed.), *Transition Metal Clusters*, Wiley, New York, 1980, p. 471.
48. Lauher, J. W., *J. Amer. Chem. Soc.*, **108**, 1521 (1986).
49. Clare, B. W., Favas, M. C., Kepert, D. L., and May, A. S., *Adv. Dynamic Stereochem.*, **1**, 1 (1985).

7

Condensed and High-Nuclearity Cluster Compounds

7.1 INTRODUCTION

The rapid development of transition metal cluster chemistry has revealed a degree of structural complexity and variety which is unparalleled in main group chemistry. Thus although borohydrides are generally pseudospherical deltahedra, the situation is more complex for transition metal clusters. In particular, although examples of metal carbonyl clusters with spherical geometries are known, there are also a large number of structures which are distinctly nonspherical. Indeed, many of the structures are not convex, but concave, and are better described as condensed polyhedra based on triangular, tetrahedral, trigonal prismatic, and octahedral components. In this context "condensed" is taken to mean that two distinct cluster skeletons can be discerned in the structure, sharing one or more common vertex atoms. Some examples of condensed transition metal clusters have been discussed in Chapter 1, and Rule 6 in Chapter 2 provides an empirical basis for understanding the electronic requirements of these clusters. The theoretical justification of this rule will be given in this chapter. This analysis also provides a basis for explaining apparent exceptions to the rule.

7.2 GEOMETRIC AND THEORETICAL ASPECTS[1-3]

Although the vertex-sharing cluster illustrated in Figure 7.1 can be described in terms of the condensation principle, this is by no means a unique description. The condensed structure can also be fragmented into two parts, the first containing the common vertex, and therefore consisting of a complete closed skeleton, and the second a fragment of a complete skeleton with one vertex missing. The latter corresponds to a *nido* deltahedron if the complete skeleton is deltahedral. In this description the second *nido* structure can be regarded as acting as a ligand toward the first *closo* skeleton. Alternatively, the condensed cluster can be fragmented into three components with two *nido* "ligands" and the central atom corresponding to the shared vertex. These alternative descriptions must of course lead to the same conclusions regarding the number of bonding orbitals in the condensed cluster. The fragment descriptions are more convenient for analyzing the bonding because the electronic requirements of the isolated fragments are well defined as a result of the analyses in the previous chapters using TSH theory. The only aspect which needs clarification

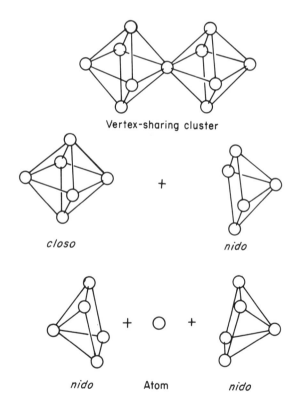

Figure 7.1 Alternative modes of fragmenting a vertex-sharing cluster.

is the number of orbitals involved in the donor-acceptor interactions between the *nido* skeleton and either the *closo* cluster or a single metal atom. The tensor surface harmonic methodology provides the necessary information because it defines the number of accessible and inaccessible molecular orbitals in the cluster as well as crucial information about the nodal characteristics and localization of the frontier orbitals.

A condensed cluster with a total of $n + m - 1$ vertex atoms, consisting of two *closo* deltahedra with n and m atoms which share a common vertex, can be partitioned into a *closo* cluster (C) with n atoms and a *nido* cluster (N) with $m - 1$ atoms. For a *closo* deltahedral transition metal cluster, the bonding orbitals are precisely filled by $14n + 2$ valence electrons. Furthermore, the positions of the ligands and their number do not influence this requirement so long as there are enough of them and their distribution is reasonably even. Indeed, the characteristic feature of the deltahedron is that it has $2n - 1$ inaccessible orbitals which are not used for bonding by virtue of their strongly antibonding metal-metal character and their inward hybridization. Therefore, we conclude that even when some of the ligands are replaced by a *nido* metal fragment, these $2n - 1$ orbitals will remain and the *closo* skeleton will still be characterized by $14n + 2$ electrons.

The *nido* cluster fragment with $m - 1$ atoms (N) is characterized by a total of $14(m - 1) + 4$ electrons when it is a separate entity. If this *nido* fragment has C_{nv} ($n \geq 3$) symmetry, then it has a nonbonding e pair of orbitals, with significant amplitude in the open face, and a lower-lying, a_1-type orbital also localized in this face (see Chapter 5). These molecular orbitals are illustrated on the right-hand side of Figure 7.2. They can interact with out-pointing orbitals of the deltahedral cluster (C) with matching symmetry characteristics as indicated in the figure. The latter orbitals are capable of accepting six electrons, usually from a set of three carbonyl ligands or a cyclopentadienyl ligand, for example. Consequently the *nido* fragment functions as a six-electron donor toward the *closo* skeleton.

The electron requirements for the condensed polyhedron are therefore:

closo	*nido*	Electrons donated
$14n + 2$	$14(m - 1) + 4$	-6

The number of electrons donated is included as -6, since three empty orbitals of the *closo* cluster must be available for accepting them from the *nido* cluster. To avoid double counting, we must subtract those electrons responsible for binding the fragments together. Alternatively, we could simply note that three more inaccessible orbitals are produced when the two fragments interact. The total electron count is therefore $14n + 14m - 14$. This expression may be rewritten as a condensation process involving two deltahedra with m and n vertex atoms, respectively: $(14n + 2) + (14m + 2) - 18$. In other words the total electron count is equal to the sum of the electron counts for the parent

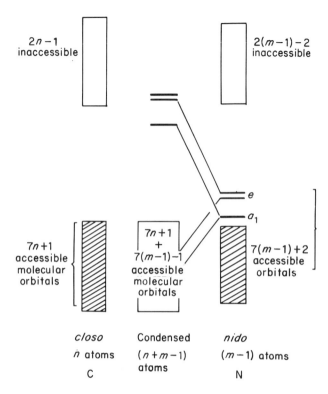

Figure 7.2 Important orbital interactions which occur when a *nido* cluster acts as a ligand toward a deltahedral cluster.

polyhedra $(A + B)$, minus the electron count characteristic of the atom common to both polyhedra (18 if the central atom conforms to the 18-electron rule).

Exactly the same conclusion is reached if the problem is analyzed in terms of two *nido* clusters, with $n - 1$ and $m - 1$ atoms, which act as "ligands" toward a common central atom. Again the relevant orbital interactions lead to the achievement of the inert gas rule for the latter. In terms of the condensation rules, this important conclusion can be interpreted using our knowledge of the number of inaccessible molecular orbitals. The two *nido* fragments have $2n - 1$ and $2m - 1$ inaccessible orbitals. If they are brought together and interact with a common transition metal atom, then nine more inaccessible orbitals result. Therefore the condensed cluster has

$$(2n - 1) + (2m - 1) + 9 = (2n + 2m) + 7 \quad \text{inaccessible orbitals.}$$

Hence, for the condensed cluster, all the accessible orbitals are filled when there are $14n + 14m - 14$ electrons. Some examples of clusters conforming to this generalization are illustrated in Figure 7.3. The procedure can also be

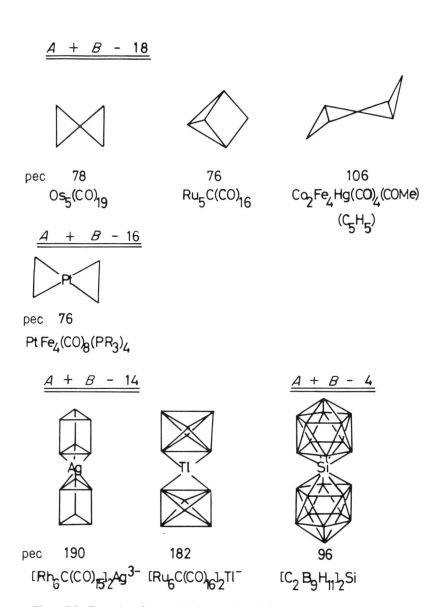

Figure 7.3 Examples of vertex-sharing condensed clusters.

used for condensed clusters sharing two vertices, e.g., $Ru_5C(CO)_{16}$, where a 62-electron "butterfly" cluster and an open triangular cluster (50 electrons) condense through the wingtip atoms.

The analysis based upon partitioning into *closo* + *nido* fragments rests on the assumption that six orbitals are available on the common atom and can interact with the a_1 and e frontier orbitals of the *nido* fragments. If the common atom has only $6-x$ acceptor orbitals, then the corresponding condensed electron count is $A + B - (18 - 2x)$. Some examples of condensed clusters with $x = 0$, 1, and 2 are illustrated in Figure 7.3. The platinum metals, which commonly form square-planar complexes, in which the central metal atom may be formally associated with 16 electrons, provide examples with $x = 1$, since they have only five acceptor orbitals; i.e., the condensed electron count is $A + B - 16$. The d^{10} ions Ag^+, Au^+, Tl^{3+}, and Hg^{2+} have $x = 2$, since they have two inaccessible p^π orbitals, and the corresponding condensed electron count is $A + B - 14$. A condensed cluster with a common main group atom is characterized by $A + B - 4$ electrons because there are only four acceptor orbitals on the shared atom (see Exercise 7.1). The fundamental principle which transcends the numerology of the condensation rules is that a knowledge of the available orbitals and the frontier orbitals of the donating cluster fragments leads to a clear account of the electronic requirements of the condensed cluster.

Before moving on to consider examples of clusters where edge and face sharing are important, it is instructive to draw attention to a large class of vertex-sharing clusters based on carboranes, where the common atom is a transition metal. Examples of such molecules are illustrated in Figure 7.4. The development of this field owed much to Hawthorne, who recognized that the $C_2B_9H_{11}^{2-}$ ligand, which is a *nido* icosahedron with an open pentagonal face, behaves analogously to the more common $C_5H_5^-$ cyclopentadienyl ligand, donating six electrons to a transition metal atom and forming sandwich compounds.[4] The most stable molecules of this type have 18 electrons around

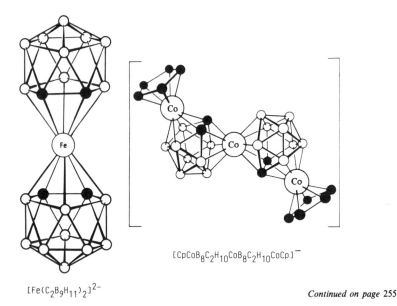

$[Fe(C_2B_9H_{11})_2]^{2-}$

$[CpCoB_8C_2H_{10}CoB_8C_2H_{10}CoCp]^-$

Continued on page 255

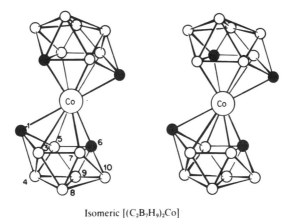

Isomeric [(C₂B₇H₉)₂Co]⁻

Figure 7.4 Examples of vertex-sharing *commo* carboranes with a central transition metal atom.

the central metal atom and are therefore analogous to ferrocene, $Fe(C_5H_5)_2$. In common with sandwich compounds there are also examples of vertex-sharing molecules with fewer than 18 electrons around the central atom and an incompletely filled set of nonbonding orbitals on the metal. There are also carborane sandwich compounds in which the central metal atom is associated with 20 or more valence electrons—the additional electrons occupy metal–ligand antibonding molecular orbitals. In these species the structure reacts to the antibonding interactions by undergoing a slip distortion. The detailed bonding analysis of such distortions is beyond the scope of this book,[5-7] but a specific example is illustrated in Figure 7.5.

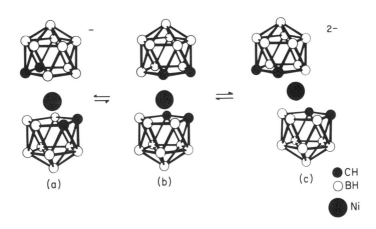

Figure 7.5 Structural consequences of reducing $Ni^{IV}(C_2B_9H_{11})_2$, which has the predicted electron count for a pair of vertex-sharing icosahedra (102).

Sec. 7.2 Geometric and Theoretical Aspects   **255**

7.3 EDGE-SHARING CONDENSED CLUSTERS

Some examples of edge-sharing condensed clusters are illustrated in Figure 7.6. They may be viewed either as an *arachno* cluster with $m - 2$ atoms acting as a "ligand" toward an n-vertex *closo* cluster, or as two *arachno* clusters, with $n - 2$ and $m - 2$ atoms, which act as "ligands" toward a central dimeric fragment (Figure 7.7). The latter is the common edge of the condensed molecule. From Chapter 5 we know that an *arachno* transition metal cluster with p vertices has a closed-shell stable electronic configuration when there are $14p + 6$ valence electrons. Furthermore, there are five bonding or nonbonding frontier molecular orbitals which have large amplitudes in the open face. These are made up of two approximately nonbonding "e" pairs (which will actually be split due to the low point group symmetry) and the fully in-phase bonding combination of the out-pointing tangential π orbitals. Another way of looking at this is to consider a Hückel problem for the out-pointing π orbitals of the open face. The five lowest-lying orbitals include one with no nodes in the face, two with one node, and two with two nodes. The *nido* problem above can be described similarly, except that with a smaller open face the two orbitals with two nodes across the face are significantly higher-lying in energy (see Chapter 5). Hence, such an *arachno* fragment is capable of donating a total of 10 electrons to another molecule if there are empty orbitals with matching symmetry characteristics. An edge-sharing condensed cluster will therefore be associated with the following total number of valence electrons:

closo fragment	*arachno* fragment	Electrons denoted
$14n + 2$	$14(m - 2) + 6$	-10

Total electron count = $14n + 14m - 30$

If reexpressed as the condensation of two deltahedra with n and m atoms, respectively, we find an electron count of

$$(14n + 2) + (14m + 2) - 34.$$

Since a dimeric transition metal cluster with a single bond between the metal atoms is associated with 34 valence electrons, e.g., $Mn_2(CO)_{10}$, the number of accessible orbitals of the condensed cluster corresponds once again to those associated with the *closo* clusters minus those of the fragment common to both polyhedra in the condensed cluster. Figure 7.6 illustrates how this conclusion can be used to account for the electronic requirements of a wide range of edge-shared condensed clusters.

The vast majority of examples in Figure 7.6 involves the condensation of a triangle and a deltahedron, and the net effect corresponds to edge bridging

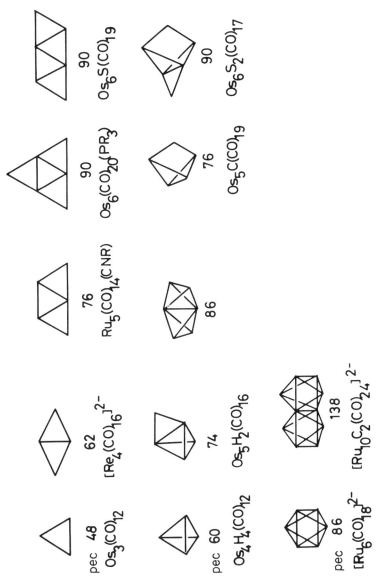

Figure 7.6 Some examples of edge-sharing clusters which conform to the condensation rules.

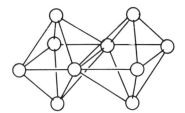

Edge-sharing cluster

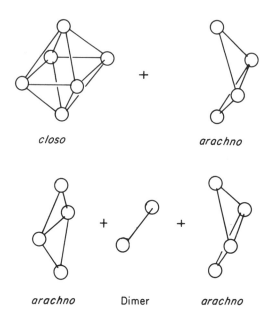

closo   *arachno*

arachno   Dimer   *arachno*

Figure 7.7 Alternative ways of fragmenting edge-sharing clusters.

by a single metal atom of the parent polyhedron. It is apparent from the condensation rule that this edge-bridging process leads to an increment in the electron count of 14 (assuming that the metal triangle is associated with 48 electrons). Successive edge bridging provides a mechanism for generating layers of close-packed metal atoms producing the "raft" structures illustrated in Figure 7.6.

If the common edge has fewer acceptor orbitals, then the condensation rule is modified accordingly. For example, in platinum chemistry the prototypical dimeric molecule is $[Pt_2(CNR)_6]^{2+}$, which is based on two platinum square planar fragments linked by a metal-metal bond and has 30 valence electrons. Consequently, the appropriate condensation rule in platinum cluster chemistry is $A + B - 30$. Examples of this generalization are given in

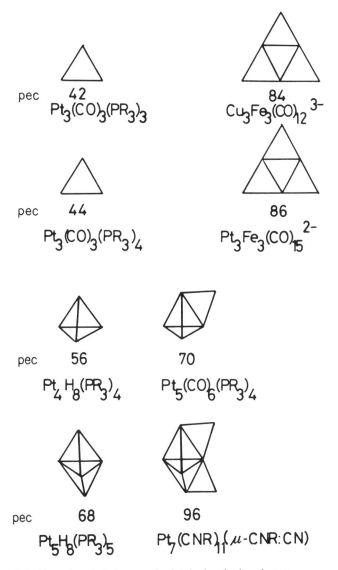

Figure 7.8 Examples of platinum and related edge-sharing clusters.

Figure 7.8. The net effect of edge bridging by such fragments is, however, also to increase the number of valence electrons by 14, as in the previous metal carbonyl examples. The examples of edge-sharing borane clusters discussed in Chapter 2 have a similar deficiency of acceptor orbitals and are characterized by $A + B - 12$ valence electrons rather than $A + B - 14$ as anticipated by a naive application of the inert gas rule.

Sec. 7.3 Edge-Sharing Condensed Clusters

7.4 FACE-SHARING CLUSTERS

A simple extension of the principles developed above to condensed clusters based on face sharing suggests that:

(1) If a triangular face is shared, then the electron count is the sum of the electron counts of the isolated clusters minus 48, because that is the electron count associated with a triangular cluster such as $Os_3(CO)_{12}$.

(2) If a square face is shared, then the electron count is the sum of the electron counts of the isolated polyhedra minus 64, this being the number of electrons associated with a square cluster with single metal-metal bonds.

While these generalizations hold for a large number of examples, they have to be used with caution. In a condensed cluster, where there is a shared triangular face, it is necessary to partition one of the fragments as a *hypho* cluster, that is, a deltahedron with three adjacent vertices missing, and make the assumption that this fragment is capable of donating 14 electrons. The departures from spherical symmetry are becoming so gross in this situation that the model begins to break down.

To take a simple main group example, which has been analyzed previously, consider the octahedral cluster $B_6H_6^{2-}$. If three adjacent vertices are lost from the octahedron to generate a *hypho*-B_3H_3 fragment, then an extrapolation of the *closo, nido, arachno* generalizations suggests that the triangular fragment should be associated with $4n + 8$ valence electrons; i.e., it would have the formula $B_3H_3^{8-}$ and would be capable of donating 14 electrons to the complete polyhedron. However, the triangular B_3H_3 fragment is clearly related to the more familiar cyclopropenyl systems which were discussed in Chapter 3. The aromatic $C_3H_3^+$ molecule has $4n + 2$ electrons and is capable of donating a total of eight electrons to a second fragment, assuming that the C—C σ and π molecular orbitals overlap sufficiently well with the acceptor orbitals. The $C_3H_3^{3-}$ anion has four additional electrons in the π antibonding set and a total of $4n + 6$ electrons. The maximum number of electrons which can be donated by this fragment is therefore 12. If the *hypho* approximation is valid, then the following partitioning holds:

closo fragment	*hypho* fragment	Electrons donated
$14n + 2$	$14(m - 3) + 8$	-12

Total electron count $= (14n + 2) + (14m + 2) - 48$

If the *hypho* approximation is not valid and there is one fewer accessible orbital in the *hypho* fragment, then we have:

closo fragment	hypho fragment	Electrons donated
$14n + 2$	$14(m - 3) + 6$	-12

Total electron count $= (14n + 2) + (14m + 2) - 50$

It can be appreciated that the *closo*, *nido*, *arachno* pattern is beginning to break down because we are deviating so far from the parent cluster that there is little memory of the *closo* electron count left.

If the mode of analysis developed above is followed through to its logical conclusion for the generation of condensed clusters by triangular face sharing, then two possible electron counts for the shared face must be considered:

(1) 48, which is the electron count associated with a triangular metal cluster such as $Os_3(CO)_{12}$
(2) 50, which corresponds to the electron count for a triangular cluster with an additional electron pair occupying an a'_2 antibonding molecular orbital (i.e., $\bar{P}^{\pi}_0$ as discussed in Section 3.3.)

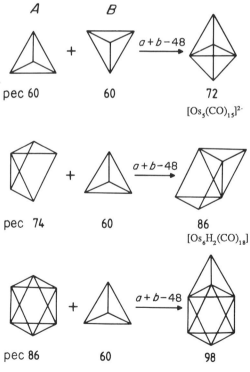

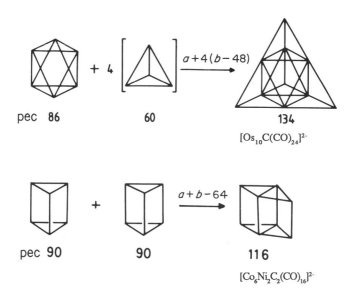

Figure 7.9 Examples of face-sharing clusters where the common face is associated with an electron count of 48 (triangular face) or 64 (square face).

The former electron count is the more usual, and some of the many examples of its application are illustrated in Figure 7.9. The alternative possibility usually arises only for columnar clusters based on the octahedron, some examples of which are illustrated in Figure 7.10. A similar duality occurs for species which share square faces; the corresponding condensation rules are $A + B - 62$ (Figure 7.10) and $A + B - 64$ (Figure 7.9), the latter pertaining to a square-planar transition metal cluster and the former to an organometallic analogue of $C_4H_4^{2-}$

Face sharing often produces tetrahedra or square pyramids and results in capping of the original cluster skeleton. This condensation mode leads to an increment in the total valence-electron count of the cluster of 12; the electronic basis of this result is discussed further in Section 7.5.

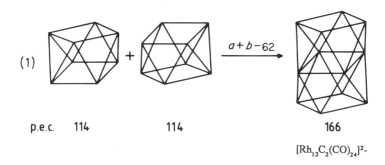

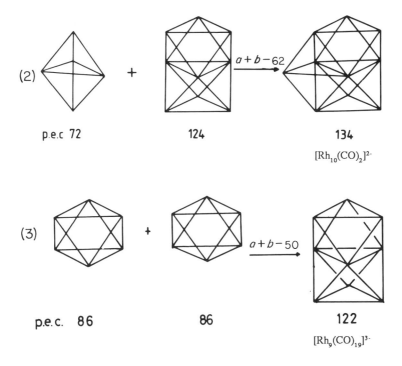

Figure 7.10 Examples of face-sharing clusters where the common face is associated with an electron count of 50 (triangular face) or 62 (square face).

7.5 THE CAPPING PRINCIPLE

Many of the condensed clusters described above can also be viewed as capped and edge-bridged structures, and the process may be repeated to build up a second sphere of atoms. Such clusters are more accurately described as "bispherical." The solution of the particle on-a-sphere problem does not allow for the description of the bonding in such clusters directly, but the application of group theoretical and perturbation theory ideas to the problem can lead to some interesting insights and account for some apparent exceptions to the condensation rules.[8-10]

The starting point of the analysis is a main group pseudospherical deltahedral cluster, $M_n H_n$, with S_{in}^σ, $P_{in}^{\pi/\sigma}$, and $(n-3)\ L_{in}^\pi$ $(L > 1)$ bonding skeletal molecular orbitals and $(n-1)\ L_{in}^\sigma$ $(L \neq 0)$ and $n\ \bar{L}_{in}^\pi$ inaccessible molecular orbitals. In this case the subscript "in" identifies the orbitals of the inner cluster skeleton. m capping atoms lie on a second and larger sphere, where the vertices are well separated from each other because $m < n$. This means that the overlap integrals between orbitals on different capping atoms are much smaller than between atoms in the inner sphere. The capping atoms are located over the faces and/or edges of the inner cluster, and their orbitals interact most strongly with the $2n + 1$ bonding skeletal molecular orbitals of this framework. Their interactions with the $2n - 1$ inaccessible orbitals are relatively weak, because the latter are inwardly hybridized and high in energy. The in-pointing radial σ valence hybrids on the caps produce m L_{out}^{σ} orbitals, e.g., S_{out}^{σ} and P_{out}^{σ} when $m = 4$, which will mix with the S_{in}^{σ} and $P_{in}^{\sigma/\pi}$ skeletal bonding orbitals of the inner shell of atoms. The subscript "out" is used to denote linear combinations of orbitals on atoms of the outer shell. These interactions can be described by perturbation theory; the net result is the creation of m additional inaccessible orbitals from the out-of-phase combinations.

The tangential L_{out}^{π} and $\bar{L}_{out}^{\pi}$ linear combinations of p_x and p_y orbitals of the outer sphere atoms also interact primarily with the bonding skeletal molecular orbitals of the inner sphere. The L_{out}^{π} molecular orbitals generally find matching partners of the same symmetry in the latter set and mixing simply gives a further m inaccessible orbitals (see Figure 7.11). The net effect of the m $\bar{L}_{out}^{\pi}$ orbitals depends critically on the symmetry of the cluster, and the following limiting situations can be identified:

(1) If the irreducible representations spanned by the $\bar{L}_{out}^{\pi}$ orbitals are matched in number and type by the bonding skeletal orbitals (excluding those which we have already counted as matching the L_{out}^{π} set), then no additional bonding orbitals will be generated and m additional inaccessible orbitals result. This simply follows from the fact that we have counted all the bonding skeletal orbitals already; although mixing with orbitals of the outer shell produces more strongly bonding combinations, there are no new accessible orbitals. Mixing with $\bar{L}_{in}^{\pi}$ orbitals of the inner shell is not usually sufficient to produce another accessible orbital if there is a matching skeletal bonding orbital. All the capped structures in Figure 2.20 of Chapter 2 conform to this generalization.

(2) Alternatively, one or more members of the $\bar{L}_{out}^{\pi}$ set may not be matched by the skeletal molecular orbitals of the inner sphere, and an additional accessible orbital is then created. Most commonly this situation results from the absence of a bonding skeletal molecular orbital of the appropriate symmetry. Such orbitals may be stabilized by mixing with members of the $\bar{L}_{in}^{\pi}$ set. This stabilization will not generally be very large, because the overlap between the outer-shell and inner-shell orbitals (which point inward) is small. The numbers of accessible and inaccessible molecular orbitals in some capped clusters are summarized in Table 7.1.

The group theoretical aspects of the capping principle can be illustrated using two specific examples based on the trigonal prism. The symmetries of the nine skeletal bonding orbitals of this cluster are those contained in the representation spanned by the edges of the prism (Chapter 2): $S^{\sigma}(A_1')$,

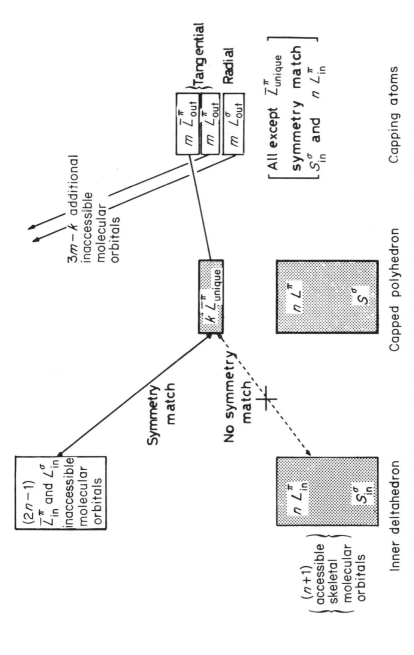

Figure 7.11 Qualitative energy-level diagram for the interaction between the orbitals of a deltahedral cluster and a surrounding shell of capping atoms.

TABLE 7.1 SUMMARY OF ACCESSIBLE AND INACCESSIBLE MOLECULAR ORBITALS IN CAPPED DELTAHEDRAL CLUSTERS

Molecular orbitals	Central polyhedron	No. of $\bar{L}^\pi_{out}$ not matched by L^π_{in}	m capped polyhedron
Accessible			
Main group	$2n + 1$	k	$(2n + 1) + m + k$
Transition metal	$7n + 1$	k	$(7n + 1) + 6m + k$
Skeletal	$n + 1$		$(n + 1 + k)$
Inaccessible			
Main group and transition metal	$2n - 1$		$(2n - 1) + 3m - k$
Total numbers of orbitals			
Main group	$4n$		$4n + 4m$
Transition metal	$9n$		$9n + 9m$

$P^{\sigma/\pi}(A_2'', E')$, $D^\pi(A_1', E'')$, and $\bar{D}^\pi(E')$. The introduction of two capping atoms on the triangular faces results in no change in point group (D_{3h}), and the frontier orbitals of the caps are:

L^σ_{out}: a_1' and a_2'' (radial)
L^π_{out}: e' (tangential)
$\bar{L}^\pi_{out}$: e'' (tangential)

All of these linear combinations find partners with the same symmetries in the set of bonding skeletal molecular orbitals, and therefore no additional accessible orbitals are generated. Hence the in-pointing hybrids of the caps all correlate formally with inaccessible orbitals. A main group trigonal prism capped in this fashion is therefore characterized by 9 occupied skeletal bonding molecular orbitals and 15 unoccupied skeletal orbitals. In a molecule such as $B_8H_8^{x-}$ with this geometry, all the skeletal bonding orbitals are filled when $x = 2$.

The tricapped trigonal prism, with the capping atoms on the square faces, also has D_{3h} point group symmetry and the in-pointing hybrids of the caps are now:

L^σ_{out}: a_1' and e' (radial)
L^π_{out}: a_2'' and e' (tangential)
$\bar{L}^\pi_{out}$: a_2' and e'' (tangential)

All the linear combinations except the a_2' orbital are matched by skeletal bonding orbitals of the prism, and consequently eight additional inaccessible orbitals result. The tricapped trigonal prism is therefore characterized by a total of nine skeletal bonding orbitals with a roughly nonbonding a_2' orbital (Figure 7.12). The latter is stabilized by mixing with an a_2' orbital belonging to the $\bar{L}_{in}^{\pi}$ set. The TSH pairing principle also suggests that there is a relatively low-lying a_2'' orbital resulting from antibonding combination of the L_{in}^{π} and L_{out}^{π} orbitals which are the partners of the above odd π orbitals.

Hence the following possibilities arise for the tricapped trigonal prism:[11]

[Trigonal prism orbitals] $(a_2')^0(a_2'')^0$ e.g., B_9Cl_9 with 9 occupied skeletal orbitals

[Trigonal prism orbitals] $(a_2')^2(a_2'')^0$ e.g., $B_9H_9^{2-}$ with 10 occupied skeletal orbitals

[Trigonal prism orbitals] $(a_2')^2(a_2'')^2$ e.g., Bi_9^{5+} with 11 skeletal molecular orbitals

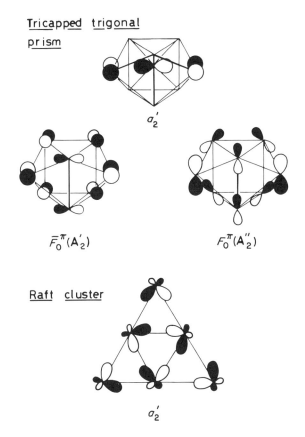

Figure 7.12 The a_2' orbital which lies in the frontier region in the tricapped trigonal prism and the D_{3h} edge-bridged triangle. The corresponding a_2'' orbital is generated by rotating each hybrid by 90°, all in the same sense.

Paramagnetic Transition Metal Carbonyl Clusters

Although the vast majority of metal carbonyl cluster compounds are diamagnetic, there are some examples of paramagnetic compounds. In addition, odd electron species can, at times, be generated in electrochemical experiments. Detailed esr experiments on such compounds can lead to the identification of the symmetry and localization of the partially occupied HOMO. Such data when combined with single-crystal structural studies on the parent cluster and related molecules with either 1 less or 1 more electron can also establish whether the HOMO is metal-metal bonding or antibonding. Strouse and Dahl's classic study* of $[Co_3(CO)_9S]$ and related triangular C_{3v} clusters with 48 valence electrons established that the additional electron resides in a molecular orbital which is cobalt-cobalt antibonding because the metal-metal bond lengths are 0.24 Å longer in the 49-electron cluster. Furthermore, a detailed analysis of the single-crystal esr spectrum of $[Co_3(CO)_9S]$ diluted in a diamagnetic host matrix of $[FeCo_2(CO)_9S]$ established that the unpaired electron resides in an orbital of A_2 symmetry which is localized on the cobalt atoms and results from a linear combination of cobalt d_{xz} orbitals.

Robinson and Vahrenkamp† have extended these esr measurements to isoelectronic triangular clusters with one, two, and three cobalt atoms (^{59}Co, $I = \frac{7}{2}$). The spectra shown in the figure have 8, 15, and 22 lines, respectively, resulting from hyperfine coupling to the cobalt atoms. In all cases the hyperfine coupling constant is approximately equal, suggesting comparable electron delocalization. In addition, no coupling to the capping ^{31}P nucleus is observed, confirming the localization of the molecular orbital in the metal triangular plane. The A_2 symmetry of the HOMO in these C_{3v} clusters ensures that there is no contribution from the s and p valence orbitals of the capping atoms, which transform as $2A_1 + E$.

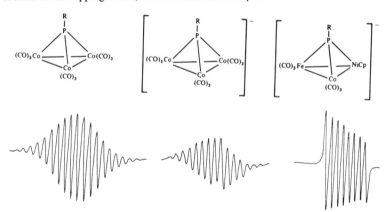

The LUMO in 48-electron triangular clusters such as $Os_3(CO)_{12}$ is closely related and has A'_2 symmetry. It is this molecular orbital which is stabilized by bridging metal atoms in the raft clusters, which are discussed in this chapter. This molecular orbital can also be stabilized by bridging π-acid and π-donor ligands and accounts for the occurrence of 42- and 44-electron triangular clusters of platinum, e.g., $[Pt_3(\mu_2\text{-}SO_2)_3(PCy_3)_3]$ and $[Pt_3(\mu_2\text{-}SO_2)_2(\mu_2\text{-}Cl)(PCy_3)_3]^-$.

*Discuss. Farad. Soc., **47**, 93 (1969).
†Reproduced with permission from Proc. Roy. Soc. A, **308**, 17 (1982).

Differences in the nodal characteristics of the frontier molecular orbitals account for differences in the bond lengths within the triangles of the trigonal prism and between the triangles. These structural details were given as Exercise 5.3 at the end of Chapter 5.

Edge bridging of triangular clusters with D_{3h} symmetry to generate raft clusters, which also have D_{3h} symmetry, generates a related "capping" problem. The three edge-bridging groups produce a linear combination of A'_2 symmetry which is not matched by any of the bonding skeletal molecular orbitals of the triangle, but only by an antibonding skeletal orbital. Therefore, raft clusters such as $Os_6(CO)_{21}$ can accommodate 90 electrons (48 for the triangle plus 3×14 for the edge bridges). The additional low-lying a'_2 orbital, illustrated in Figure 7.12, is not occupied in this cluster. Electrochemical studies on such species have indeed confirmed the presence of a low-lying molecular orbital,[12] although additional esrd experiments would be required to confirm its identity.

$[Fe_3Pt_3(CO)_{15}]^{2-}$ is a related cluster with 86 valence electrons in total. The parent triangular Pt_3 skeleton is characterized by 42 electrons in compounds such as $Pt_3(CO)_3(PCy_3)_3$ (Cy is cyclohexyl), and the presence of three edge bridges suggests a total electron count of $42 + 3 \times 14 = 84$. Hence it appears that the additional low-lying a'_2 orbital is occupied in this cluster. In this case, electrochemical oxidation is easily achieved to give an 85-electron raft cluster. The related raft cluster $[Cu_3Fe_3(CO)_{12}]^{3-}$ has 84 valence electrons, and in this case the a'_2 orbital is vacant.

In summary, whenever a pseudospherical cluster has a ring of edge-bridging atoms, there is a strong possibility that an additional molecular orbital (of A'_2 symmetry if there is a threefold axis) will be utilized to accommodate an additional electron pair.

In higher-symmetry bispherical clusters with more complete outer shells of capping atoms (rather than only toroids as above), degenerate sets of additional accessible molecular orbitals can be generated by the capping process. For example, in a hexacapped cube there is a combination of the tangential molecular orbitals of the capping atoms with T_{1g} symmetry which does not match the symmetries of any of the skeletal bonding orbitals of the cube. Consequently, the additional accessible molecular orbitals have T_{1g} symmetry and the hexacapped cube can be associated with either 12 (characteristic of the inner cube alone) or 15 skeletal electron pairs. $[Rh_{15}(CO)_{30}]^{3-}$ has such a structure and 15 skeletal electron pairs.

Some illustrative examples of the capping principle are given in Table 7.2 for transition metal cluster compounds. The rule works very well in general, and there are not many instances where additional skeletal molecular orbitals are occupied. The presence of d orbitals in such clusters can prevent the appearance of an additional skeletal bonding orbital because the outpointing d orbitals of the inner shell of atoms generally correlate with accessible orbitals. Mixing with these orbitals can therefore produce no additional bonding combinations.

TABLE 7.2 EXAMPLES OF BISPHERICAL CLUSTERS (CAPPED STRUCTURES)*

Description	Example	Ref.	N_i	N_o	N_t	Skeletal electron pairs	Extra electron pairs
Octahedron	$[Os_6(CO)_{18}]^{2-}$	a	6	0	6	7 $(N_i + 1)$	—
Capped octahedron	$[Os_7(CO)_{21}]$	b	6	1	7	7 $(N_i + 1 = N_t)$	0
Bicapped octahedron	$[Os_8(CO)_{22}]^{2-}$	c	6	2	8	7 $(N_i + 1)$	0
Tetracapped octahedron	$[Os_{10}C(CO)_{24}]^{2-}$	d	6	4	10	7 $(N_i + 1)$	0
	$[Pd_{10}(CO)_{12}(PBu_3)_6]$	e	6	4	10	8 $(N_i + 2)$	1
Hexacapped octahedron	$[Pd_6Fe_6(CO)_{23}H]^{3-}$		6	6	12	8 $(N_i + 2)$	1
Octahedron†	$[Cu_6(PPh_3)_6H_6]$	f	6	0	6	6 (N_i)	—
Tetracapped octahedron	$[Cu_6Fe_4(CO)_{16}]^{2-}$	g	6	4	10	6 (N_i)	0
	$[Hg_6Rh_4(PMe_3)_{12}]$	h	6	4	10	6 (N_i)	0
Cube	$[Ni_8(PPh)_6(CO)_8]$	i	8	0	8	12 $(3N_i/2)$	—
Pentacapped cube (centered)	$[Rh_{14}(CO)_{25}]^{4-}$	j	8	5	13	12 $(3N_i/2)$	0
	$[Rh_{14}(CO)_{26}]^{2-}$	k	8	5	13	12 $(3N_i/2)$	0
Rhombic dodecahedron/ hexacapped cube (centered)	$[Rh_{15}(CO)_{30}]^{3-}$		8	6	14	15 $[(3N_i/2) + 3 = N_t + 1]$	3
Pentagonal prism	$C_{10}H_{10}$	l	10	0	10	15 $(3N_i/2)$	—
Tetracapped pentagonal prism (centered) $(2\diamondsuit + 2\square)\ddagger$	$[Rh_{15}C_2(CO)_{28}]^-$	m	10	4	14	16 $[(3N_i/2) + 1]$	1
Square antiprism§	$[Co_8C(CO)_{18}]^{2-}$	n	8	0	8	9 $(N_i + 1)$ (closo)	—
Tetracapped square antiprism (4△)	$[Co_3Ni_9C(CO)_{20}]^{3-}$	o	8	4	12	10 $(N_i + 2)$	1

TABLE 7.2 (cont.)

Description	Example	Ref.	N_i	N_o	N_t	Skeletal electron pairs	Extra electron pairs
Square antiprism	$[Ni_8C(CO)_{16}]^{2-}$	o	8	0	8	11 $(N_i + 3)$ (arachno)	—
	$[Bi_8]^{2+}$	p	8	0	8	11 $(N_i + 3)$ (arachno)	—
Trigonal prism	C_6H_6	q	6	0	6	9 $(3N_i/2)$	—
	$[Rh_6C(CO)_{15}]^{2-}$	r	6	0	6	9 $(3N_i/2)$	
Capped (and edge-bridged) trigonal prism (1□)	$[Rh_8C(CO)_{19}]^{2-}$	s	6	1	7	9 $(3N_i/2)$	0
Bicapped trigonal prism (2△)	$[Cu_2Rh_6C(CO)_{15}(NCMe)_2]$	t	6	2	8	9 $(3N_i/2)$	0
Tricapped trigonal prism (3□)	$[B_9H_9]^{2-}$	u	6	3	9	10 $[(3N_i/2) + 1 = N_t + 1]$	1
	$[TlSn_8]^{3-}$	v					
	$[Bi_9]^{5+}$ ¶	w				11 $[(3N_i/2) + 2 = N_t + 2]$	2
Square-face-sharing fused trigonal prisms (A)	$[Co_6Ni_2C_2(CO)_{16}]^{2-}$	x	8	0	8	10	—
Bicapped A (2□)	$[Ni_{10}C_2(CO)_{16}]^{2-}$	y	8	2	10	11	1
Twinned cuboctahedron (centered)	$[Rh_{13}(CO)_{24}H_{5-8}]^{n-}$	z	12	0	12	13 $(N_i + 1)$	—
Cuboctahedron			12	0	12	13 $(N_i + 1)$	—
Distorted tetracapped cuboctahedron (4□)	$[Ni_{16}C_4(CO)_{23}]^{4-}$	aa	12	4	16	17 $(N_i + 5 = N_t + 1)$	4
Hexacapped (and edge-bridged) cuboctahedron (centered) (6□)	$[Pd_{23}(CO)_{22}(PEt_3)_{10}]$	bb	12	6	18	15 $(N_i + 3)$	2

TABLE 7.2 (cont.)

*N_i = number of inner-sphere atoms (central core); N_o = number of outer-sphere (capping) atoms (not counting edge-bridging groups which form essentially localized two-center two-electron bonds to the inner-sphere polyhedron); N_t = total number of cluster atoms (not counting central/interstitial atoms or edge-bridging atoms, which donate all their valence electrons to the cluster).

†Polyhedral clusters of the group 1B and 2B metals generally have one skeletal electron pair less than normal. Octahedral clusters of this type are therefore characterized by 84 cluster valence electrons. Another example of a bispherical cluster based on an 84-electron octahedron is [Ag$_6${Fe(CO)$_4$}$_3${(Ph$_2$P)$_3$CH}], which has three faces capped by Fe(CO)$_4$ groups and the other capped by a tripodal phosphine ligand (Briant, C. E., Smith, R. G., and Mingos, D. M. P., J. Chem. Soc., Chem. Comm., 586 (1984)).

‡The numbers in parentheses refer to the number and type of face which is capped, in those cases where the inner-sphere polyhedron possesses more than one type of face.

§The square antiprism may be regarded as either a four-connected polyhedron (in which case it should possess $N + 1 = 9$ sep's as for a *closo* cluster) or an *arachno* deltahedron (characterized by $N + 3 = 11$ sep's).

¶[For the [Bi$_9$]$^{5+}$ cluster both $3F_0^\pi$ and $3P_0^\pi$ are occupied ($3P_0^\pi$ is a fairly low-lying LUMO for [B$_9$H$_9$]$^{2-}$); Wade, K., and O'Neill, M. E., *Polyhedron*, **2**, 963 (1983).

[a] McPartlin, M., Eady, C. R., Johnson, B. F. G., and Lewis, J., *J. Chem. Soc., Chem. Comm.*, 883 (1976).

[b] Eady, C. R., Johnson, B. F. G., Lewis, J., Mason, R., Hitchcock, P. B., and Thomas, K. M., *J. Chem. Soc., Chem. Comm.*, 385 (1977).

[c] Jackson, P. F., Johnson, B. F. G., Lewis, J., and Raithby, P. R., *J. Chem. Soc., Chem. Comm.*, 60 (1980).

[d] Jackson, P. F., Johnson, B. F. G., Lewis, J., Nelson, W. J. H., and McPartlin, M., *J. Chem. Soc., Dalton Trans.*, 2099 (1982).

[e] Mednikov, E. G., Eremenko, N. K., Gubin, S. P., Slovokhotov, Yu. L., and Struchkov, Yu. T., *J. Organomet. Chem.*, **239**, 401 (1982).

[f] Churchill, M. R., Bezman, S. A., Osborn, J. A., and Wormald, J., *Inorg. Chem.*, **11**, 1818 (1972).

[g] Doyle, G., Heaton, B. T., and Ochiello, E., *Organometallics*, **4**, 1224 (1985).

[h] Jones, R. A., Real, F. M., Wilkinson, G., Galas, A. M. R., and Hursthouse, M. B., *J. Chem. Soc., Dalton Trans.*, 126 (1981).

TABLE 7.2 (*cont.*)

[i]Lower, L. D., and Dahl, L. F., *J. Amer. Chem. Soc.*, **98**, 5046 (1976).

[j]Ciani, G., Sironi, A., and Martinengo, S., *J. Organomet. Chem.*, **192**, C42 (1980).

[k]Martinengo, S., Ciani, G., and Sironi, A., *J. Chem. Soc., Chem. Comm.*, 1140 (1980).

[l]Eaton, P. E., Or, Y. S., and Branca, S. J., *J. Amer. Chem. Soc.*, **103**, 2134 (1981).

[m]Albano, V. G., Sansoni, M., Chini, P., Martinengo, S., and Strumulo, D., *J. Chem. Soc., Dalton Trans.*, 970 (1976).

[n]Albano, V. G., G, Chini, P., Ciani, G., Martinengo, S., and Sansoni, M., *J. Chem. Soc., Dalton Trans.*, 463 (1978).

[o]Longoni, G., Ceriotti, A., Della Pergola, R., Manassero, M., Perego, M., Piro, G., and Sansoni, M., *Philos. Trans. Roy. Soc. London, A*, **308**, 47 (1982).

[p]Corbett, J. D., *Inorg. Chem.*, **7**, 198 (1968); Krebs, B., Hucke, M., and Brendel, C. J., *Angew. Chem., Int. Ed. Engl.*, **21**, 445 (1982).

[q]Katz, T. J., and Acton, N., *J. Amer. Chem. Soc.*, **95**, 2738 (1973).

[r]Albano, V. G., Braga, D., and Martinengo, S., *J. Chem. Soc., Dalton Trans.*, 717 (1981).

[s]Raithby, P. R., in Johnson, B. F. G. (ed.), *Transition Metal Clusters*, Wiley, New York, 1980, chap. 2, p. 63.

[t]Albano, V. G., Braga, D., Martinengo, S., Chini, P., Sansoni, M., and Strumulo, D., *J. Chem. Soc., Dalton Trans.*, 52 (1980).

[u]Klanberg, F., and Muetterties, E. L., *Inorg. Chem.*, **5**, 1955 (1966).

[v]Burns, R. C., and Corbett, J. D., *J. Amer. Chem. Soc.*, **104**, 2804 (1982).

[w]Herschaft, A., and Corbett, J. D., *Inorg. Chem.*, **2**, 979 (1963); Friedman, R. M., and Corbett, J. D., *Inorg. Chim. Acta*, **7**, 525 (1973).

[x]Arrigoni, A., Ceriotti, A., Della Pergola, R., Longoni, G., Manassero, M., Masciocchi, N., and Sansoni, M., *Angew. Chem., Int. Ed. Engl.*, **23**, 322 (1984).

[y]Ceriotti, A., Longoni, G., Manassero, M., Masciocchi, N., Resconi, L., and Sansoni, M., *J. Chem. Soc., Chem. Comm.*, 181 (1985).

[z]Ciani, G., Sironi, A., and Martinengo, S., *J. Chem. Soc., Dalton Trans.*, 519 (1981).

[aa]Ceriotti, A., Longoni, G., Manassero, M., Masciocchi, N., Piro, G., Resconi, L., and Sansoni, M., *J. Chem. Soc., Chem. Comm.*, 1402 (1985).

[bb]Mednikov, E. G., Eremenko, N. K., Slovokhotov, Yu. L., and Struchkov, Yu. T., *J. Organomet. Chem.*, **301**, C35 (1986).

7.6 HIGH-NUCLEARITY SPHERICAL CLUSTERS[13–15]

The discussion developed above has shown that there are simple relationships between the electron counts of capped clusters and those of the parent constituents. When the degree of capping is extensive, the structures are no longer easily related to simple cluster skeletons; it then becomes more convenient to describe them in terms of those close-packed arrangements which are characteristic either of bulk metals or metal crystallites. Extensive condensation of tetrahedral and octahedral clusters leads to packing arrangements related to hexagonal close-packed (hcp) and cubic close-packed (ccp) systems. In addition, arrangements with fivefold symmetry are possible for metal particles, although they cannot form the basis of infinite structures. Furthermore, as the cluster size increases, the energy differences between alternative close-packed arrangements become smaller and it is no longer possible to associate a particular electron count with a specific geometry.

The observed close-packed arrangements of clusters with 13 to 44 metal atoms are summarized in Table 7.3, and it is apparent that there are representative examples of all the important packing modes. These clusters have either a single atom or a group of atoms at the center and the remaining atoms form successive concentric shells around the single atom or groups (Figure 7.13). The molecular orbital pattern for each shell can be defined individually, and then the interactions between the layers may be evaluated using a perturbation theory approach analogous to that developed above for capped structures. In capped structures, each cap results in an increment in the total electron count of 12 unless there are additional molecular orbitals which are symmetry-isolated.

In gold clusters, the L^π orbitals of the peripheral atoms are very high-lying in energy and the bonding interactions are dominated by the radial orbitals of the individual $Au(PR_3)$ fragments. The resultant L^σ cluster orbitals in order of increasing energy are $S^\sigma, P^\sigma, D^\sigma, F^\sigma$, etc. In a perturbational analysis the cluster molecular orbitals associated with S^σ, P^σ, and D^σ mix with the $5d, 6s,$ and $6p$ valence orbitals of the central gold atom and generate nine low-lying molecular orbitals. All nine of these orbitals are usually filled in centered spherical gold cluster compounds; hence, the clusters $Au[Au(PR_3)]_n{}^{x+}$ are characterized by $12n + 18$ valence electrons. The $12n$ component corresponds to the filled d orbitals and radial metal-phosphine bond for each vertex of the outer shell of atoms.

Let us now consider a hypothetical situation where we have n_s $Au(PR_3)$ fragments surrounding an interstitial cluster fragment in a spherical fashion (the subscript s stands for "surface"). The peripheral gold fragments would generate a set of $S^\sigma, P^\sigma, D^\sigma, F^\sigma$, etc., molecular orbitals in the same way as that described above for a cluster with a single interstitial atom. However, the energy spread of these cluster orbitals is likely to be smaller because the gold atoms on average will be separated by larger internuclear distances. If the

TABLE 7.3 EXAMPLES OF HIGH-NUCLEARITY CLUSTER COMPOUNDS WHERE RADIAL BONDING INTERACTIONS PREDOMINATE

Compound	Reference	n_i	n_s	Structure	Electron count Obs.	Electron count Calc.	
$[Au_9(PPh_3)_8]^+$	a	1	8	bcc	114	114	
$[Au_{11}I_3(PPh_3)_7]$	b	1	10	bcc/icp	138	138	$12n_s + 18$
$[Au_{13}Cl_2(PMePh_2)_{10}]^{3+}$	c	1	12	icp	162	162	
$[Pt_{19}(CO)_{22}]^{4-}$	d	2	17	icp	238	238	$12n_s + 34$*
$[Rh_{22}(CO)_{35}H_{5-q-m}]^{q-}$	e	2	20	fcc/bcc	$273 + m$	274	
$[Au_{13}Ag_{12}Cl_6(PPh_3)_{12}]^{m+}$	f	3	22	icp	$317 - m$	314	$12n_s + 50$*
$[Pt_{26}(CO)_{32}]^{2-}$	g	3	23	hcp	326	324 (326)	$12n_s + 48$ (50)†
$[Ni_{38}Pt_6(CO)_{48}H_{6-n}]^{n-}$	h	6	38	fcc	542	542	
$[Pt_{38}(CO)_{44}H_m]^{2-}$	g	6	32	fcc	$470 + m$	470	$12n_s + 86$‡

*Linear M_3 interstitial moiety.
†Triangular M_3 interstitial moiety. Although isolated M_3 clusters are characterized by 48 valence electrons, the presence of bridging metal atoms can lead to the stabilization of an a'_2 molecular orbital and a valence electron count of 50 (Evans, D. G., and Mingos, D. M. P., *Organometallics*, **2**, 435 (1983)).
‡Octahedral M_6 moiety.

[a] van der Linden, J. G. M., Paulissen, M. L. H., and Schmitz, J. E. J., *J. Amer. Chem. Soc.*, **105**, 1903 (1983).
[b] Smits, J. M. M., Beurskens, P. T., van der Velden, J. W. A., and Bau, J. J., *J. Cryst. Spectrosc. Res.*, **13**, 373 (1983).
[c] Briant, C. E., Theobald, B. R. C., White, J. W., Bell, L. K., and Mingos, D. M. P., *J. Chem. Soc., Chem. Comm.*, 201 (1981).
[d] Washecheck, D. M., Wucherer, E. J., Dahl, L. F., Ceriotti, A., Longoni, G., Manassero, M., Sansoni, M., and Chini, P., *J. Amer. Chem. Soc.*, **101**, 6110 (1979).
[e] Vidal, J. L., Schoening, R. C., and Troup, J. M., *Inorg. Chem.*, **20**, 227 (1989).
[f] Teo, B. K., and Keating, K., *J. Amer. Chem. Soc.*, **106**, 2224 (1984).
[g] Ceriotti, A., Chini, P., Longoni, G., Washecheck, D. M., Murphy, M. A., Nagaki, D. A., Montag, R. A., and Dahl, L. F., personal communication.
[h] Ceriotti, A., Demartin, F., Longoni, G., Manassero, M., Marchionna, M., Piva, G., and Sansoni, M., *Angew. Chem.*, in press.

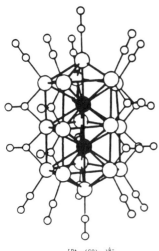

$[Pt_{19}(CO)_{22}]^{4-}$
238 valence electrons

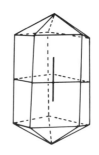

(a)

17 surface atoms (12 x 17)	204 electrons
Diatomic bonded interstitial moiety	34 electrons
Total	238 electrons

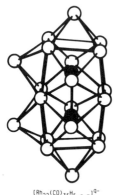

$[Rh_{22}(CO)_{35}H_{5-q-m}]^{q-}$
273 + m valence electrons

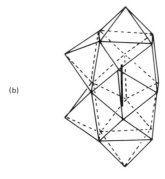

(b)

20 surface atoms (20 x 12)	240 electrons
Diatomic interstitial moiety	34 electrons
Total	274 electrons

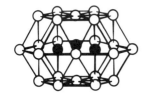

$[Pt_{26}(CO)_{32}]^{2-}$
326 valence electrons

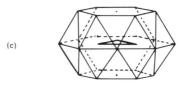

(c)

23 surface atoms (23 x 12)	276 electrons
Triangular interstitial moiety	48 electrons
Total	324 electrons

N.b. triangular $[Os_3(CO)_{12}]$ has 48 valence electrons

But edge bridging of clusters leads to the stabilisation of an extra molecular orbital of a_2' symmetry, $[Os_6(CO)_{15}(P(OMe)_3)_6]$ undergoes a reversible 2 electron reduction

Continued on p. 277

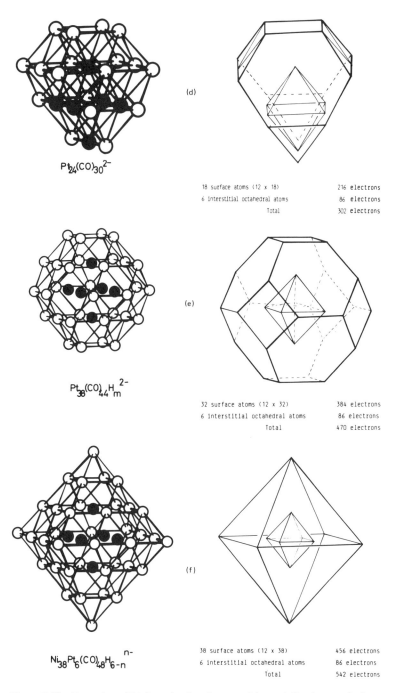

Figure 7.13 Examples of high-nuclearity clusters with partially close-packed arrangements of atoms. The figures on the right emphasize the outer polyhedron and the interstitial moiety. The interstitial atoms are shaded.

Sec. 7.6   High-Nuclearity Spherical Clusters

central cluster moiety is an m-vertex deltahedron, it will generate a set of $7m + 1$ cluster orbitals which are either metal-metal bonding or capable of overlapping with molecular orbitals on the outer sphere of metal atoms. If this were an isolated deltahedral cluster, then these molecular orbitals would act as acceptors from ligand-based orbitals, but in a spherical cluster it is the outer sphere of atoms which emulate the ligand sphere. The central deltahedral fragment also has $2m - 1$ inaccessible molecular orbitals, which do not interact effectively with an outer shell of atoms because they are either inwardly hybridized or too high-lying in energy.[16]

It is apparent that such a cluster will be characterized by a total of $12n_s + (14m + 2)$ electrons (Figure 7.14). Expressed in more general terms, the number of valence electrons associated with a multispherical cluster where radial bonding predominates is $12n_s + \Delta_i$, where Δ_i is the number of valence electrons associated with the interstitial moiety. The same arguments apply to other multispherical clusters where the tangential bonding interactions are weak. The high-nuclearity clusters in Table 7.3 all fit in with this generalization. When there is a single transition metal atom at the center of the cluster, the characteristic electron count for the interstitial moiety is 18. When two atoms are present, the characteristic interstitial electron count is 34, corresponding to a dimer such as $Mn_2(CO)_{10}$. For three metal atoms, the naive interpretation of the generalization would suggest an interstitial electron count of 48 corresponding to a metal triangular cluster such as $Os_3(CO)_{12}$. However, it was apparent from the capping principle arguments developed above that when a triangular cluster is bridged by metal atoms an additional a'_2 orbital is stabilized and becomes accessible. Therefore, such clusters are characterized by $12n_s + 50$ electrons; e.g., $[Pt_{26}(CO)_{32}]^{2-}$. Table 7.3 also gives some examples of high-nuclearity clusters with interstitial octahedral moieties which are characterized by $12n_s + 86$ valence electrons; their structures are illustrated in Figure 7.13.

The example developed above assumed that the interactions between the surface atoms were rather weak and that the L^π_{out} cluster orbitals were too high in energy to be important in bonding. At the other extreme we could imagine a situation where there are a relatively large number of metal atoms on the surface which interact very strongly to generate not only L^σ but also low-lying L^π bonding molecular orbitals. In such a situation the outer sphere of atoms would be associated with $7n_s + 1$ accessible orbitals and $2n_s - 1$ inaccessible orbitals. In such a situation the interaction between the outer shell of atoms and the interstitial moiety could lead to a cluster where the electron count is dominated by the requirement of the outer sphere. The preferred electron count would then be $14n_s + 2$ for a deltahedral outer shell. There are numerous examples of such clusters with a single interstitial atom; some are given in Table 7.4a.

An intermediate situation could also pertain when the outer sphere of atoms results in some components of the L^π set being low enough in energy

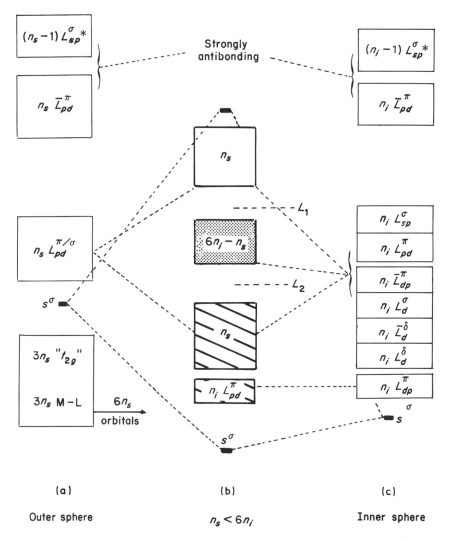

Figure 7.14 Qualitative energy-level diagram for the interaction between the orbitals of a deltahedral cluster enclosed by a shell of metal atoms. Filling of the molecular orbitals up to level L_1 corresponds to a total of $6n_s + 7n_i + 1$ accessible orbitals, i.e., $12n_s + \Delta_i$ electrons, where Δ_i is characteristic of the inner deltahedron.

to be occupied, but not all of them. In Table 7.4b a series of rhodium clusters which have a single interstitial atom and provide examples of this behavior are summarized. Their structures are based on body centered cubic, face centered cubic, and hexagonal close packing, and they share a common valence-electron count of $12n_s + 24$. This strongly suggests that the S^σ, P^σ, and D^σ cluster orbitals associated with the outer shell correlate with occupied orbitals, as for

Sec. 7.6   High-Nuclearity Spherical Clusters

TABLE 7.4a EXAMPLES OF HIGH-NUCLEARITY CLUSTERS WHERE RADIAL AND TANGENTIAL EFFECTS ARE IMPORTANT.*

Compound	Reference	n_i	n_s	Structure	Electron count Obs.	Calc.	
$[Rh_{13}(CO)_{24}H_5]$	a	1	12	hcp anticuboctahedron	170	170	
$[Rh_{15}(CO)_{30}]^{3-}$	b	1	14	bcc deltahedron	198	198	$12n_s + (2n_s + 2)$
With nonmetal interstitial atoms							
$[Ru_6C(CO)_{17}]$		1	6	Octahedron	86	86	
$[Rh_{10}S(CO)_{22}]^{2-}$		1	10	Bicapped square antiprism	142	142	
$[Rh_{12}Sb(CO)_{27}]^{3-}$		1	12	Icosahedron	170	170	

*The clusters summarized in this table have either deltahedral or four-connected polyhedral geometries. *nido* and *arachno* clusters derived from them have electron counts of $14n_s + 4$ and $14n_s + 6$, respectively. Condensed clusters derived from these polyhedra have electron counts governed by the condensation rules given in refs. 1 and 5.

TABLE 7.4b EXAMPLES OF HIGH NUCLEARITY CLUSTERS WHERE THE TANGENTIAL BONDING INTERACTIONS MAKE A PARTIAL CONTRIBUTION

Compound	Reference	n_i	n_s	Structure	Electron count Obs.	Electron count Calc.	
$[Rh_{14}(CO)_{25}]^{4-}$	c	1	13	bcc	180	180	
$[Rh_{14}(CO)_{26}]^{2-}$	d	1	13	bcc	180	180	
$[Rh_{15}(CO)_{27}]^{3-}$	e	1	14	bcc/hcp	192	192	$12n_s + 24$
$[Rh_{17}(CO)_{30}]^{3-}$	f	1	16	hcp	216	216	
$[Rh_{22}(CO)_{37}]^{4-}$	g	1	21	fcc/hcp	276	276	

[a] Albano, V. G., Ceriotti, A., Chini, P., Martinengo, S., and Ankler, W. M., *J. Chem. Soc., Chem. Comm.*, 859 (1975).
[b] Vidal, J. L., Kapieak, L. A., and Troup, J. M., *J. Organomet. Chem.*, **215**, C11 (1981).
[c] Ciani, G., Sironi, A., and Martinengo, S., *J. Organomet. Chem.*, **192**, C42 (1980).
[d] Martinengo, S., Ciani, G., and Sironi, A., *J. Chem. Soc., Chem. Comm.*, 1140 (1980).
[e] Martinengo, S., Ciani, G., Sironi, A., and Chini, P., *J. Amer. Chem. Soc.*, **100**, 7096 (1978).
[f] Ciani, G., Magni, A., Sironi, A., and Martinengo, S., *J. Chem. Soc., Chem. Comm.*, 1280 (1981).
[g] Martinengo, S., Ciani, G., and Sironi, S., *J. Amer. Chem. Soc.*, **102**, 7564 (1980).

the gold clusters above. Three additional L^π (presumably F^π) cluster orbitals resulting from tangential rhodium-rhodium interactions are also occupied.

EXERCISES

7.1 Consider a vertex-sharing deltahedral main group cluster with $n + m - 1$ vertices which may be partitioned into an n-vertex *closo* deltahedron and an $m - 1$ vertex *nido* deltahedron. Using this viewpoint, show that the total valence-electron count of the condensed cluster is $(4n + 2) + (4m + 2) - 4$ (note that the *closo* fragment has only one suitable accepter orbital, not three). (The partitioning in terms of *nido* deltahedron plus *nido* deltahedron plus single atom is not as convenient here, because the single atom cannot match all six acceptor orbitals of the *nido* fragments, unlike a transition metal atom. An analysis in terms of inaccessible orbitals of the *nido* fragments is also not so straightforward in this case.)

7.2 Derive the cluster orbitals generated by the three in-pointing hybrids per capping vertex (one σ-type and two π-type) in the tricapped trigonal prism (square faces capped) and the edge-bridged triangle. Show that in each case an a_2' orbital is contained in the $\bar{L}^\pi$ set which does not mix with any of the skeletal bonding orbitals of the capped or bridged species. How do the $\bar{L}^\pi$ orbitals of the inner shell of atoms affect the resulting pattern of energy levels?

7.3 Perform the same analysis as in Exercise 7.2 for the hexacapped cube. How many additional accessible orbitals may arise in this case? What are the most likely electron counts for this structure?

7.4 Throughout this chapter there is a pervading theme, namely, that when atoms on two different spherical shells interact, the accessible skeletal orbitals which result are those generated by the largest set of atoms. Explain how this principle arises from a simple perturbational argument, and show how it applies to (a) capping by isolated vertices, (b) capping by rings of atoms, and (c) bispherical clusters where the number of atoms in the outer shell is either greater than or less than the number in the inner shell. How do "extra" accessible orbitals arise in these cases?

7.5 Chihari, T., et al. (*J. Chem. Soc., Chem. Commun.*, 886 (1988)) have reported the synthesis and structural characterization of $[Ru_8H_2(CO)_{21}]^{2-}$ from $Ru_4H_4(CO)_{12}$ and sodium. The structure has been described as an octahedron fused to a trigonal bipyramid through a triangular face. Is this consistent with the condensation rule? At $-20°C$, the ^1H nmr spectrum shows resonances at $\delta = 6.88$ and -12.87 ppm with relative intensities 1:1. What does this suggest about the locations of the hydrido ligands?

7.6 Lewis, J., et al. (*J. Chem. Soc., Chem. Comm.*, 1358 (1988)) have described the characterization of $Os_6(CO)_{18}(O_2CCF_3)_2$, and they have described the skeletal geometry as a ladder of triangles. How does this structure relate to that described in this chapter for the "raft" cluster $Os_6(CO)_{21}$? Is the condensation rule equally applicable to both clusters? (CF_3CO_2 is a three-electron bridging ligand.)

7.7 Examples of condensed clusters with transition metal and main group vertex atoms are illustrated on page 283. Confirm that they obey the condensation rule,

Rule 6 of Chapter 2. (*Hint*: Calculate the electron count as if all the atoms were transition metal atoms and then use Rule 3 to introduce the main group atoms.) In this Figure, pec stands for "polyhedral electron count."

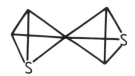

M_5A_2: pec = 82;
e.g., $Os_5S_2(CO)_4H_2$

M_7A_2: pec = 108;
e.g., $Os_7S_2(CO)_{20}$

M_4A_2: pec = 66;
e.g., $Mo_2Fe_2S_2(Cp)_2(CO)_8$

M_3A_4: pec = 58;
e.g., $Co_3(Cp)_3B_4H_4$

M_3A_4: pec = 60;
e.g., $Fe_3(CO)_8(C_2Ph_2)_2$

M_4A_3: pec = 66;
e.g., $Mo_2Co_2S_3(Cp)_2(CO)_4$

M_3A_2 p.e.c 54;
e.g., $Mn_3(CO)_{10}B_2H_7$

REFERENCES

1. Mingos, D. M. P., *J. Chem. Soc., Chem. Comm.*, 206 (1983).
2. McPartlin, M., and Mingos, D. M. P., *Polyhedron*, **3**, 1321 (1984).
3. Mingos, D. M. P., *Acc. Chem. Res.*, **17**, 311 (1984).
4. Callahan, K. P., and Hawthorne, M. F., *Adv. Organomet. Chem.*, **14**, 145 (1978).
5. Forsyth, M. I., and Mingos, D. M. P., *J. Organometallic Chem.*, **C37**, 146 (1978).
6. Forsyth, M. I., Mingos, D. M. P., and Welch, A. J., *J. Chem. Soc., Dalton Trans.*, 1363 (1978).
7. Calhorda, M. J., Mingos, D. M. P., and Welch, A. J., *J. Organometallic Chem.*, **309**, 228 (1982).
8. Mingos, D. M. P., and Forsyth, M. I., *J. Chem. Soc., Dalton Trans.*, 610 (1977).
9. Thomas, K. M., Mason, R., and Mingos, D. M. P., *J. Amer. Chem. Soc.*, **95**, 3802 (1973).
10. Johnston, R. L., and Mingos, D. M. P., *J. Aer. Chem. Soc., Dalton Trans.*, 1445 (1985).
11. Wade, K., and O'Neill, M. E., *Polyhedron*, **2**, 963 (1983).
12. Evans, D. G., and Mingos, D. M. P, *Organometallics*, **2**, 435 (1983).
13. Mingos, D. M. P., *J. Chem. Soc., Chem. Comm.*, 1352 (1985).
14. Mingos, D. M. P., *Chem. Soc. Rev.*, **15**, 31 (1986).
15. Mingos, D. M. P., and Zhenyang, L., *J. Chem. Soc., Dalton Trans.*, 1657 (1988).
16. Mingos, D. M. P., and Zhenyang, L., *J. Organometallic Chem.*, **341**, 523 (1988).

8

Clusters Where δ Orbitals Must Be Considered Explicitly

8.1 INTRODUCTION

The bonding theories developed for transition metal clusters in Chapters 5 and 7 generally assumed that the molecular orbitals derived from the δ orbitals were completely occupied. While this is usually the case for clusters with π-acceptor ligands (such as carbon monoxide), there are some other important classes of cluster with π-*donor* ligands. Chlorine, bromine, sulfur, and alkoxy (OR, where R = alkyl group) are ligands of this type which stabilize clusters of the earlier transition metals such as niobium, tantalum, molybdenum, and tungsten. These metals are more electropositive than the "later" transition metals, considered in the previous chapters, and form clusters where the metal atoms are in higher formal oxidation states. π-donor ligands can help to stabilize such clusters. Although π-donor and π-acceptor ligands have complementary electronic functions in clusters of early and late transition metals, there are a number of similarities between them. For example, both types of ligand form bridges across the edges of metal cluster skeletons, and although fluxionality is not common in halide clusters, it has been observed in alkoxy clusters.

The d orbitals of the earlier transition metals overlap more strongly because they are less contracted and consequently the separations between bonding and antibonding molecular orbitals are larger. Furthermore, the absence of π-acceptor ligands means that the antibonding metal-metal molecular orbitals are not stabilized by metal-ligand backbonding effects.

8.2 OCTAHEDRAL CLUSTERS

Many of the clusters with π-acceptor ligands have octahedral metal atom geometries. These octahedral clusters may in turn be divided into two classes: those with edge-bridging ligands and those with face-bridging ligands. $[M_6(\mu_2\text{-Cl})_{12}L_6]^{2+}$ (M = Nb or Ta) are examples of edge-bridged octahedra, while $[M_6(\mu_3\text{-Cl})_8L_6]^{4+}$ (M = Mo or W) are face-bridged octahedra (Figure 8.1). In each case, the six additional donor ligands, L, are terminally bound and define the vertices of a larger octahedron. These ligands can be lone pairs from halides on adjacent clusters leading to three-dimensional linked arrangements in the solid state. An example of a face-bridged molybdenum cluster with alkoxy ligands is provided by $Mo_6(\mu_3\text{-OMe})_8(OMe)_6^{2-}$.[1] Discrete ions of both types are also known; for example, $(NH_4)_2[Mo_6Cl_{14}]\cdot H_2O$ and $K_4[Nb_6Cl_{18}]$, which have 8 face-bridging and 12 edge-bridging ligands, respectively, in addition to the 6 terminally bound chlorine atoms.[2-3]

The bonding in both types of cluster has been described in terms of localized two-center or three-center bonds in Chapter 2.[4] The delocalized molecular orbital model developed below provides additional insights into the structural and electronic properties of these clusters.[5-9] In the localized description, five hybrids at each metal atom are directed toward the five ligands, which are arranged in a square-pyramidal fashion. The remaining four hybrids per vertex point into the faces or along the edges of the octahedron when the bridging atoms lie along the edges or in the faces, respectively. If we include the five metal-ligand σ bonds for each vertex, then there are 38 and 42 strongly bonding orbitals for the edge-bridged and face-bridged clusters, respectively; i.e., 76 and 84 valence electrons are associated with these clusters.

These electron counts are both lower than the $14n + 2 = 86$ valence electrons associated with octahedral transition metal cluster carbonyls. Furthermore, clusters with π-donor ligands also exhibit different electronic requirements when vertices are lost to give *nido* and *arachno* species.[10]

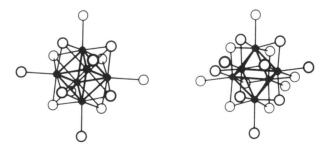

Figure 8.1 Octahedral $M_6X_8L_6$ and $M_6X_{12}L_6$ clusters with face-bridging (*left*) and edge-bridging (*right*) π-donor ligands. The metal atoms are shaded.

In the usual TSH hybridization scheme, three inwardly hybridized orbitals (one σ-type and two π-type) are used for skeletal bonding, and for the octahedron, $2n - 1 = 11$ inaccessible skeletal molecular orbitals result. Specifically, for the octahedron these inaccessible orbitals consist of $P_{x,y,z}^{\sigma/\pi}(T_{1u})$, $D_{x^2-y^2,z^2}^{\sigma/\pi}(E_g)$, $\bar{P}_{x,y,z}^{\pi}(T_{1g})$, and $\bar{D}_{xz,\,yz,\,xy}^{\pi}(T_{2u})$. These inaccessible orbitals are also present in the clusters with π-donor ligands, but there are some additional inaccessible orbitals in these molecules derived from the d^δ orbitals.

We choose local axes so that the $d_{x^2-y^2}$ orbitals point along the edges of the metal octahedron and the d_{xy} orbitals lie over the faces. Any given vertex and the five nearest neighbor ligands (in either a face- or an edge-bridging cluster) are invariant under a subset of the operations of the point group O_h. This subset is actually the subgroup C_{4v}. The representation spanned by the five ligands in C_{4v} is easily shown to be $2A_1 \oplus B_1 \oplus E$ for a face-bridged cluster and $2A_1 \oplus B_2 \oplus E$ for an edge-bridged cluster. It is also easy to show that a metal σ orbital transforms as A_1, two π orbitals as E, and two δ orbitals as $B_1 \oplus B_2$. In fact, this can be achieved simply by reading off the symmetry species of the appropriate Cartesian representations of the atomic orbitals from the C_{4v} character table. Since the inwardly hybridized orbitals include one of σ symmetry and two π orbitals, it is clear that all the remaining metal hybrids except one of the δ orbitals are needed to match the ligand orbitals. For a face-bridged cluster, the $d_{x^2-y^2}^\delta$ orbitals are "left over," while for an edge-bridged cluster the d_{xy}^δ orbitals remain. The $d_{x^2-y^2}$ and d_{xy} functions represent two complementary subsets of the complete L^δ set as shown in Figure 8.2.

The twelve δ orbitals therefore interact in a complementary fashion with sets of face- and edge-bridging ligands. We can easily work out the symmetries of the δ cluster orbitals and divide them into linear combinations of either $d_{x^2-y^2}^\delta$ or d_{xy}^δ functions:

	$[Mo_6Cl_8L_6]^{4+}$	$[Ta_6Cl_{12}L_6]^{2+}$	
	$d_{x^2-y^2}^\delta$ functions	d_{xy}^δ functions	
Accessible	$D^\delta(E_g)$	$\bar{D}^\delta(E_u)$	Inaccessible
Accessible	$\bar{D}^\delta(T_{2u})$	$D^\delta(T_{2g})$	Inaccessible
Inaccessible	$\bar{F}^\delta(A_{2g})$	$F^\delta(A_{2u})$	Accessible

These orbitals can be correlated with occupied or unoccupied levels by taking into account mixing with the other cluster orbitals and noting that even δ cluster orbitals are generally bonding in character, while odd δ orbitals are antibonding.[11] For clusters with face-bridging ligands, the $\bar{F}^\delta(A_{2g})$ orbital is the only inaccessible member of the above set because the mixing between the antibonding $\bar{D}^\delta(T_{2u})$ and $\bar{D}^\pi(T_{2u})$ functions results in bonding (accessible) and antibonding (inaccessible) combinations. The occurrence of this additional inaccessible δ orbital $\bar{F}^\delta(A_{2g})$ leads to a reduction in the total electron count

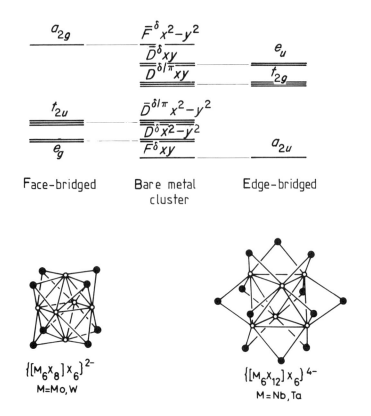

Figure 8.2 The relationship between the δ cluster orbitals formed from $d^\delta_{x^2-y^2}$ functions in a face-bridged cluster (*left*) and d^δ_{xy} functions in an edge-bridged cluster (*right*).

from 86 to 84. On the other hand, for an edge-bridged cluster only the $F^\delta(A_{2u})$ orbital remains accessible after mixing, because we have already counted a set of $D^\pi(T_{2g})$ skeletal bonding orbitals. Hence, $D^\delta(T_{2g})$ correlates with an inaccessible set in this case. Such clusters are therefore characterized by 76 valence electrons and have five additional inaccessible δ orbitals, $\bar{D}^\delta(E_u)$ and $D^\delta(T_{2u})$.

If the complete L^δ set were available for skeletal bonding in these octahedral clusters, six bonding and six antibonding orbitals would be generated before mixing. In $M_6Cl_8L_6$ and $M_6Cl_{12}L_6$, the division of the available L^δ orbitals into unequal occupied and unoccupied subsets leads to the observed electron counts of 84 and 76.

A great advantage of the delocalized description is that it can equally well describe molecules with variable electron counts. For example, $K_4[Nb_6Cl_{18}]$ has a total of 76 valence electrons filling all the skeletal bonding orbitals of an edge-bridged octahedron, and the Nb—Nb bond length is

2.91 Å. The existence of the related ions $[Nb_6Cl_{18}]^{3-}$ and $[Nb_6Cl_{18}]^{2-}$ can be associated with the depopulation of the $F^\delta(A_{2u})$ orbital (see Figure 8.3). The metal-metal bond lengths increase on oxidation to 2.97 and 3.03 Å, respectively, confirming the weakly bonding nature of the a_{2u} orbital. Furthermore, the fact that only $[Nb_6Cl_{18}]^{3-}$ is paramagnetic ($\mu_{eff} = 1.7$ Bohr magnetons, indicating one unpaired electron) confirms that the frontier molecular orbital is singly degenerate.[12]

A very pleasing description of these orbitals is possible using the TSH theory pairing principle.[13] This relationship is clearly seen in the δ orbital combinations tabulated above—the orbitals of one set can be obtained from those of the other set by applying the parity operator, just as the $d^\delta_{x^2-y^2}$ functions can be obtained from the d^δ_{xy} functions. (Recall from Chapter 3 that

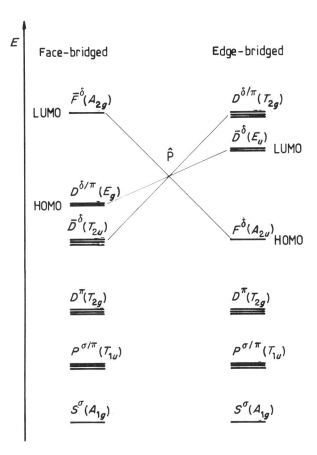

Figure 8.3 Qualitative energy-level diagrams for the face-bridged octahedron (*left*) and the edge-bridged octahedron (*right*), and the parity relation between the frontier orbitals.

Sec. 8.2 Octahedral Clusters

Luminescence and Redox Photochemistry of $[M_6X_{14}]^{2-}$ Clusters (M = Mo or W; X = Cl or Br)

Solids and solutions containing the ions $[Mo_6Cl_{14}]^{2-}$, $[Mo_6Br_{14}]^{2-}$, and $[W_6Cl_{14}]^{2-}$ are yellow or orange, since their absorption spectra are dominated by intense bands in the near ultraviolet. The emission spectra of these ions show a band at about 800 nm. The phosphorescence lifetimes obtained from emission decay kinetics are amongst the longest known for transition metal complexes and have been attributed to the occurrence of a long-lived phosphorescent excited state $[M_6X_{14}]^{2-*}$. The close similarities in the emission spectra for $[Mo_6Cl_{14}]^{2-}$ and $[Mo_6Br_{14}]^{2-}$ strongly suggest that the electronic transition from the phosphorescent excited state to the ground state involves molecular orbitals localized almost exclusively on the metal atoms.

Cyclic voltammetry on the ions in CH_3CN indicates reversible ion electron oxidation processes, e.g., for $[Mo_6Cl_{14}]^{2-}$ at 1.29 V vs. Ag/0.1 M $AgNO_3$. The combination of luminescence and redox properties makes these ions ideal as photoreceptors for light-induced chemical reactions. Their chemical stability under a variety of conditions and their ability to undergo electron transfer processes rapidly, both in their ground and excited states, are also important in this context.

With electron acceptors such as methyl viologen (A), the luminescence of $[Mo_6Cl_{14}]^{2-*}$ is quenched according to the following reaction:

$$[Mo_6Cl_{14}]^{2-*} + A \longrightarrow [Mo_6Cl_{14}]^- + A^-$$

The resultant ion $[Mo_6Cl_{14}]^-$ is a powerful oxidant. Therefore, these cluster ions function in a manner analogous to that reported for $[Ru(bipy)_3]^{2+}$, which has been investigated in great detail as part of photochemical energy storage systems.

Source: Gray, H. B., et al., *J. Amer. Chem. Soc.*, **105**, 1878 (1983).

---

for the δ orbitals the parity operation corresponds to a rotation through 45° about each radius vector, all in the same sense.) As shown in Figure 8.3, the parity relationship is not greatly disturbed by mixing with L^π orbitals.[13]

In order to emphasize the close electronic relationships between π-donor and π-acceptor clusters, the total number of skeletal bonding orbitals associated with the octahedral clusters has been stressed. The significant electronegativity difference between the metal atoms and the ligands in π-donor clusters and the absence of back donation also provides reasonable grounds for describing the electron distribution in these clusters in more localized terms. In particular, in both $[Mo_6Cl_8L_6]^{4+}$ and $[Nb_6Cl_{12}L_6]^{2+}$, there are 30 metal-ligand bonds (6 terminal and 24 bridge bonds) localized predominantly on the ligands—these clusters differ in the number of electrons accommodated in metal-metal bonding orbitals (24 in the former and 16 in the latter). Hence we can assign formal oxidation numbers to the metal atoms in these clusters.

8.3 *nido* AND *arachno* π-DONOR CLUSTERS

A *nido* octahedron is, of course, a square pyramid, as in $[Mo_5(\mu_3\text{-Cl})_4(\mu_2\text{-Cl})_4Cl_5]^{2-}$ (Figure 8.4a).[14] An *arachno* octahedron, however, could have either a square or a butterfly of metal atoms, depending upon which two vertices are "lost." The molybdenum atoms in $Mo_4(\mu_2\text{-OPr}^i)_8Cl_4$ are square planar (Figure 8.4b), while those in $Mo_4(OPr^i)_8Br_4$ (Figure 8.4c) define a butterfly.[15] All these compounds are related to the *closo* octahedron with face bridges rather than edge bridges. The number of valence electrons associated with these structures is 69, 52, and 52, respectively; i.e., the electron count does not remain constant at 84.

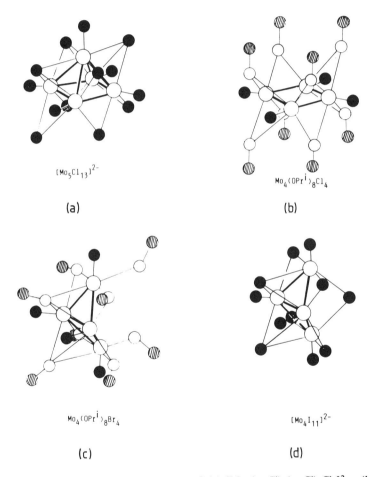

Figure 8.4 The molybdenum skeletons of (a) $[Mo_5(\mu_3\text{-Cl})_4(\mu_2\text{-Cl})_4Cl_5]^{2-}$, (b) $Mo_4(\mu_2\text{-OPr}^i)_8Cl_4$, (c) $Mo_4(OPr^i)_8Br_4$, and (d) $[Mo_4I_{11}]^{2-}$. The structure in part is a *nido* face-bridged octahedron, and the others are *arachno* face-bridged octahedra.

> ### Colors of π-donor clusters
>
> π-donor clusters which can be isolated as separated anions are generally colored; some examples are given below. Replacement of halide ligands by S, Se, etc., generally leads to a darker coloration.
>
> | | |
> |---|---|
> | $[Mo_6Cl_8Cl_6]^{2-}$ | Yellow |
> | $[Mo_6Br_8Br_6]^{2-}$ | Yellow |
> | $[W_6Cl_8Cl_6]^{2-}$ | Yellow |
> | $Pb[Mo_6Cl_8Cl_6]$ | Yellow |
> | Mo_6S_8 | Black |
> | $[Re_3Cl_{12}]^{3-}$ | Red ⎫ Two bands at |
> | $Re_3Cl_9(PPh_3)_3$ | Purple ⎬ around 500–550 nm |
> | $Re_3Cl_9(OPPh_3)_3$ | Red ⎭ and 750–800 nm |
> | $[Re_2Cl_8]^{2-}$ | Royal blue |
> | Re_6X_{14} | Red (X = Cl, Br, I) |
> | $[Re_6Se_6Cl_8]^{2-}$ | Red |
> | $Re_6S_8Br_4$ | Black |
> | $[Nb_6Cl_{12}Cl_6]^{2-}$ | Green-black |
> | $[Nb_6Cl_{12}Cl_6]^{4-}$ | Green |
> | Nb_6Cl_{14} | Black |
>
> π-donor clusters with infinite structures linked by metal-metal bands are generally black if they are semiconductors or metallic conductors; the latter also have a metallic luster.

When a metal vertex is lost, the four ligands which were face-bridging around the missing vertex become edge-bridging around the new square face. Hence they change from five- to three-electron donors. This behavior should be contrasted with that of the carbonyl ligand, which generally functions as a two-electron donor regardless of the number of nearest-neighbor metal atoms.

Detailed calculations for the above species[5] show that three skeletal bonding orbitals are lost for the removal of each MoCl vertex from the octahedral cluster. Since five Mo—Cl σ bonds are also lost, this means that the number of strongly bonding orbitals decreases by 8 for each vertex lost. Hence, 68 and 52 valence electrons are expected for *nido* and *arachno* species, respectively, in reasonable agreement with the electron counts of the three clusters described above. For clusters with edge-bridging ligands, on the other hand, the loss of an M—L vertex is predicted to leave the number of skeletal bonding orbitals unchanged,[5] so that 10 valence electrons (associated with the 5 M—L σ bonds) are "lost" on the removal of each M—L vertex. Hence, the *closo*, *nido*, *arachno* progression for edge-bridged clusters of this type should be associated with 76, 66, and 56 valence electrons, respectively.

The above results can actually be understood more easily by considering the effect of decapping on the two-center two-electron edge bonds (face-bridged

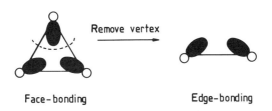

Figure 8.5 Removing one of the metal vertices from a triangular face converts a face-bonding orbital (*left*) to an edge-bonding orbital (*right*).

cluster) or the three-center two-electron face bonds (edge-bridged cluster).[5] Removing a metal vertex from an octahedron gives a square-based pyramid with four basal edges corresponding to the four triangular faces which are lost. The effect of decapping on the four face-bonding orbitals localized in the faces surrounding the missing vertex is to produce four essentially edge-bonding orbitals (Figure 8.5). Hence the number of skeletal bonding orbitals in an edge-bridged π-donor cluster is unaffected by decapping. Now consider the effect of decapping on the four edge-bonding orbitals connected to the critical vertex in a face-bridged cluster. Four linear combinations may be formed from the four resulting out-pointing orbitals (Figure 8.6): one is bonding, two are nonbonding, and one is antibonding. Only the fully in-phase bonding orbital is sufficiently low-lying to be occupied, and hence the number of occupied skeletal bonding orbitals decreases by 3.

If two trans M—L vertices are removed from an edge-bridged cluster of this type, then a square-planar metal skeleton remains. From the above analysis we would expect to find two edge bonds for each edge of the

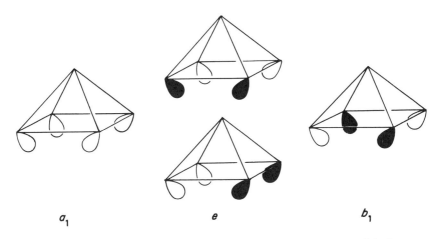

Figure 8.6 Bonding, nonbonding, and antibonding linear combinations of the four orbitals which result when an M—L vertex is removed from a face-bridged π-donor cluster.

Sec. 8.3 *nido* and *arachno* π-Donor Clusters

remaining skeleton, corresponding to the two face-bonding orbitals localized in the two triangular faces above and below each edge in the octahedron. This description is equivalent to a set of four metal-metal double bonds in the ring. $[Re_3(\mu_2\text{-}Cl)_6Cl_6]^{3-}$ can also be described in such terms,[15] i.e., as a *trans-arachno* trigonal bipyramid or as a triangular ring of rhenium atoms with three formal metal-metal double bonds.

8.4 THREE-DIMENSIONAL LINKED CLUSTERS

There is a very wide range of clusters based upon octahedra of early transition metals with varying numbers of halide or chalcogenide (sulfur, selenium, tellurium, etc.) ligands and different patterns of interconnection between the structural units.[16] This family of clusters has attracted a great deal of interest because of the remarkable range of electrical properties they possess. These properties are determined by the number of electrons and ligands associated with each octahedron of metal atoms.

There are a variety of ways in which adjacent clusters can be linked, as illustrated schematically in Figure 8.7.[16] For reasons of clarity, only a few adjacent clusters are shown. The X^a ligands are located at the apices of the octahedron and X^i ligands above the faces. Compounds such as $K_4Nb_6Cl_{18}$ ($K_4Nb_6Cl_{12}^i Cl_6^a$) and $HgMo_6Cl_{14}$ ($HgMo_6Cl_8^i Cl_6^a$) have all the co-ordination sites filled, and no intercluster linking is present. In contrast, $NaMo_6Cl_{13}$ ($NaMo_6Cl_8^i Cl_4^a Cl_{2/2}^{a-a}$) has the octahedra linked through the *trans*-apical halides to give an infinite network. In this case the $Cl_{2/2}^{a-a}$ ligands are

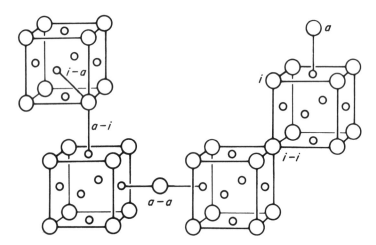

Figure 8.7 Different kinds of cluster linkages via X atoms and the symbols used to describe them. The cube of X^i atoms is emphasized, and the metal-metal bonds are not shown.

Alkoxide clusters: A bridge between π-donor and π-acceptor ligands

Chisholm and his coworkers have characterized a range of alkoxide clusters of the early transition metals.* Many of these compounds have structures which are related to those exhibited by clusters with π-donor ligands. For example, $[Mo_6(\mu_3\text{-}OMe)_8(OMe)_6]^{2-}$ is isoelectronic to $[Mo_6Cl_8L_6]^{4+}$ and has the same octahedral metal geometry with face-capping ligands. Some alkoxide clusters also bear a striking resemblance to the cluster subunits found in the solid state structures of reduced metal oxides. For example, $W_4(OEt)_{16}$ contains the same $M_4(\mu_3\text{-}O)_2(\mu_2\text{-}O)_4O_8$ unit seen in $B_{1.14}Mo_8O_{16}$. There are in addition some interesting similarities between alkoxide and carbonyl clusters. For example, the following pairs of compounds

$Co_3(\mu_3\text{-}CR)(CO)_9$ $W_3(\mu_3\text{-}CR)(OR)_9$

$Fe_3(\mu_3\text{-}NH)(CO)_{10}$ $W_3(\mu_3\text{-}NH)(OR)_{10}$

have triangular metal frameworks capped by either CR or NH ligands. Similarly, the carbido clusters $Fe_4C(CO)_{13}$ and $W_4C(\mu_2\text{-}O)(OPr^i)_{12}$ both have butterfly arrangements of metal atoms with the carbido ligand between the wingtips.

Chisholm has proposed that the CO and OR ligands lead to similar cluster geometries because of the complementary nature of their bonding interactions with the metal frameworks. The total number of valence electrons in the two types of cluster is the critical factor, and it appears that electron-rich metal fragments stabilized by π-acceptor ligands such as $Co(CO)_3$ can result in the same geometric preference as electron-poor fragments with π-donor ligands such as $Mo(OR)_3$. Furthermore, both fragments have frontier orbitals of the same local symmetry occupied by three valence electrons, and hence they are isolobal.

*Chisholm, M. H., Clark, D. L., and Hampden-Smith, M. J., *Angew. Chem., Int. Ed.*, **28**, 432 (1989).

---

apical and are shared between two octahedra. The electronic balance and the degree of coverage of the M_6 units by the X ligands (X may be halide, chalcogenide, etc.) determines the physical properties of the compounds. A total coverage always results either in semiconducting or insulating properties even when the metal-metal bonding orbitals are partially occupied. An increasing degree of cluster linkage leads in a stepwise fashion to increased electrical conductivity because the M_6 octahedra move closer together. The transition from insulator to metallic behavior is recognizable in the molybdenum cluster compounds given in Table 8.1. When the halide ligands are successively replaced by dianionic chalcogen atoms (denoted by $\hat{X}$), the formal metal oxidation state is maintained by reduction in the $(X, \hat{X})$/metal ratio. The different patterns of interconnection with decreasing $(X, \hat{X})$/Mo ratio are illustrated schematically in Figure 1.7 in Chapter 1.

The family of clusters $Mo_6\hat{X}_8$ ($\hat{X}$ = S, Se, Te) have structures in which

TABLE 8.1 HALIDES, HALIDE CHALCOGENIDES, AND CHALCOGENIDES WITH Mo_6X_8 CLUSTERS: STRUCTURES AND PROPERTIES. VALUE OF z INDICATES THE NUMBER OF ELECTRONS IN M—M BONDING STATES

Compounds		Structural principle	z	Properties
$A_2Mo_6X_{14}$	$A = Li, K, Rb, Cs, Cu$ $X = Cl$ $Cs_2Mo_6Cl_8Br_6$	$Mo_6X_8^i X_6^a$	24	Insulators
AMo_6X_{14}	$A = V—Ni, Zn—Hg, Sn,$ $Pb, Mg—Ba, Eu, Yb$ $X = Cl$	$Mo_6X_8^i X_6^a$	24	Insulators
AMo_6X_{13}	$A = Na, Ag; X = Cl$	$Mo_6X_8^i X_4^a X_{2/2}^{a-a}$	24	Insulators
Mo_6X_{12}	$X = Cl, Br, I$	$Mo_6X_8^i X_2^a X_{4/2}^{a-a}$	24	Insulators
$Mo_6X_{10}\tilde{X}$	$X = Cl, Br; \tilde{X} = S—Te$	$Mo_6(X_7\tilde{X})_8^i X_{6/2}^{a-a}$	24	Insulators
$Mo_6X_8\tilde{X}_2$	$X = I; \tilde{X} = Se, Te$ $X = Br; \tilde{X} = S$	$Mo_6(X_5\tilde{X})^i X_{2/2}^{i-i} X_{6/2}^{a-a}$	24	Insulators
$Mo_6X_6\tilde{X}_3$	$X = I; \tilde{X} = S, Se$ $Mo_6Br_6S_3$	$Mo_6X_4^i X_{2/2}^{i-a} \tilde{X}_{2/2}^{i-a} X_{4/2}^{a-a} \tilde{X}_{2/2}^{a-i}$	24	Narrow band gap semiconductor $(0.02$ eV at $T < 100$ K$)$
$Mo_6X_2\tilde{X}_6$	$X = Br, I; \tilde{X} = S$	$Mo_6X_2^i X_{6/2}^{i-a} \tilde{X}_{6/2}^{a-i}$	22	Metals, superconductors $T_c \approx 14$ K
$Mo_6\tilde{X}_8$	$\tilde{X} = S, Se, Te$	$Mo_6\tilde{X}_2^i \tilde{X}_{5/2}^{i-a} \tilde{X}_{6/2}^{a-i}$	20	Metals, superconductors
$AMo_6\tilde{X}_7$	$Cs_{0.6}Mo_6S_7$	$Mo_6\tilde{X}_{2/2}^{i-i} \tilde{X}_{6/2}^{i-a} \tilde{X}_{6/2}^{a-i}$	22	Metal, superconductor $T_c \approx 8$ K

each molybdenum octahedron is surrounded by eight face-bridging ligands and are metallic conductors. These face-bridging ligands also act as bridges between different Mo_6 octahedra (Figure 8.8). Each metal octahedron is formally associated with 20 electrons in metal-metal bonding molecular orbitals. (The easiest way to see this is to formulate the cluster as $(Mo_6)^{16+}(\hat{X}^{2-})_8$ and write all the metal-ligand bonds as dative bonds from the ligands.) This is obviously less than the 24 electrons associated with $[Mo_6(\mu_3\text{-}Cl)_8Cl_6]^{2-}$ clusters discussed above. The metallic conduction arises because there are unoccupied accessible skeletal orbitals and the metal octahedra interact significantly with one another. Furthermore, the structure adopted in the solid state (Figure 8.8) is the same as that of the superconducting Chevrel phases.[17]

The chemical and physical properties of the Chevrel phase compounds have been studied in some detail because $PbMo_6\hat{X}_8$ becomes superconducting at $T_c \approx 14$ K and set a new record for upper critical magnetic fields ($H_{c2} \approx 60$ tesla). Bonding studies of these compounds have indicated that there are strong interactions between the closely spaced clusters. The transition from the local bonding scheme for the "molecule" $Mo_6\hat{X}_8$ to the band structure for the infinite solid leads to a broadening of all levels. In addition, the special arrangement of clusters in the $Mo_6\hat{X}_8$ structure (Figure 8.8) allows

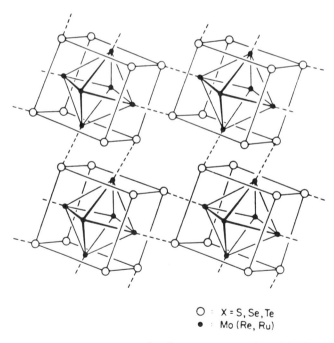

○ : X = S, Se, Te
● : Mo (Re, Ru)

Figure 8.8 The structure of $Mo_6\hat{X}_8$ ($\hat{X}$ = S, Se, Te) involves eight face-capping ligands some of which are shared with four neighboring $Mo_6\hat{X}_8$ units.

each molybdenum to form a donor-acceptor bond with an $\hat{X}$ atom (type $\hat{X}^{i-a}$) of an adjacent cluster. The Mo—$\hat{X}$ bonds which link octahedra are essentially as strong as those involving only a single cluster unit, and the intercluster metal-metal bonding is about one-tenth as strong as the intracluster metal-metal bonding. Band structure calculations on Chevrel phase compounds show that large changes in the conduction band occur when the metal octahedron is formally associated with between twenty and twenty-three d electrons. Depletion of the conduction band reduces the repulsive interactions between the clusters, and the intercluster distances therefore decrease. Consequently the width of the conduction band is very sensitive to lattice vibrations which can couple to the occupation of the conduction states. This strong electron-phonon coupling is thought to be essential for the superconducting properties.[18-19]

8.5 π-DONOR CLUSTERS WITH INTERSTITIAL ATOMS

In Chapter 1 (Section 1.7.2), some metal carbonyl clusters with interstitial atoms such as H, C, N, Si, and P were described. The π-donor clusters of the earlier transition metals are also able to incorporate interstitial atoms such as Be, B, C, and N; some examples are given in Table 8.2.[16] In this table there are also some examples of clusters with first-row transition metals in octahedral interstitial sites, e.g., [Th$_6$Br$_{15}$Fe] (Th$_6$Br$^i_{12}$Br$^{a-a}_{6/2}$Fe).

Recently Corbett and his coworkers[20] have described many examples of halide clusters with interstitial main group atoms (B, C, N, Al, Si, Ge, S), transition metal atoms (Mn, Fe, Co, Ni, Ru, Rh, Pd, Re, Os, Ir, Pt), and atoms from groups I and II (H, K, Be). The following represents a typical synthesis:

$$4CsX + 11Zr_{powder} + 13ZrX_4 + 4C \xrightarrow[\text{2 weeks}]{850°C} 4CsZr_6X_{14}C \quad (X = Cl, Br).$$

The following series of isoelectronic compounds illustrates the range of species currently accessible with main group interstitial atoms:

Zr$_6$Cl$_{12}$Be	Zr$_6$Cl$_{13}$B	Zr$_6$Cl$_{14}$C	Zr$_6$Cl$_{15}$N
Sc[Sc$_6$Cl$_{12}$N]	K[Zr$_6$Cl$_{13}$Be]	K[Zr$_6$Cl$_{14}$B]	Na[Zr$_6$Cl$_{15}$Be]
Zr$_6$Cl$_{15}$N	K[Zr$_6$Cl$_{15}$C]	K$_2$[Zr$_6$Cl$_{15}$B]	K$_3$[Zr$_6$Cl$_{15}$Be]
Zr$_6$Cl$_{13}$B	K[Zr$_6$Cl$_{14}$B]	K$_2$[Zr$_6$Cl$_{15}$B]	Cs$_3$[Zr$_6$Cl$_{16}$B] Rb$_5$[Zr$_6$Cl$_{18}$B]

All the compounds are based on the M$_6$X$_{12}$ octahedral edge-bridged structure and are linked through halide bridges to give infinite structures (see Table 8.2). All of the species listed above have a total of 14 electrons involved in metal-metal and metal-interstitial bonding. Using the molecular orbital scheme developed in this chapter for M$_6$X$_{12}$L$_6$, this corresponds to the

TABLE 8.2 EXAMPLES OF HALIDE CLUSTERS WITH INTERSTITIAL ATOMS

Parent cluster type	Centered cluster	Intercluster connections
$[Nb_6I_{11}]$	$[Nb_6I_{11}H]$	$M_6X_8^i X_{6/2}^{a-a}$
$Cs[Nb_6I_{11}]$	$[Sc_7Cl_{12}Y]$, $Y = B, N$	$M_6X_6^i X_{6/2}^{i-a} X_{6/2}^{a-i}$
	$[Sc_7I_{12}M']$, $M' = Co, Ni$	
	$[Zr_6Cl_{12}Be]$	$M_6X_6^i X_{6/2}^{i-a} X_{6/2}^{a-i}$
Nb_6Cl_{14}	$[Zr_6Cl_{14}Y]$, $Y = B, C$	$M_6X_{10}^i X_{2/2}^{i-a} X_{2/2}^{a-i} X_{4/2}^{a-a}$
$[Nb_6F_{15}]$	$[Th_6Br_{15}M']$, $M' = Mn, Fe, Co$	$M_6X_{12}^i X_{6/2}^{a-a}$

Chemical Shifts of Interstitial Atoms

The encapsulated atoms in metal clusters can be studied using high-resolution nmr techniques. If the atoms are at the center of the cluster, they generally show resonances at low fields (see the table below), i.e, they are very deshielded. This should not be interpreted in terms of highly positively charged interstitial atoms, but rather a reflection of the different diamagnetic and paramagnetic effects experienced by a nucleus in such an unusual chemical environment.

NMR DATA ON ENCAPSULATED ATOMS

Compound	Encapsulated atom	Geometric environment	Chemical shift (ppm)	NMR Ref.
$[Rh_{13}(CO)_{24}H_3]^{2-}$	Rh	hcp	+3547	3.16 MHz
$[Rh_{13}(CO)_{24}H_2]^{3-}$	Rh	hcp	+4954	3.16 MHz
$[Rh_{13}(CO)_{24}H]^{4-}$	Rh	hcp	+6370	3.16 MHz
$[Rh_6(CO)_{13}C]^{2-}$	C	Octahedron	+338	TMS
$[Rh_6(CO)_{15}C]^{2-}$	C	Trigonal prism	+264	TMS
$[Co_6(CO)_{15}C]^{2-}$	C	Trigonal prism	+332.8	TMS
$[Co_8(CO)_{18}C]^{2-}$	C	Square antiprism	+388	TMS
$[Rh_9(CO)_{21}P]^{2-}$	P	Square antiprism	+282.3	(H_3PO_4)
$[Co_6(CO)_{15}H]^-$	H	Octahedron	+23.2	TMS
$[Ru_6(CO)_{18}H]^-$	H	Octahedron	+16.4	TMS
$[Ni_{12}(CO)_{21}H_2]^{2-}$	H	Distorted octahedron	−18	TMS
$[Ni_{12}(CO)_{21}H]^{3-}$	H	Distorted octahedron	−24	TMS
$[Rh_{13}(CO)_{24}H_3]^{2-}$	H	Square pyramid fluxional*	−29.3	TMS
$[Rh_{13}(CO)_{24}H_2]^{3-}$	H	Square pyramid fluxional†	−26.7	TMS

*At −90°C two signals in the ratio 1:2 are observed at −28.3 and −30.8 ppm.
†At −90° two signals in the ratio 1:1 are observed at −26.9 and −27.8 ppm.

> The "normal" high-field chemical resonances observed for the interstitial hydrogen atoms in $[Ni_{12}CO)_{21}H_{4-n}]$ and $[Rh_{13}(CO)_{24}H_{5-n}]^{n-}$ have been attributed to the fact that hydrogen is no longer in a high-symmetry environment and experiences anisotropic paramagnetic chemical shift effects.
>
> The ^{15}N chemical shifts for nitrido clusters are not as deshielded as a simple extrapolation from the carbido clusters would suggest:
>
> | | δ rel. to NH_3 | | δ |
> | --- | --- | --- | --- |
> | $[Co_6N(CO)_{16}]^-$ | 196.2 | $[Co_6C(CO)_{15}]^{2-}$ | 332.8 |
> | $[Rh_6N(CO)_{15}]^-$ | 107.8 | $[RhC(CO)_{15}]^{2-}$ | 264 |
> | $[Ru_6N(CO)_{16}]^-$ | 538   ref. NH_4^+ | $[Ru_6C(CO)_{16}]^{2-}$ | 461.2 (rel. to CH_4) |
>
> It has been suggested that the ^{15}N chemical shifts are more sensitive to site symmetry effects and to the excitation energy term which mixes in the paramagnetic contributions.
>
> The first example of an interstitial boride metal carbonyl cluster has recently been described: $[Fe_4Rh_2B(CO)_{16}]^-$. The compound exists in *cis* and *trans* isomeric forms in solution, and the ^{11}B nmr spectrum shows resonances at 211 and 205.5 ppm with respect to $BF_3 \cdot OEt_2$. These large deshielded chemical shifts and the narrow linewidths of the resonances (12 Hz) confirm that the ^{11}B is in a high-symmetry interstitial site.
>
> *Fehlner, T. P., et al., *J. Amer. Chem. Soc.*, **111**, 1877 (1989).

occupation pattern (see Figure 8.3): $S^\sigma(A_{1g})$, $P^{\sigma/\pi}(T_{1u})$, and $D^\pi(T_{2g})$, with $F^\delta(A_{2u})$ unoccupied. For example, $Zr_6Cl_{12}Be \equiv [Zr_6Be]^{12+}12Cl^-$ has $6 \times 4 + 2 - 12 = 14$ electrons occupying these orbitals. When the interstitial atom is a transition metal, there is an additional accessible e_g pair of orbitals originating from the central atom d functions. The t_{2g} set is not formally available, because it interacts with the $D^\pi(T_{2g})$ occupied skeletal orbitals. Therefore, such species are associated with a maximum of 18 skeletal bonding electrons. Hence, for $Zr_6I_{14}Fe \equiv [Zr_6Fe]^{14+}14I^-$, we have $6 \times 4 + 8 - 14 = 18$ electrons occupying the above orbitals.

The bonding in all of these clusters can be understood using the principles developed in Chapter 4 (Section 4.4) for radial clusters.[21-28] The interstitial main group atom has s and p valence orbitals which match the S^σ and P^σ cluster orbitals, and the resulting interactions are strong because the overlap integrals are large. The p orbitals of the interstitial main group atom also overlap strongly with tangential P^π cluster orbitals as shown in Figure 8.9. Therefore the interstitial atom experiences both strong σ- and π-type interactions with the d orbitals of the transition metal. This "multiple bonding" results in an effective contraction of the radius of the interstitial atom and accounts for the small radius ratios noted in Table 1.11 in Chapter 1. The orbital interaction diagram for an octahedral metal carbonyl cluster with an interstitial carbon atom, which is illustrated in Figure 8.10, confirms that the electronic requirements of the cluster are unaffected by the introduction of the

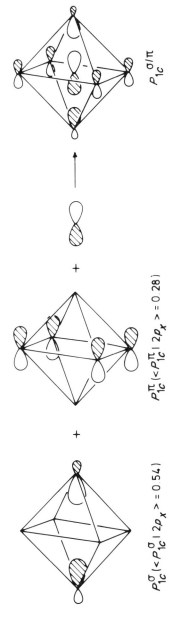

Figure 8.9 The overlap of the p orbitals of an interstitial main group atom with the P^π and P^σ cluster orbitals.

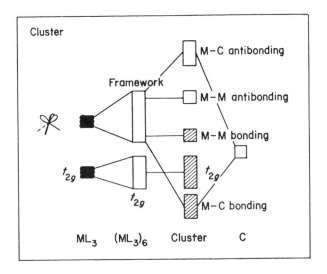

Figure 8.10 Schematic energy-level diagram for an octahedral metal carbonyl cluster with an interstitial main group atom.

main group atom. Hence both $[Ru_6(CO)_{18}]^{2-}$ and $[Ru_6C(CO)_{18}]^{2+}$ are characterized by 86 valence electrons.

The delocalized nature of the bonding in these clusters ensures that the interstitial atoms attain approximate electroneutrality, and an anionic description such as C^{4-}, N^{3-}, etc., is not appropriate. Simple semiempirical molecular orbital calculations on octahedral metal carbonyl clusters have indicated that the charge on the interstitial carbon atom is close to zero and becomes more negative when the carbon is in an exposed position, e.g., in square-pyramidal $Fe_5C(CO)_{15}$ and in the "butterfly" cluster $[Fe_4C(CO)_{12}]^{2-}$ shown in Figure 8.11.[28] In these clusters the π interactions between the out-pointing orbital on the carbon, which has a high proportion of p character, and the cluster framework orbitals have interesting geometric effects. For example, in the square-pyramidal cluster the carbon lies only about 0.1 Å below the basal plane, and in the "butterfly" cluster it lies on the line connecting the wingtip atoms. In this way the π-bonding interactions illustrated in Figure 8.11 are maintained. Previously, we have always assumed that a main group atom on the surface of a cluster occupies a conventional vertex position and has an out-pointing radial nonbonding orbital. In these clusters this orbital is not nonbonding, but enters into a strong π interaction with the adjacent cage atoms. In addition to the geometric consequences of this interaction, the nucleophilicity of the carbon atoms in these clusters is lower than would have been anticipated.

These clusters with either interstitial or exposed atoms are useful models for binary carbides, nitrides, etc., and also for metal surfaces which are

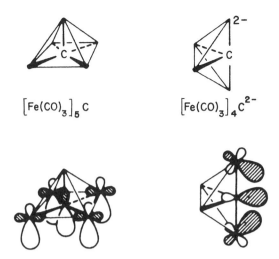

Figure 8.11 The square-pyramidal $Fe_5C(CO)_{15}$ and "butterfly" $[Fe_4C(CO)_{12}]^{2-}$ clusters and the relevant π-type interactions which occur between Fe and C.

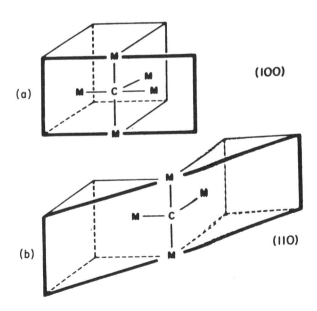

Figure 8.12 The (100) face of a body-centered cubic metal (a) and the (110) face of a face-centered cubic metal (b). Reproduced with permission from Wijeyesekara, S. D., et al., *Organometallics*, **3**, 949 (1984).

implicated in dissociative chemisorption processes. For example, the bonding in NbC, which has an NaCl structure, and WC, which has a hexagonal close-packed structure, can be related to that in carbido-metal clusters with octahedral and trigonal prismatic geometries.[22] In addition, it is thought that surface carbides, which are present in Fischer-Tropsch heterogeneous catalytic processes, may be structurally related to the square-pyramidal and butterfly carbido structures described above.[29] In particular, if surface carbides are stabilized when the carbon atom lies approximately in the plane of the surface atoms, then the fourfold site of a (100) face of a body-centered cubic metal (Figure 8.12) is relevant since it creates a square-pyramidal environment for carbon. Carbon in a (110) face of a face-centered cubic metal (Figure 8.12) resembles the butterfly arrangement observed in $[Fe_4C(CO)_{12}]^{2-}$.

For larger clusters, the cavity size becomes too large for an atom such as carbon, and the interactions with both the σ and π cluster orbitals become weaker. Simple semiempirical molecular orbital calculations suggest that it might be possible to stabilize CH_x fragments within the cluster in some circumstances. In these clusters it has proved possible experimentally to stabilize C—C (ethanido) fragments with a wide range of C—C distances.[30-31]

8.6 COLUMNAR CLUSTERS

The tensor surface harmonic methodology developed above is based on a free-electron model for a particle on a sphere, and consequently it is not as good a description of clusters whose shapes are more closely related to a cylinder. The solutions of the Schrödinger equation for a particle on-a-cylinder problem have been developed to account for the bonding in such clusters. Approximate wavefunctions are obtained by taking the products of the eigenfunctions for a particle on a ring and those of a particle in a box. The resultant wavefunctions are characterized by two quantum numbers k and λ which define the number of phase changes along the length of the cylinder and perpendicular to the cylindrical face, respectively. Only an outline of the conclusions will be given here.[32]

For a main group cluster, the number of skeletal bonding orbitals for a cylindrical geometry is the same as that for a pseudospherical deltahedral geometry (i.e., $n + 1$), and there are $4n + 2$ valence electrons in total. Furthermore, the nodal characteristics of the radial and tangential molecular orbitals are related to those of a spherical cluster, and the two sets of molecular orbitals can be easily correlated. Molecular orbital calculations on $B_9H_9^{2-}$ clusters with spherical and cylindrical topologies have shown that the former is substantially more stable, because the sphere maximizes the radial and tangential bonding interactions and leads to a more even distribution of electron density.

In a roughly spherical transition metal carbonyl cluster, the L^δ molecular orbitals derived from d_{xy} and $d_{x^2-y^2}$ are all filled, but their repulsive effects are not large, because the overlap between d^δ orbitals on adjacent atoms is small. Most important, the backbonding effects associated with the π-acceptor carbonyl ligands mitigate the antibonding nature of some of the δ orbitals. In a cylindrical cluster, the d^δ overlap integrals are larger for purely geometric reasons, and those d^δ orbitals associated with inner shells of atoms experience particularly strong interactions. The antibonding interactions between d^δ orbitals of the inner layers of metal atoms cannot be mitigated either by hybridization with the remaining valence orbitals (see Figure 8.13) or by

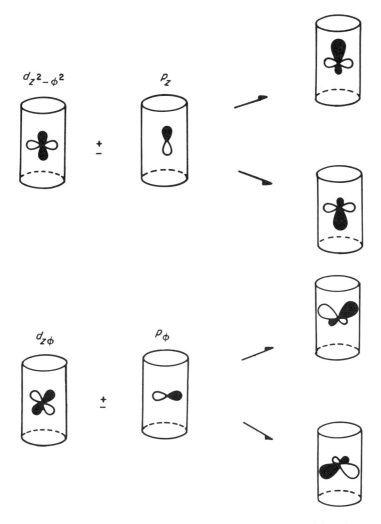

Figure 8.13 Schematic representation of the possible orbital mixings between $d^\delta_{z^2-\phi^2,z\phi}$ and $p^\pi_{z,\phi}$ atomic orbitals for atoms in a central cluster shell. It is noteworthy that this mixing does not produce out-pointing hybrids.

bridging carbonyls (because in a cylindrical cluster the number of carbonyls which can bond to the metal atoms is geometrically constrained). Therefore a cylindrical cluster is characterized by $2n - 1$ inaccessible molecular orbitals, with a high proportion of metal p character, and some additional inaccessible δ cluster orbitals resulting from out-of-phase combinations of d^δ functions from the inner rings of atoms. This situation is therefore reminiscent of that encountered for π-donor clusters of the earlier transition metals.

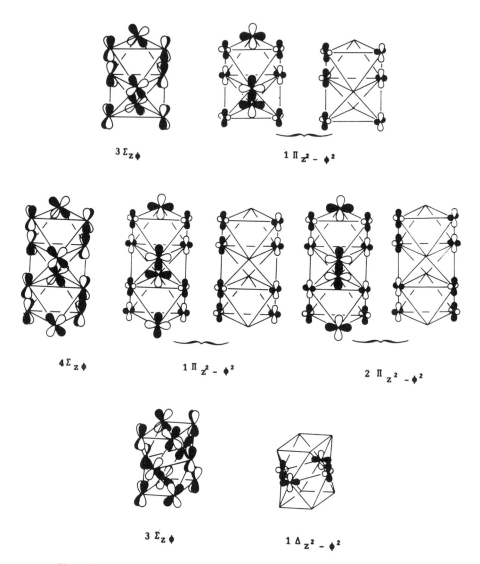

Figure 8.14 The inaccessible orbitals of some columnar clusters consisting of stacks of metal triangles with three and four layers.

The symmetries of the additional inaccessible molecular orbitals can be derived quite easily for columnar clusters based upon stacks of metal triangles, and follow a definite pattern. The relevant molecular orbitals are illustrated in Figure 8.14 for columnar clusters with three and four layers. There are

$$3(n_z - 2) - [(n_z - 2)/2]$$

inaccessible orbitals for clusters of this type, where n_z is the number of layers. The square brackets denote that the integer part is to be taken if $n_z - 2$ is odd. We are therefore left with the following number of accessible orbitals:

$$7n + 1 - 3(n_z - 2) + [(n_z - 2)/2].$$

This analysis helps to explain the electron counts of the following columnar rhodium clusters:

$Rh_6(CO)_{16}$	octahedral ($n_z = 2$), 86 valence electrons
$[Rh_9(CO)_{19}]^{3-}$	face-sharing octahedra ($n_z = 3$), 122 valence electrons
$[Rh_{12}(CO)_{25}]^{2-}$	three face-sharing octahedra ($n_z = 4$), 160 valence electrons

The structures of these clusters are shown in Figure 8.15.

In the platinum clusters $[Pt_3(CO)_6]_n^{2-}$, the bonding occurs primarily within the triangles, and the intertriangle bonding is rather weak. In these circumstances the overlap between the d^δ orbitals of the ring atoms is small and no additional inaccessible δ orbitals are generated. The platinums have a d^{10} configuration which leads to a cancellation of the bonding effects associated with the d shell. All the clusters have a total of $42n_z + 2$ valence electrons corresponding to the formation of three strong bonds within each triangle

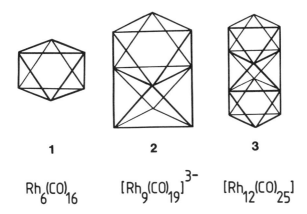

Figure 8.15 Some examples of columnar rhodium carbonyl clusters.

TABLE 8.3 COMPARISON BETWEEN CARBONYL AND SULFIDO COLUMNAR CLUSTERS BASED ON STAGGERED METAL TRIANGLES

Compound	Electron counts	Unavailable orbitals $2n - 1 + $ (extra d^δ)	Compounds	Electron counts				
				d electrons	Valence electrons	+	n electron pairs*	Unavailable orbitals† $2n - 2 + $ (extra d_δ)
$[Rh_6(CO)_{16}]$	86	$2 \times 6 - 1$	$[Mo_6S_8]^{4-}$	24	72	+	$(12) = (84)$	$2 \times 6 - 2$
$[Rh_9(CO)_{19}]^{3-}$	122	$2 \times 9 - 1 + (3)$	$[Mo_9S_{11}]^{4-}$	36	102	+	$(18) = (120)$	$2 \times 9 - 2 + (3)$
$[Rh_{12}(CO)_{25}]^{2-}$	160	$2 \times 12 - 1 + (5)$	$[Mo_{12}S_{14}]^{6-}$	50	134	+	$(24) = (158)$	$2 \times 12 - 2 + (5)$

*The sp hybridized out-pointing orbital of each metal which is available for one terminal ligand is unoccupied in this series of clusters.
†There is one more unavailable orbital ($\bar{F}_d^\delta$) in $Mo_6S_8L_6$.

(formed from three inwardly directed hybrids), the occupation of the remaining six hybrids per vertex, and a single molecular orbital which is bonding between triangles.

Although the arguments developed above have concentrated primarily on carbonyl clusters, they can easily be extended to halide and sulfide clusters of the earlier transition metals. For example, the inaccessible a_{2u} orbital in $[Mo_6X_8L_6]^{4+}$ is joined by $3(n_z - 2) - [(n_z - 2)/2]$ additional inaccessible δ orbitals in the columnar clusters $Mo_9S_{11}^{4-}$ and $Mo_{12}S_{14}^{6-}$. These conclusions are summarized in Table 8.3.[8]

EXERCISES

8.1 What are the formal oxidation states of the molybdenum and niobium atoms in the clusters $[Mo_6Cl_8L_6]^{4+}$ and $[Nb_6Cl_{12}L_6]^{2+}$? How do these clusters differ from transition metal carbonyl clusters of later transition metals?

8.2 Work out which σ, π, and δ cluster orbitals are required in an octahedron and reduce the representation spanned by the ligand donor orbitals in $[Mo_6(\mu_3\text{-}Cl)_8Cl_6]^{2-}$ (face-bridged) and $[Nb_6(\mu_2\text{-}Cl)_{12}Cl_6]^{4-}$ (edge-bridging). Hence, deduce which metal valence orbitals remain after forming the skeletal bonding orbitals from the three inwardly hybridized orbitals and the metal-ligand σ bonds. Verify that mixing with orbitals of the same symmetry then generates the energy-level spectra described in the text.

8.3 Derive a qualitative energy-level diagram for the metal-metal bonding and antibonding orbitals in $[Re_3Cl_{12}]^{3-}$. (Each rhenium atom contributes one σ, two π, and one δ orbital for skeletal bonding. The σ and π interactions are similar to those discussed for $C_3H_3^+$ in Chapter 2.)

8.4 Predict the electrical conductivity (i.e., insulator or semiconductor vs. metallic conductor) of the compounds containing $Re_6(\mu_3\text{-}\hat{X})_8\hat{X}_6$ structural units for clusters with different modes of interlinking (i.e., different $Re_6\hat{X}_y$ stoichiometries) (see reference 17 for some possibilities).

8.5 Haupt H.-J., et al.[34] (*Angew. Chem., Int. Ed.*, **27**, 263 (1988)) have reported the following bond lengths for triangular rhenium cluster compounds:

$Re_3(CO)_9(\mu_2\text{-}PPh_2)_3$	2.914 (4) Å
$Re_3(CO)_9(PPh_3)(\mu_2\text{-}H)_4$	2.797 (1), 3.193 (1), and 3.234 (1) Å
$Re_3(CO)_6(\mu_2\text{-}H)_2(\mu_2\text{-}PPh_2)_3$	2.730 (1) Å

The authors have suggested that these compounds represent a transition between triangular clusters formed by π-acceptor ligands, such as $Os_3(CO)_{12}$, and those formed with π-donor ligands, such as $[Re_3Cl_9]^{3-}$. To what extent can this suggestion be supported by localized bonding arguments? Can you develop the molecular orbital picture for $[Re_3Cl_9]^{3-}$ to provide a delocalized account of these bond length variations?

8.6 Gray H. B., et al.[35] (*J. Amer. Chem. Soc.*, **105**, 1878 (1983)) have described the esr spectrum of $[Mo_6Cl_{14}]^-$ in a frozen solution of CH_2Cl_2 at 10 K. The anion was produced electrochemically from $[Mo_6Cl_{14}]^{2-}$. The spectrum has been interpreted in terms of an axially symmetric system with one unpaired spin and $g_\perp = 2.10$, $g_\parallel = 2.00$. By reference to the molecular orbital diagrams presented in this chapter, suggest why the ion might undergo a tetragonal distortion. What are the possible point groups and ground states of the ion?

REFERENCES

1. Chisholm, M. H., Clark, D. L., Hampden-Smith, M. J., and Hoffman, D. M., *Angew. Chem., Int. Ed.*, **28**, 432 (1989).
2. Schäffer, H., and Schnering, H. G., *Angew. Chem., Int. Ed.*, **76**, 833 (1964).
3. Corbett, J. D., *Adv. Chem. Series (Solid State Chemistry)*, **186**, 329 (1980).
4. Kettle, S. F. A., *Theo. Chim. Acta*, **3**, 211 (1965).
5. Johnson, R. L., and Mingos, D. M. P., *Inorg. Chem.*, **25**, 1661 (1986).
6. Cotton, F. A., and Haas, T. E., *Inorg. Chem.*, **8**, 2041 (1969).
7. Bursten, R. E., Cotton, F. A., and Stanley, G. G., *Israel J. Chem.*, **19**, 132 (1980).
8. Hughbanks, T., and Hoffmann, R., *J. Amer. Chem. Soc.*, **105**, 1150 (1983).
9. Smith, J. D., and Corbett, J. D., *J. Amer. Chem. Soc.*, **107**, 5702 (1985).
10. Chisholm, M. H., Errington, R. J., Folting, K., and Huffman, J. C., *J. Amer. Chem. Soc.*, **104**, 2025 (1982).
    Stensvad, S., Helland, B. J., Bakrich, M. W., Jacobson, R. A., and McCarley, R. E., *J. Amer. Chem. Soc.*, **100**, 6257 (1978).
11. Stone, A. J., *Molec. Phys.*, **41**, 1339 (1980).
12. Perrin, A., and Sergent, M., *New J. Chem.*, **12**, 337 (1988).
13. Fowler, P. W., and Ceulemans, A., *Inorg. Chim. Acta*, **75**, 105 (1985).
14. Jodden, K., Schnering, H. G., and Schäffer, H., *Angew. Chem., Int. Ed.*, **14**, 570 (1975).
15. Stensvad, S., Hellend, B. J., Bakich, M. W., Jacobson, R. A., and McCarley, R. E., *J. Amer. Chem. Soc.*, **100**, 6257 (1978).
16. Simon, A., *Angew. Chem., Int. Ed.*, **27**, 159 (1988).
17. Chevrel, R., and Sergent, M., *Superconductivity in Ternary Compounds*, Topics in Current Physics, Springer-Verlag, Berlin, Heidelberg, New York, 1982.
18. Nohl, H., and Anderson, O. K., *Supercond. $d-f$ Band Met. Conf.*, **4**, 161 (1982).
19. Anderson, O. K., Klose, W., and Nohl, H., *Phys. Rev.*, **B17**, 1209 (1978).
20. Ziebarth, R. P., and Corbett, J. D., *J. Amer. Chem. Soc.*, **110**, 1132 (1988).
21. Kollis, J. W., Basolo, F., and Shriver, D. F., *J. Amer. Chem. Soc.*, **104**, 5626 (1982).
22. Wijeyesekera, S. D., Hoffmann, R., and Wilker, C. N., *Organometallics*, **3**, 949, 962 (1984).
23. Harris, S, and Bradley, J. S., *Organometallics*, **3**, 1984 (1086).

24. Housecroft, C. E., *J. Organometallic Chem.*, **276**, 297 (1984).
25. Halet, J. F., Saillard, J.-Y., Lissilour, R., McGlinchey, M. J., and Jaouen, G., *Organometallics*, **5**, 139 (1986).
26. Brint, P., O'Cuill, K., and Spalding, T. R., *Polyhedron*, **5**, 1791 (1986).
27. Chisholm, M. H., Clark, D. L., Huffman, J. C., and Smith, C. A., *Organometallics*, **6**, 1280 (1987).
28. Halet, J.-F., Evans, D. G., and Mingos, D. M. P., *J. Amer. Chem. Soc.*, **110**, 87 (1988).
29. Bradley, J. S., *Adv. Organometal. Chem.*, **22**, 1 (1983).
30. Ceriotti, A., Longoni, G., Piro, G., Manaserro, N., Masciocchi, N., and Sansoni, M., *New J. Chem.*, **12**, 501 (1988).
31. Halet, J.-F., and Mingos, D. M. P., *Organometallics*, **7**, 51 (1988).
32. Mingos, D. M. P., and Zhenyang, L., *J. Organometallic Chem.*, **339**, 367 (1988).

Index

Actinide clusters, 31
Alkali metal clusters, 15, 154, 166, 170, 173, 179
Alkaline-earth metal clusters, 15, 173
Alkoxide clusters, 20, 295
Alternant hydrocarbons, 146
Anticuboctahedron, 86
Arachno clusters:
   boranes, 59–61, 90
   electron counts, 87, 195–97, 206
   pi-donor ligands, 291–94
   rearrangements, 230
   transition metal, 92
Arsenido clusters, 44
Avoided crossing, 221

Benzene, Hückel treatment, 123
Bicapped square antiprism, 45, 91
Bipolar clusters, 208–9
Bipyramidal clusters, 210
Bispherical clusters, 270–71
Bond enthalpies, 6
Bond lengths:
   boranes and carboranes, 87
   boron halide clusters, 57
   hydrocarbon clusters, 63
   metal carbonyl clusters, 33–34
   metal-metal, 2
   naked clusters, 59

Bonding:
   arachno clusters, 87, 182, 195–97
   bispherical clusters, 263–73
   boron hydrides, 183–86
   capped clusters, 263–73
   closo clusters, 87, 144, 182
   edge-localized, 85, 100, 293
   edge-sharing clusters, 256–59
   face-localized, 85, 99–100, 293
   face-sharing clusters, 260–63
   four-connected clusters, 200–201
   gold clusters, 154
   high-nuclearity clusters, 274–83
   localized vs. delocalized, 81
   metal-metal, 26–30
   multi-center, 51
   multiple, 26–31, 102
   nido clusters, 87, 144, 182, 195–97
   partial involvement of tangential orbitals, 212
   platinum clusters, 213–16
   radial, 154–79
   three-connected clusters, 198–99
   transition metal clusters, 201–7
   vertex-sharing clusters, 250–55
   pi-donor clusters, 285–94
   nido and *arachno* pi-donor clusters, 291–94
   columnar clusters, 304–9
Bond order, 5

Boranes, 58
　bonding, 183–86
　bond lengths, 87
　charge distributions, 191–92
　HOMO-LUMO gaps, 186–90
　infrared spectroscopy, 228
　nmr spectroscopy, 228
　physical properties, 56
　skeletal rearrangements, 227–32
　structures, 60, 186–90
　synthesis, 56
Boron halides:
　physical properties, 57
　synthesis, 57
Butadiene ring closure, 221

Capped square antiprism, 45, 91, 224
Capping Ib metal atoms, 236
Capping principle, 98, 183, 227, 230, 263–73
Carbido clusters, 44, 300
　infrared data, 205
Carbonyl clusters, *see* Transition metal carbonyl clusters
Carbonyl ligands:
　infrared spectroscopic data, 205
　rearrangements, 240–45
Carboranes:
　bonding, 183–86
　bond lengths, 87
　shapes, 186–90
Cavity radii, 42
Chalcogenide clusters, 296
Characteristic electron counts, table, 206
Charge distributions, in boranes, 191–92
Chemical shifts of interstitial atoms, 299
Chevrel-Sergent compounds, 22, 25, 297
Close-packed structures, 37–38
Closo, nido, arachno nomenclature, 62
Closo clusters:
　boranes, 59–61
　electron counts, 87, 144, 206
　rearrangements, 219, 230
　transition metals, 92
Cluster cone angle, 11–12
Cluster orbitals:
　delta, 148
　pi, 132–38
　sigma, 126–30
Cohesive energy, 6
Colors:
　naked clusters, 59

pi-acid clusters, 202
pi-donor clusters, 292
Columnar metal carbonyl clusters, 38, 41, 304–9
Commo carboranes, 255
Condensation principle, 95
Condensed clusters, 38–41, 249–73
　edge-sharing, 256–59
　electron counting rules, 95, 230
　face-sharing, 260–63
　main group, 96
　with pi-donor ligands, 22, 260–63
　platinum, 259
　transition metal, 13, 97–98
　vertex-sharing, 250–55
Conservation of orbital symmetry, 220–22
Coulson-Rushbrooke pairing theorem, 146
Cubane, 62, 76
　iron-sulfur analogue, 24–27
Cuboctahedron, 200
Cuneane, 62, 76
Cyclobutene ring opening, 221
Cyclopentadienyl clusters, 55
Cylindrical clusters, 304

Debor principle, 183
Degenerate edge, definition, 219
Delocalized bonding schemes vs. localized bonding schemes, 54, 81
Delta cluster orbitals, 148–50, 285–309
Deltahedra, 56, 61, 87
　capped, 266
　electron-counting rules for, 87, 144
Diamond-square-diamond rearrangement, 219–20
　frontier orbitals, 225
Dissociation energies, 9
Dodecahedral clusters, 211
Dodecahedrane, 75–76
Donor characteristics:
　interstitial atoms, 77
　ligands, 75

Edge-cleavage process, 229–31
Edge-localized bonding, 85, 100, 293
Edge-sharing condensed clusters, 256–59
Effective Atomic Number Rule, 73
Electrical and magnetic properties, 22
Electron-counting rules, 73–113, 206
　closo, nido and *arachno* clusters, 87
　condensed clusters, 95

effective atomic number rule, 73
four-connected clusters, 86
replacing main group by transition metal atoms, 74
ring clusters, 73
symmetry-forced deviations, 207–11
three-connected clusters, 57
Electron-deficient clusters, 57
Electron-delocalized clusters, 58–62
  definition, 57
Electronic closed-shell requirements for clusters, 72, 206, 207–11
Electron-precise clusters:
  definition, 56
  tetrahedral, 64
Electron-rich clusters, 105–8
  definition, 57
Enthalpies of formation:
  metal atoms, 8
  metal carbonyls, 7
Equilateral triangle, TSH treatment of, 138
Esr:
  metal carbonyls, 268
  sodium clusters, 170
Euler's rule, 216

Face-capping growth sequence, 110
Face-localized bonding schemes, 85, 99–100, 293
Face-sharing clusters, 260–63
Fast atom bombardment mass spectroscopy, 211
Ferredoxins, 24–27
Fluxionality:
  carbonyl ligands, 240–45
  cluster skeleton, 218–40
Four-connected clusters, 86
  bonding, 200–201
  electron count, 86, 206
Frontier orbitals:
  for the DSD process, 225
  of some isolobal fragments, 79

Gold clusters:
  bonding, 154, 157, 167
  hemispherical, 177
  pseudospherical, 174–77
  skeletal rearrangements, 234–36
  solid-state nmr, 235
  structures, 48

synthesis, 49
toroidal, 174–77
Gold-silver clusters, 46, 48, 178

Halide clusters, 100–103, 296
  synthesis, 19
Hemispherical gold clusters, 177
Heterometallic clusters, 39
High-nuclearity clusters, 36, 38, 46, 274–83
HOMO-LUMO gaps, in boranes, 186–90
Hückel energy levels:
  benzene, 123
  octahedron, 129
Hückel theory, 119
  cyclic polyenes, 121
  linear polyenes, 119
Hybridization schemes, 80
Hydrido clusters, 49–54
  infrared spectroscopic data, 204
  location of ligands, 208
  multi-center bonding, 51–52
  structures, 51
Hydrocarbon clusters, 63
Hypho fragments, 260

Icosahedron, 45, 91, 110, 174
Infrared spectroscopy:
  boranes, 228
  carbido clusters, 205
  hydrido clusters, 205
  metal carbonyls, 205
Insulating properties, 296
Interstitial atoms:
  beryllium, 300
  bonding, 168–70
  boron, 300
  carbon, 300
  in carbonyl clusters, 42–46
  donor characteristics, 77
  as models of surfaces, 304
  nitrogen, 300
  nmr chemical shifts, 299
  in pi-donor clusters, 298–304
  in radially bonded clusters, 168–70
  structures, 45
  synthesis, 44
  transition metal, 86, 298
Intrinsic nodes, 142
Iron-sulfur cubane clusters, 24–27
  redox properties, 27

Isocyanide clusters, 55
Isolobal principle, 79
  fragments, 79, 213
Isostructural main group and transition metal clusters, 93–94

Jellium model, 66, 179

Kinetic factors, 10

Lanthanide clusters, 31–32
Legendre polynomials, 157
Ligands:
  donor characteristics, 75
  fluxionality, 240
  steric influence, 11
Linear combination of atomic orbitals (LCAO) method, 117–18
Linked clusters, 294–98
Lipscomb's diamond-square-diamond process, 219
Lithium clusters, 15, 166
Localized bonding schemes vs. delocalized bonding schemes, 54, 81
Location of hydrido ligands, 208
Luminescence and redox photochemistry of pi-donor clusters, 290

Magic angle sample spinning, 235
Magic numbers, for sodium clusters, 65
Magnesium clusters, 173
Main group clusters, 55–64, 76, 94
  condensed, 95–96
  and isostructural transition metal clusters, 78, 94
  three-connected, 62–64
Mass spectroscopy:
  for ligated clusters, 211
  for sodium clusters, 65
McCarley's compound, 31
Metal carbonyl clusters, 32–46
  bond lengths, 33–34
  columnar, 41
  enthalpies of formation, 7
  high resolution solid state nmr, 235
  infrared spectroscopy, 205
  with interstitial atoms, 42–46
  ligand rearrangements, 240–45
  in molecular beams, 108–12
  paramagnetic, 268
  structures, 33–34, 45
  synthesis, 35
  ^{13}C nmr of carbonyl ligands, 242
Metal hydrido clusters, 49
  structures, 51
Metallic radii, 3
Metallocarboranes, 254
  slipped, 255
Metal-metal bonding, 4, 26, 30, 53
Metal-metal bond lengths, 2
  carbonyl clusters, 33–34
Metal phosphine clusters, 46–49
  structures, 48
  synthesis, 49
Molecular beams, 64–66, 108–12, 166, 179
Molecular orbitals, for radially bonded clusters, 163
Mössbauer spectroscopy, 172
Multi-center bonding, 51–52
Multiple bonding, 26–31, 102
Multispherical clusters, 179

Naked clusters, 14, 58–59, 89, 91
  synthesis, 59
Nickel clusters, in molecular beams, 108–12
Nido clusters:
  boranes, 59–61, 90
  electron counts, 87, 195–97
  pi-donor clusters, 291–94
  rearrangements, 228, 231–32
  transition metal, 92
Nitrido clusters, 44, 300
Nitrosyl clusters, 55
Nmr:
  boranes, 228
  carbonyl ligands, 242
  interstitial atoms, 299
  platinum clusters, 237–38
  solid-state, 235
  variable-temperature, 243
Nodes:
  of D^σ atomic orbitals, 157
  intrinsic, 142
Nondegenerate edge, definition, 219
Nonpolar clusters, 208–9

Oblate and prolate clusters, 164–68
Octahedral halide clusters, 21, 100–103, 286–94

Octahedron, 86, 91, 200, 239
  local axes, 127
  pi cluster orbitals, 140
  sigma cluster orbitals, 128, 162
One-electron atom, 124
Organometallic clusters, 14
Oxidative coupling, 35
Oxide cluster, 20

Pairing principle:
  boranes, 192–95
  Coulson-Rushbrooke, 146
  tensor surface harmonic theory, 145, 151–52, 192–95
  TSH theory, simple applications, 146–47
Palladium clusters, skeletal rearrangements, 236–40
Paramagnetic transition metal carbonyl clusters, 268
Partial involvement of tangential orbitals in bonding, 212–16
Particle in-a-box, 120
Particle on-a-ring, 121
Particle on-a-sphere, 126
Phosphido clusters, 44
Phosphine clusters, *see* Metal phosphine clusters
Photoelectron spectra, 82
  $Mo_4S_4(\eta\text{-}C_5H_4Pr^i)_4$, 204
  P_4, 194
  $Re_3H_3(CO)_{12}$, 83
Physical properties:
  boranes, 56, 228
  boron halides, 57
  electrical and magnetic, 22
  from esr, 170
  from fast atom bombardment mass spectroscopy, 211
  hydrocarbon clusters, 63
  infrared spectroscopy of ligands, 205–6
  luminescence of pi-donor clusters, 290
  from mass spectroscopy, 65, 211
  from Mössbauer spectroscopy, 172
  from nmr, 235, 237, 242
  paramagnetism, 268
  photoelectron spectra, 83, 194, 204
  from X-ray diffraction, 5
Pi-acid clusters, 14, 32–46
  colors, 202
Pi cluster orbitals, 132–38
  for $B_6H_6^{2-}$, 141

  for boranes, 189
  choosing coefficients for, 133
  sketching, 136
  splitting pattern in boranes, 190
Pi-donor clusters, 14, 18
  bonding, 285–94
  colors, 292
  condensed chain clusters 22
  with interstitial atoms, 298–304
Platinum clusters, 40, 50, 215
  condensed, 259
  high-nuclearity, 276
  with hydrido ligands, 50
  nmr, 237–38
  skeletal rearrangements, 236–40
  synthesis, 49
Polar clusters, 208–9
Polyhedral clusters, 14
Polyhedral Skeletal Electron Pair Theory (PSEPT), 72
Post-transition metal clusters, 89
Prismane, 76
Prolate and oblate clusters, 164–68
  skeletal rearrangements, 234–36
Pseudospherical gold clusters, 174–77
  skeletal rearrangements, 234–36

Radial bonding, 154–79
Rearrangements:
  carbonyl ligands, 240–45
  cluster skeleton, 218–40
Redox properties:
  condensation syntheses, 35, 39
  iron-sulfur clusters, 27
  luminescence of pi-donor clusters, 290
Ring compounds, electron counts, 73, 206

Secular equations, 118
Selenido clusters, 26–27, 29
Semibridging effects, 242
Semiconductivity, 296
Sigma cluster orbitals, 126–30, 163
  energies, 160
  sketching, 136
  trigonal prism, 161
Skeletal rearrangements in clusters, 218–46
  application to boranes, 227
  clusters with capping Ib metals, 236
  and electron counting rules, 227
  and the pairing principle, 223

Skeletal rearrangements in clusters (*cont.*)
  platinum and palladium clusters,
    236–40
  radially bonded clusters, 234–36
  square-diamond, diamond-square
    rearrangement, 231–32
  transition metals, 232–34
  TSH theory analysis, 222–27
Sketching cluster orbitals, 136
Slipped metallocarboranes, 255
Sodium clusters, 16, 65, 173
  esr, 170
  magic numbers, 65
  mass spectroscopy, 65
Spherical harmonics, 123–26
  transformation, 131–32
  vector spherical harmonics, 135
Square antiprism, 45, 86, 91, 100
Square-diamond, diamond-square
  rearrangement, 231–32, 246
Stereochemical nonrigidity, 235
Steric effects:
  in ligand fluxionality, 241
  in ligand packing, 11
Stibinido clusters, 44
Styx method, 103–5
Suboxide clusters, 14, 16
Sulphido clusters, 20, 27, 29, 44
Sulfur dioxide clusters, 55
Superconductivity, 22, 296–97
Symmetries of molecular orbitals, 84
Symmetry-forced deviations from electron
  counting rules, 207–11
Synthesis:
  boranes, 56
  boron halide clusters, 57
  clusters with interstitial atoms, 44
  gold phosphine clusters, 49
  heterometallic clusters, 39
  metal carbonyls, 35
  metal halide clusters, 19
  naked clusters, 59
  platinum phosphine clusters, 49
  thermodynamic and kinetic aspects,
    9–13

Tensor solid harmonic theory, 179
Tensor surface harmonic theory, 117–53
  pairing principle, 145–47, 151–52,
    192–95
  and skeletal rearrangements, 222–27
Tetrahedral electron-precise clusters, 64
Tetrahedrane, 75
Thermal condensation, 35
Thermochemistry, 6–10
Three-connected clusters, 62–72, 76–77
  bonding, 198–99
  electron counts, 73, 206
  symmetries of molecular orbitals, 84
Tolman's cone angle, 11
Toroidal gold clusters, 174–77
Transformation of spherical harmonics,
  131
Transition metal, interstitial, 86, 298
Transition metal clusters, 17–26, 94,
  201–207
  bonding, 201–7
  closo, *nido* and *arachno*, 92
  condensed, 97–98
  electron-rich, 108
  and isostructural main-group clusters,
    78, 94
  skeletal rearrangements, 232–40
Tricapped trigonal prismatic clusters, 91,
  210, 224, 267
Trigonal bipyramid, 91, 161, 224
Trigonal prism:
  rearrangements, 239
  sigma cluster orbitals, 161

Vector spherical harmonics, 135
Vertex-sharing clusters, 250–55

X-ray diffraction, 5

Zeolite-Y, 16

Summary of the PSEPT Rules

(1) MG/TM ring compounds have a total of $6n/16n$ valence electrons.
(2) MG/TM three-connected cluster compounds have $5n/15n$ valence electrons.
(3) If a transition metal atom occupying a vertex position is replaced by a main group atom, then the characteristic number of valence electrons is reduced by 10.
(4) MG/TM four-connected clusters are characterized by a total of $4n + 2/14n + 2$ valence electrons, so long as the n vertex atoms lie approximately on a spherical surface.
(5) MG/TM *closo*, *nido*, and *arachno* deltahedral molecules are characterized by $4n + 2/14n + 2$, $4n + 4/14n + 4$, and $4n + 6/14n + 6$ valence electrons.
(6) The total electron count of a condensed cluster is equal to the sum of the electron counts for the parent clusters minus the electron count characteristic of the common atom, pair of atoms, or face of atoms. These characteristic electron counts are 18 for a shared metal vertex, 34 for a shared edge of metal atoms, 48 for a shared metal triangle, and 62 for a shared metal square.

MG = main group   TM = transition metal n = number of vertices

Edge bridging results in an increment in the valence electron count by 14 (TM). Face capping results in an increment in the valence electron count by 12 (TM). The number of skeletal electron pairs (SEPs) in a cluster is related to the total electron count as follows:

SEP	Total number of electrons (pec)
$n + 1$	$4n + 2/14n + 2$
$n + 2$	$4n + 4/14n + 4$
$n + 3$	$4n + 6/14n + 6$
$3n/2$	$5n/15n$